T0201892

MULTILAYER NETWORKS

Multilayer Networks

Structure and Function

Ginestra Bianconi

OXFORD
UNIVERSITY PRESS

OXFORD
UNIVERSITY PRESS

Great Clarendon Street, Oxford, OX2 6DP,
United Kingdom

Oxford University Press is a department of the University of Oxford.
It furthers the University's objective of excellence in research, scholarship,
and education by publishing worldwide. Oxford is a registered trade mark of
Oxford University Press in the UK and in certain other countries

© Ginestra Bianconi 2018

The moral rights of the author have been asserted

First published 2018
First published in paperback 2022

Impression: 1

All rights reserved. No part of this publication may be reproduced, stored in
a retrieval system, or transmitted, in any form or by any means, without the
prior permission in writing of Oxford University Press, or as expressly permitted
by law, by licence or under terms agreed with the appropriate reprographics
rights organization. Enquiries concerning reproduction outside the scope of the
above should be sent to the Rights Department, Oxford University Press, at the
address above

You must not circulate this work in any other form
and you must impose this same condition on any acquirer

Published in the United States of America by Oxford University Press
198 Madison Avenue, New York, NY 10016, United States of America

British Library Cataloguing in Publication Data
Data available

Library of Congress Cataloging in Publication Data
Data available

ISBN 978–0–19–875391–9 (Hbk.)
ISBN 978–0–19–286554–0 (Pbk.)

Printed and bound by
CPI Group (UK) Ltd, Croydon, CR0 4YY

Links to third party websites are provided by Oxford in good faith and
for information only. Oxford disclaims any responsibility for the materials
contained in any third party website referenced in this work.

To Christoph

Preface

Multilayer networks are formed by several networks that evolve and interact with each other. These networks are ubiquitous and include social networks, financial markets, multimodal transportation systems, infrastructures, climate networks, ecological networks, molecular networks and the brain. The multilayer structure of these networks strongly affects the properties of dynamical and stochastic processes defined on them, which can display unexpected characteristics. For example, interdependencies between different networks of a multilayer structure can cause cascades of failure events that can dramatically increase the fragility of these systems; spreading of diseases, opinions and ideas might take advantage of multilayer network topology and spread even when its single layers cannot sustain an epidemic when taken in isolation; diffusion on multilayer transportation networks can significantly speed up with respect to diffusion on single layers; finally, the interplay between multiplexity and controllability of multilayer networks is a problem with major consequences in financial, transportation, molecular biology and brain networks.

In the last twenty years, considerable attention has been devoted to the study of single networks. It has been found that despite their different functions, many biological, social or technological systems can share similar properties when they are studied from the network perspective. Recently, the multiplexity of many networks has been identified as an important aspect of networked systems that needs to be addressed to improve our understanding of biological and man-made networks. The subject is currently raising great scientific attention, and several important new results have been obtained. This book will present a comprehensive account of this emerging field.

The book includes three parts:

- PART I: SINGLE AND MULTILAYER NETWORKS
  This part (chapter 1) outlines the main research questions that have been driving the research on multilayer network structure and function.
- PART II: SINGLE NETWORKS
  This part provides an introduction to the main results obtained in Network Science for the characterization of the structure (chapter 2) and the function (chapter 3) of single networks. This part constitutes the reference point for appreciating the results that hold for multilayer networks.
- PART III: MULTILAYER NETWORKS
  This part constitutes the core of the book and discusses the main properties of multilayer network structure and function.

Three initial chapters (chapters 4–6) set the stage for the rest of book. They discuss the relevance of the multilayer networks framework for a variety of applications (chapter 4), provide the mathematical definitions of multilayer networks (chapter 5) and introduce their basic structural properties (chapter 6).

Subsequently, several chapters are devoted to the characterization of the structure of multilayer networks and extraction of relevant information using their built-in correlations (chapter 7), their mesoscale community structure (chapter 8) and structural properties determining the nodes' and layers' centralities (chapter 9).

Bridging between the chapters focusing on multilayer network structures (chapters 5–9) and the chapters focusing on multilayer network dynamics (chapters 11–16), novel modelling frameworks especially tailored to multilayer networks are presented in chapter 10, together with randomization algorithms.

The active research activity on the dynamics and function of multilayer networks is covered in chapters 11–16. These chapters provide a general perspective on the major dynamical processes, including: percolation and avalanches (chapters 11–12), epidemic spreading (chapter 13), diffusion (chapter 14), dynamical systems and synchronization (chapter 15) and finally opinion dynamics and game theory models (chapter 16).

- APPENDICES
A series of appendices providing more detailed mathemetical discussion of some of the major results in multilayer networks complements the material presented in the main body of the book.

Our aim has been to provide an overview of the field which could guide the reader in understanding the recent literature on multilayer networks. Given the fast pace at which new results are continuously published on the subject, it has become impossible to cover entirely the rapidly growing literature in the field. Our aim is to provide a pedagogical presentation and an in-depth discussion of the main results on multilayer networks, allowing students and reseachers to be quickly introduced to the field. We have therefore made some choices based on our perception of what is more relevant to cover in the book. This does not imply that the work not covered here is less valuable and we apologize in advance to the Authors of the papers not cited here.

This book will be of interest for graduate students and researchers in Network Science working at the interface between two or more disciplines such as: physics, mathematics, statistics, economy, engineering, computer science, neuroscience and cell biology. While the book will provide a theoretical introduction to the main results on multilayer networks, at the same time it will remain widely accessible to the general interdisciplinary reader.

Ginestra Bianconi
London, 31 October 2017

Acknowledgements

This book is the result of multiple interactions with the network scientists working in multilayer networks and with many interested students to whom I gave lectures on the topic. I am most grateful to all of them for their shared passion for multilayer network structure and function.

Special thanks go to my precious multilayer network collaborators: A. Arenas, A. Barrat, A. Baronchelli, M. Barthelémy, G. J. Baxter, S. Boccaletti, F. Battiston, D. Cellai, R. A. da Costa, L. Dall' Asta, R. Criado, C. I. Del Genio, S. N. Dorogovtsev, J. P. Gleeson, J. Gómez-Gardeñes, A. Halu, M. Karsai, J. Iacovacci, V. Latora, E. López, J. F. F. Mendes, G. Menichetti, R. Mondragón, S. Mukherjee, V. Nicosia, P. Panzarasa, M. A. Porter, F. Radicchi, C. Rahmede, D. Remondini, M. Romance, I. Sendina-Nadal, J. Stéhle, Z. Wang, Z. Wu, M. Zanin, K. Zhao, J. Zhou among which are the coauthors of an influential review article on multilayer networks that has been the starting point for this book: S. Boccaletti, R. Criado, C. I. Del Genio, J. Gómez-Gardeñes, M. Romance, I. Sendina-Nadal, Z. Wang, M. Zanin.

I thank the Physics Department at Seoul National University, the London Mathematical Society, the 2016 NetSci School, the Mathematics Department of the Politecnico of Torino, the Como School for Advanced Studies, and the Short Course on Complex Networks at Oxford University for hosting my courses and lectures on multilayer networks that have been very useful in shaping this book. These events would not have been possible without the support of my great friends and colleagues: A. Arenas, J. Coon, B. Kahng, S. Majid, Y. Moreno, M. A. Porter, F. Vaccarino.

In writing this book I benefited from great discussions with S. Havlin on the robustness of interdependent networks, with A. Arenas and Y. Moreno on epidemic spreading and diffusion processes on multilayer networks and with S. Boccaletti on the synchronization of multilayer networks.

Sönke Adlug and Ania Wronski from Oxford University Press have been invaluable too, for their excellent editorial work that has motivated me throughout the writing of the book. I also thank the School of Mathematical Sciences at Queen Mary University of London that has allowed me to make this book a reality.

Finally, this book would not have seen the light were it not for the enthusiasm, support and encouragement of my husband Christoph. I am most grateful to him for being the first reader of this book and giving me many relevant comments and suggestions.

Contents

Part I

Single and Multilayer Networks

1

Complex Systems as Multilayer Networks

1.1   What are multilayer networks?

The fundamental idea behind Network Science is that important information about a complex system can be gained by studying its underlying network structure. This simple yet powerful point of view has provided the tools for gaining unprecedented knowledge on the rich interplay between the structure and function of complex systems.

The field of Network Science has been flourishing in the last decades, where we have witnessed a Big Data explosion in social science, biology and engineering. Network Science is a highly interdisciplinary field that combines tools and techniques coming from physics, mathematics, statistics, biology, engineering and computer science. Now almost twenty years since the beginning of the field, we have reached understanding of complex networks and their universal topological properties and we have revealed the rich interplay between structure and dynamics in complex network architectures.

In the last few years it has been pointed out by several researchers that our understanding of complex networks has so far had an important limitation. In fact, rarely do networks work in isolation. From infrastructures and transportation systems to cells and the brain, most networks are multilayer, i.e. they are formed by several interacting networks. For example, in modern society different infrastructures are related by a complex web of interdependencies and a failure in the power grid can trigger failures in the Internet, the financial market and transportation. When commuting to the workplace, the inhabitants of large cities usually take more than one means of transportation including bus, metropolitan trains and underground. In the cell, the protein–protein interaction network, signalling networks, metabolic networks and transcription networks are not isolated but interacting, and the cell is not alive if any one of these networks is not functioning. In the brain, understanding the relation of functional and structural networks forming a multilayer network is of fundamental importance.

Multilayer networks have been first introduced in the context of social sciences to describe different types of social ties. Up to now, social networks remain one of the typical

Multilayer Networks. Ginestra Bianconi, Oxford University Press (2018).
© Ginestra Bianconi. DOI: 10.1093/oso/9780198753919.001.0001

examples of multilayer networks. Nevertheless, multilayer networks have attracted a significant interdisciplinary interest only in the last few years, because it has become clear that characterizing multilayer networks is fundamental to understanding most complex networks including cellular networks, the brain, complex infrastructures and economical networks in addition to social networks (chapter 4).

Interestingly, the framework of multilayer networks (chapters 5–6) can also be applied to temporal networks, i.e. networks that change over time. Temporal networks can be used to describe a large variety of data, ranging from contacts networks recording face-to-face social interactions to time-resolved correlations between different regions of the brain. In this case, the multilayer network is formed by temporal slices each describing the interactions occurring in a given temporal interval. It turns out that the multilayer approach for studying temporal networks can be extremely useful for advancing our understanding of the dynamical processes occurring in them, such as diffusion and epidemic spreading.

1.2   Information gain in multilayer networks

A multilayer network is not to be confused with a larger network including all the interactions. As a network ultimately is a way to encode information about the underlying complex system, there is a significant difference between considering all the interactions at the same level and including the information on the different natures of the different interactions. In a multilayer network, each interaction has a different connotation, and this property is correlated with other structural characteristics, allowing network scientists to extract significantly more information from the complex system under investigation (chapters 7–9).

A major theme of this book is the discussion of the main types of correlation that are present in multilayer network datasets. We will show how these correlations can be quantified and we will present several techniques for extracting relevant information from multilayer network datasets that cannot be found by considering networks in isolation. These include ranking algorithms aimed at assessing the relevance of nodes in multilayer networks and algorithms that aim to extract the mesoscale organization of multilayer networks in different multilayer network communities.

This field is expected to have significant impact in a variety of contexts, including most notably network medicine and brain research. In brain research, the ability to make sense of the main structural characteristics of brain data is essential to advance our understanding of the interplay between the structure of the connectome, describing the macroscopic wiring of the brain, and functional brain networks, shedding light on brain dynamics. Network medicine and personalized medicine aim at finding the best treatment for a specific patient by integrating several medical datasets that usually take the form of multilayer networks. The advance in our ability to extract relevant information from these datasets is therefore of fundamental significance for the well-being of society.

For making sense of the large set of multilayer networks we need to combine inference algorithms and techniques with null models of multilayer networks (chapter 10). The null models define well-controlled network structures that constitute the reference point to which the results obtained by investigating real datasets can be compared. Additionally, null models can be taken as benchmark structures over which we can run simulations of dynamical processes. This can allow us to test the effect of multilayer network structures on the characteristic behaviour of the dynamics taking place on them.

1.3   Overview of dynamical processes on multilayer networks

In multilayer networks, links might indicate different types of interactions. This property of multilayer networks has essential consequences on the dynamical processes (chapters 11–16) defined on such structures. From percolation to diffusion and game theory, in general the dynamical interactions between nodes in a multilayer network will take a different functional form depending on the nature of the link. For example, if we consider a multilayer transportation network formed by the airport network of flight connections, the train and the road transportation networks, we will observe that the rules determining the diffusion within each of the networks might be different, and that changing from one means of transportation (diffusion from one layer to the other of the multilayer networks) might again follow other dynamical rules. This scenario does not only apply to transportation networks but also to the diffusion of ideas and behaviours in social networks. Therefore, the fact that diffusion on multilayer networks is characterized by different rates depending on the type of link significantly changes the properties of this dynamical process and has a variety of practical consequences. Similarly, the nodes of a multilayer network might respond differently to the damage of nodes in the same layer or in another layer. For example, take an Internet router. This router might still be functional even if one of the connected Internet routers is damaged, but it might not be functional anymore if the power plant providing energy to the router is damaged. Therefore, in a multilayer network we can distinguish between connectivity links providing connectivity to the nodes of each layer and interdependency links that imply the immediate damage of one node if the other linked node is damaged. This property, common to many interconnected infrastructures, makes them more fragile than single networks. Therefore, the implications of interdependencies on the robustness of multilayer networks is essential to build more reliable and resilient global infrastructures.

   In recent years it has been shown that considering the multilayer nature of networks can significantly modify the conclusions reached by considering single networks. A number of dynamical processes, including percolation, diffusion, epidemic spreading and game theory, present a phenomenology that is unexpected if one considers the layers in isolation. Moreover, it has been shown that the structural correlations built in multilayer network structures can significantly change the dynamical properties of the multilayer network. This spectacular interplay between structure and dynamics is very likely to open new scenarios for applications and control of multilayer networks, including the design of more resilient infrastructures and transportation systems and the possibility of reprogramming cancer cells.

Part II

Single Networks

2
The Structure of Single Networks

2.1 Networks

Networks are formed by a set of nodes describing the elements of a complex system connected pairwise by several links describing their complex web of interactions. Most networks reflect in their structure a rich interplay between randomness and order. For instance, in social networks the establishment of a friendship may depend on a series of contingent events, while in the brain the connections between neurons are not all determined by genomic information. If stochasticity is ubiquitous in complex networks, these networks are not maximally random either; rather, they obey organization principles that make them functional. Network Science characterizes network structures to increase our understanding of complex systems, as it is assumed that the underlying network structure of a complex system encodes information about its function. In this respect, the effort made in biology to gather reliable and complete information on biological interactions is noticeable. This work ranges from the high-throughput experiments that aim to complete the information about the human protein interaction network to the big projects that aim to map the human connectome. Network Science includes network inference and characterization of network structure, but also goes beyond topology and aims at identifying the effects that networks have on social, technological and biological processes and at predicting the behaviour of complex systems. In this chapter we will focus on the major results obtained by studying network structure, while in the subsequent chapter we will focus on network dynamics. Our intention is here to give some relevant background on single networks which might serve as a reference to the core of the book on multilayer networks. However, given the space limitations, we will not be able in any way to give a complete account of the large literature that exists on Network Science. We suggest to the novice wanting to deepen his understanding to read the relevant monographies on single networks [14, 107, 225, 105, 184]. Conversely, the very experienced reader familiar with most of the results valid for single networks can use the material of chapter 2 and chapter 3 only as a reference for the discussion of multilayer networks presented in Part III (chapters 4–16).

Multilayer Networks. Ginestra Bianconi, Oxford University Press (2018).
© Ginestra Bianconi. DOI: 10.1093/oso/9780198753919.001.0001

2.2   Single network types

A graph $G = (V, E)$ is formed by the pair of sets V and E where V is the set of nodes (or vertices) and E is the set of links (or edges). Networks are graphs that describe real interacting systems as diverse as the brain or the Internet. Single networks come in different types depending on several aspects characterizing their interactions.

Single networks can be classified as *undirected* or *directed* networks.

Undirected networks are formed by undirected interactions, and in these networks if node i is linked to node j then automatically node j is linked to node i. For instance, Facebook is an undirected network, as the Facebook friendship indicates an undirected interaction that has been agreed to by the two involved accounts. Similarly, in biology a protein interaction network is undirected, as any protein interaction indicates whether two proteins bind together to form a protein complex.

On the contrary, directed networks are networks in which the interactions are directed, and if node i points to node j it is not automatically true that node j points to node i. The World Wide Web is a clear example of a directed network where links from one webpage to another are not typically reciprocated. Within online social networks, Twitter is a clear example of a directed network where accounts do not always follow each other.

Single networks can also be classified as *unweighted* or *weighted*.

Weighted networks are networks where a *weight* is associated with each interaction, describing typically a measure of the 'intensity' of the interaction. For instance, in the airport network formed by flight connections between airports, a weight can be associated with the links according to the traffic (in terms of number of passengers) of each connection. In networks generated from correlations between time series such as brain functional networks or financial networks between assets, weights can be associated with links where stronger weights indicate large and positive correlations.

Unweighted networks, on the contrary, are networks in which each interaction is either present or absent. Unweighted networks might correspond to networks in which the weights are disregarded or networks in which the weights are the same for every interaction.

The most fundamental types of single networks are *simple networks* that are undirected and unweighted, in which interactions exist only between different nodes.

While the above classification of single networks depends on the properties of the network interactions, it is also possible to consider networks having nodes with different properties.

Bipartite networks are networks formed by two distinct types of nodes in which interactions exist exclusively between different types of nodes. Bipartite networks include the networks between actors and movies where each actor is connected to a movie if he has acted in it, or the network between scientists and papers where each scientist is connected to a paper if he has authored it.

2.3 Basic definitions

2.3.1 Nodes and links

The most basic properties of single networks $G = (V, E)$ are the total number of nodes N (also called the *network size*) and the total number of links L with

$$N = |V|,$$
$$L = |E|, \tag{2.1}$$

where the symbol $|X|$ indicates the cardinality of the set X. We will indicate the labelled nodes of the network with $i = 1, 2, \ldots, N$. Therefore, the set of nodes V is given by

$$V = \{i | i \in \{1, 2, \ldots, N\}\}. \tag{2.2}$$

The links will be indicated as pairs of node labels (i, j) where for undirected networks the order is irrelevant, while for directed networks the order indicates that node i points to node j. Note that for undirected networks each undirected link joining two given nodes of the network is counted once, while for directed networks a link from node i to node j is counted independently of the link which might eventually connect node j to node i.

Bipartite networks, where nodes can be cast into two different sets and interactions only exist between nodes belonging to different sets, should be treated somewhat differently. In fact, a bipartite network comprises three sets: $G_B = (V, U, E)$, where the sets V and U indicate two different groups of nodes (for instance, V and U might indicate actors and movies). These two sets might have different cardinality, $|V| = N_V$ and $|U| = N_U$, indicating in our example the total number of actors and the total number of movies respectively. The elements of the set V will be indicated by Latin letters i, j etc. The elements of the set U will be indicated by Greek letters μ, ν etc. Finally, the set E indicates the set of links connecting nodes of the set V only to nodes of the set U.

2.3.2 Adjacency matrix and incidence matrix

Any single network $G = (V, E)$ is fully determined by its adjacency matrix. The adjacency matrix is an $N \times N$ matrix $\mathbf{a}$, whose elements a_{ij} indicate whether node i is linked to node j. The specific definition of the adjacency matrix depends on whether the network is directed or undirected, weighted or unweighted.

For unweighted and undirected networks the adjacency matrix elements a_{ij} are given by

$$a_{ij} = \begin{cases} 1 & \text{if node } i \text{ is linked to node } j, \\ 0 & \text{otherwise.} \end{cases} \tag{2.3}$$

Therefore, it follows that for these networks the adjacency matrix is symmetric. In this case, as long as there are no tadpoles, i.e. links starting and ending on the same node, the total number of links L can be expressed in terms of the adjacency matrix as

$$L = \frac{1}{2} \sum_{i=1}^{N} \sum_{j=1}^{N} a_{ij}. \tag{2.4}$$

For unweighted and directed networks the matrix elements a_{ij} are given by

$$a_{ij} = \begin{cases} 1 & \text{if node } i \text{ points to node } j, \\ 0 & \text{otherwise.} \end{cases} \tag{2.5}$$

Therefore, in this case the adjacency matrix is asymmetric. As long as there are no tadpoles, the total number of links L can be expressed in terms of the adjacency matrix as

$$L = \sum_{i=1}^{N} \sum_{j=1}^{N} a_{ij}. \tag{2.6}$$

For weighted networks the only difference is that the adjacency matrix elements do not just take values equal to either zero or one, but also integers or real values equal to the weights attributed to the corresponding links. Therefore, for weighted and undirected networks the adjacency matrix element a_{ij} is given by

$$a_{ij} = \begin{cases} w_{ij} & \text{if node } i \text{ is linked to node } j \text{ with weight } w_{ij} > 0, \\ 0 & \text{otherwise.} \end{cases} \tag{2.7}$$

In this case, as long as there are no tadpoles the total number of links can be expressed as

$$L = \frac{1}{2} \sum_{i=1}^{N} \sum_{j=1}^{N} \theta(a_{ij}), \tag{2.8}$$

where $\theta(x)$ is the Heaviside function with $\theta(x) = 1$ if $x > 0$ and $\theta(x) = 0$ if $x \leq 0$. Instead, in the case of weighted and directed networks the adjacency matrix has elements a_{ij} given by

$$a_{ij} = \begin{cases} w_{ij} & \text{if node } i \text{ points to node } j \text{ with weight } w_{ij} > 0, \\ 0 & \text{otherwise.} \end{cases} \tag{2.9}$$

In this case, as long as there are no tadpoles the total number of links can be expressed as

$$L = \sum_{i=1}^{N} \sum_{j=1}^{N} \theta(a_{ij}), \tag{2.10}$$

where $\theta(x)$ is the Heaviside function.

A bipartite network is usually described by incidence matrices that determine all the interactions in the network. For undirected (weighted) bipartite networks a unique $N_V \times N_U$ incidence matrix $\mathbf{b}$ can be used to represent the system. This matrix has elements

$$
b_{i\mu} = \begin{cases} w_{i\mu} & \text{if node } i \in V \text{ is connected to node } \mu \in U \text{ with weight } w_{i\mu} > 0, \\ 0 & \text{otherwise,} \end{cases}
$$

(2.11)

where in the unweighted case all the non-zero weights are equal to one, i.e. $w_{i\mu} = 1$.

In the directed (weighted) case the bipartite network is described by two incidence matrices, the $N_V \times N_U$ matrix $\mathbf{b}$ and the $N_U \times N_V$ matrix $\tilde{\mathbf{b}}$. These matrices have elements

$$
b_{i\mu} = \begin{cases} w_{i\mu} & \text{if node } i \in V \text{ points to node } \mu \in U \text{ with weight } w_{i\mu} > 0, \\ 0 & \text{otherwise,} \end{cases}
$$

$$
\tilde{b}_{\mu i} = \begin{cases} w_{\mu i} & \text{if node } \mu \in U \text{ points to node } i \in V \text{ with weight } w_{\mu i} > 0, \\ 0 & \text{otherwise.} \end{cases}
$$

(2.12)

Here we have again used the convention that for unweighted networks all the non-zero weights are equal to one.

2.3.3   Degree, degree sequence and degree distribution

Degree

In an undirected network $G = (V, E)$ the *degree* k_i of a node i is the number of links incident to it. The degree k_i of a node i can be written in terms of the adjacency matrix. For an undirected and unweighted network we have

$$
k_i = \sum_{j=1}^{N} a_{ij},
$$

(2.13)

while for undirected and weighted networks we have

$$
k_i = \sum_{j=1}^{N} \theta\left(a_{ij}\right).
$$

(2.14)

In a directed network $G = (V, E)$, the *in-degree* $k_{i,in}$ of node i and the *out-degree* $k_{i,out}$ of node i are given respectively by the number of links pointing to node i and the number of nodes to which node i points. The in/out-degree of node i can be expressed in terms of the adjacency matrix as

$$
k_{i,in} = \sum_{j=1}^{N} a_{ji},
$$

$$
k_{i,out} = \sum_{j=1}^{N} a_{ij}
$$

(2.15)

for unweighted networks and as

$$k_{i,in} = \sum_{j=1}^{N} \theta\left(a_{ji}\right),$$

$$k_{i,out} = \sum_{j=1}^{N} \theta\left(a_{ij}\right) \qquad (2.16)$$

for weighted networks.

In an undirected (weighted) bipartite network $G = (V, U, E)$ the degrees k_i of nodes $i \in V$ and the degrees k_μ of nodes $\mu \in U$ can be expressed in terms of the incidence matrix as

$$k_i = \sum_{\mu=1}^{N_U} \theta\left(b_{i\mu}\right),$$

$$k_\mu = \sum_{i=1}^{N_V} \theta\left(b_{i\mu}\right). \qquad (2.17)$$

In the case of a directed (weighted) bipartite network the in-degrees and the out-degrees of nodes $i \in V$ and $\mu \in U$ are given by

$$k_{i,in} = \sum_{\mu=1}^{N_U} \theta\left(\tilde{b}_{\mu i}\right),$$

$$k_{i,out} = \sum_{\mu=1}^{N_U} \theta\left(b_{i\mu}\right),$$

$$k_{\mu,in} = \sum_{i=1}^{N_V} \theta\left(b_{i\mu}\right),$$

$$k_{\mu,out} = \sum_{i=1}^{N_V} \theta\left(\tilde{b}_{\mu i}\right). \qquad (2.18)$$

Degree Sequence and Average Degree

In an undirected network the degree sequence is the ordered sequence of the degrees of each node of the network, i.e.

$$\{k_1, k_2, \ldots, k_N\}. \qquad (2.19)$$

The average degree $\langle k \rangle$ is given by

$$\langle k \rangle = \frac{1}{N} \sum_{i=1}^{N} k_i = \frac{2L}{N}. \tag{2.20}$$

In a directed network, we distinguish between the in- and out-degree sequences given by the ordered sequences of in and out-degrees, i.e.

$$\{k_{1,in}, k_{2,in}, \dots k_{N,in}\},$$
$$\{k_{1,out}, k_{2,out}, \dots k_{N,out}\}. \tag{2.21}$$

The average in-degree and the average out-degree are given by

$$\langle k_{in} \rangle = \frac{1}{N} \sum_{i=1}^{N} k_{i,in},$$

$$\langle k_{out} \rangle = \frac{1}{N} \sum_{i=1}^{N} k_{i,out}. \tag{2.22}$$

Using the definition of the in- and out-degree, it is apparent that the average in- and out-degree are equal, i.e.

$$\langle k_{in} \rangle = \langle k_{out} \rangle = \frac{L}{N}. \tag{2.23}$$

The extension of the above definitions to the bipartite network is straightforward. The only notable thing to mention is the relation between the average degrees of the nodes $i \in V$ and $\mu \in U$. In fact, for an undirected bipartite network we have

$$\langle k_i \rangle_{i \in V} N_V = \langle k_\mu \rangle_{\mu \in U} N_U. \tag{2.24}$$

For a directed bipartite network we have

$$\langle k_{i,in} \rangle_{i \in V} N_V = \langle k_{\mu,out} \rangle_{\mu \in U} N_U,$$
$$\langle k_{i,out} \rangle_{i \in V} N_V = \langle k_{\mu,in} \rangle_{\mu \in U} N_U. \tag{2.25}$$

Degree distribution

The degree distribution $P(k)$ of an undirected network determines the fraction of nodes of degree k, or equivalently the probability that a random node has degree k. Indicating with $N(k)$ the number of nodes of degree k the degree distribution is given by

$$P(k) = \frac{1}{N}N(k) = \frac{1}{N}\sum_{i=1}^{N}\delta(k, k_i), \tag{2.26}$$

where $\delta(x, y)$ is the Kronecker delta, i.e. $\delta(x, y) = 1$ if $x = y$ and otherwise $\delta(x, y) = 0$. For directed networks we distinguish between the in-degree distribution $P^{in}(k)$ and the out-degree distribution $P^{out}(k)$, indicating the probability that a random node has in-degree $k_{i,in} = k$ and out-degree $k_{i,out} = k$ respectively. Let us indicate with $N^{in/out}(k)$ the total number of nodes of the network with in/out-degree k. Then the in/out-degree distribution $P^{in/out}(k)$ of a directed network is given by

$$P^{in}(k) = \frac{1}{N}N^{in}(k) = \frac{1}{N}\sum_{i=1}^{N}\delta(k, k_{i,in}),$$

$$P^{out}(k) = \frac{1}{N}N^{out}(k) = \frac{1}{N}\sum_{i=1}^{N}\delta(k, k_{i,out}). \tag{2.27}$$

The extension of these definitions to bipartite networks is straightforward.

2.3.4 Clustering coefficient

The clustering coefficient [310] is a fundamental measure to evaluate the local organization of a network.

Specifically, the local clustering coefficient C_i of node i of degree k_i measures the density of triangles passing through node i or equivalently the probability that two neighbours of node i are connected. It is defined as

$$C_i = \begin{cases} \dfrac{\text{\# of triangles passing though node } i}{k_i(k_i - 1)/2} & \text{for} \quad k_i > 1, \\ 0 & \text{for} \quad k_i = 0, 1, \end{cases} \tag{2.28}$$

where $\frac{1}{2}k_i(k_i - 1)$ enumerates the number of pairs of distinct nodes which are neighbours of node i. From this definition it is apparent that the local clustering coefficient is a number between zero and one. If the clustering coefficient is one, i.e. $C_i = 1$, each pair of neighbours of node i are connected to each other, whereas if the clustering coefficient is zero, i.e. $C_i = 0$, there are no pairs of neighbours of node i that are linked.

The *global clustering coefficient* C of a network of size N is the average of the local clustering over all the nodes of the network and is given by

$$C = \frac{1}{N}\sum_{i=1}^{N}C_i. \tag{2.29}$$

This quantity evaluates the 'locality' of the interactions in a network.

Sometimes in the literature an alternative global clustering coefficient is introduced, called the 'transitivity'. The *transitivity* T of a network of size N is given by

$$T = 3 \frac{\text{\# triangles in the network}}{\text{\# distinct paths of length 2}}. \tag{2.30}$$

Compared to the global clustering coefficient, the transitivity has the advantage that in very sparse networks the transitivity is not dominated by the contributions coming from nodes of degree smaller than two which have a local clustering coefficient equal to zero.

Both the global clustering coefficient C and the transitivity T assume values between zero and one, where values close to one indicate a network with large density of triangles while values close to zero indicate network structures with a very low density of triangles that locally are similar to trees.

2.3.5   Shortest distance and diameter

Navigability is central in Network Science. This term refers to the possibility of exploring large parts of the network by following its paths. A path $\mathcal{P}$ of a network $G = (V, E)$ is a sequence of nodes $\mathcal{P} = (i_0, i_1, \ldots i_n)$ where each node is connected by a link to the subsequent node in the list, i.e. $(i_p, i_{p+1}) \in E$ for $p = 0, 1, \ldots, n - 1$. If the network is directed the path $\mathcal{P}$ is called a directed path, as links are followed along their direction. The length of a given path $\mathcal{P}$ from a node $i_0 = i$ to a node $i_n = j$ is determined by its number of links, i.e. by n.

By adopting this metric, the shortest distance d_{ij} between a node i and a node j is given by the shortest length of any path going from i to j. If node j cannot be reached from node i by following any path, we put $d_{ij} = \infty$.

Two important global measures can be extracted from the matrix of distances d_{ij}: the average shortest distance and the diameter of the network.

The average shortest distance $\bar{\ell}$ of a connected network is the average of the shortest distances between any two distinct nodes of the network. Therefore, in a connected network we have

$$\bar{\ell} = \frac{1}{N(N-1)} \sum_{i=1}^{N} \sum_{\substack{j=1 \\ j \neq i}}^{N} d_{ij}. \tag{2.31}$$

The *diameter* D of a connected network is the maximum of the shortest distances between any two nodes of the network. Therefore, we have

$$D = \max_{i,j \neq i} d_{ij}. \tag{2.32}$$

Naturally from these definitions it follows that the average shortest distance cannot be larger than the diameter

$$\bar{\ell} \leq D. \tag{2.33}$$

2.3.6   Connected components and giant component

Undirected networks are called connected if from every node we can reach any other node by following suitable paths; otherwise they are called disconnected.

Disconnected networks are formed by several connected components, i.e. connected subgraphs of the network formed by a set of nodes which are disconnected from any other node of the network.

Typically in real complex networks we desire to have connected networks because connectedness allows for the possibility of diffusing information and navigating the network. Additionally, in many cases connectivity ensures the proper function of the network, as in infrastructures. However, often the requirement that the network is connected can be very stringent, as networks might be disconnected as a result of their intrinsic nature or as an effect of the local damage of nodes or links. Among disconnected networks it is therefore useful to distinguish between networks in which the largest connected component connects a finite fraction of the nodes, and networks in which every connected component includes only a small fraction of all the nodes of the network. In the first scenario, we might still benefit from diffusing information and navigating on the largest connected component. Moreover, we can still assume that the nodes in the largest connected component are able to perform the network function. In the second scenario, the network is instead decomposed into many small components and effectively dismantled.

These considerations reveal the interest in defining the giant component of undirected networks. The *giant component* of an undirected network is the largest connected component of the network as long as it includes a finite fraction of the nodes of the network. The giant component plays a central role in percolation theory where the robustness of the network to random damage of its nodes or links is monitored by evaluating the size of the giant component that results after the inflicted damage.

The study of connectedness in directed networks is somewhat more involved. In this case the analysis can be performed either by neglecting the direction of the links or taking into account the direction of the links. The connected components of the undirected network obtained from a directed network by neglecting the direction of the links are called *weakly connected components* and the corresponding giant component is called the *giant weakly connected component*. If, instead, the direction of the links is taken into account, a much richer structure is revealed. For instance, the same giant weakly connected component can be decomposed into a giant strongly connected component, a giant in-component, a giant out-component, tendrils and tubes. The *giant strongly connected component* is formed by the set of nodes such that from any node we can reach any other node in the set by following directed paths; the *giant in-component* is formed by nodes that do not belong to the giant strongly connected component but from which it is possible to reach nodes in the giant strongly connected component by following directed paths; the *giant out-component* instead is formed by nodes that do not belong to the giant strongly connected component but that are reachable from the nodes in the giant strongly connected componentby following directed paths; finally, the tendrils and the tubes are formed by nodes that belong to the giant weakly connected component but that cannot be reached from and cannot reach nodes of the giant strongly connected component by following directed paths.

2.4   Universalities

2.4.1   Complex networks

Despite their large variety of functions, the network topology of very disparate complex systems is characterized by universal properties. These universal properties have been characterized in two ground-breaking papers from the late nineties. In their seminal paper [310] Watts and Strogatz showed that networks ranging from the neuronal network of the worm *C. elegans* to the network of collaborations between movie actors are *small world*. In their pivotal work [15] Barabási and Albert showed that the vast majority of networks are *scale-free*. Scale-free networks have a very heterogenous distribution of degrees and their dynamical behaviour is dominated by the hub nodes having degree order of magnitude larger than the average.

2.4.2   Small-world networks

Small-world networks are networks with large clustering coefficient and small-world distance property.

A network is said to have a large clustering coefficient when the average clustering is finite and independent of the network size. Real networks that have a fixed number of nodes N are said to have a large clustering coefficient when their clustering coefficient is much larger than the one observed in a network with the same number of nodes N and the same number of links L where the links have been distributed randomly between the nodes. Therefore, a large clustering coefficient implies that the connections of the network are much more clustered and local than in the random network.

A network has the small-world distance property if its diameter D is of the same order or of a lower order than the logarithm of the number of nodes N, i.e.

$$D \simeq \mathcal{O}(\ln N) \tag{2.34}$$

or

$$D \simeq o(\ln N). \tag{2.35}$$

If a network has the small-world distance property it follows also that the average shortest distance between the nodes is of the same order or of a lower order than the logarithm of the total number of nodes. This property reveals that nodes in a small-world network are in close proximity to each other, while it is known that nodes in a regular square lattice of dimension d have a much larger typical distance: $\bar{\ell} \simeq \mathcal{O}(N^d)$.

The small-world distance property implies that any two nodes of the network are connected by the shortest paths including only relatively few links. For instance, in social networks this property is well known as the 'six degrees of separation', i.e. the phenomenon that any two people in the world are only a few shaken hands apart from each other. Note that you should not attribute any fundamental role to the number six of the 'six degrees of separation', but rather focus on the fact that this number is of the same order of magnitude as the logarithm of the total number of nodes in the social network. In fact, six was the average shortest social distance found originally in 1967 by Milligram's experiment devised to quantify this phenomenon [294]. Subsequent experiments have found different values for the typical social distance between people. However, they have confirmed the small-world distance properties of social networks.

While the small-world distance property has been first observed in the context of social networks, in Ref. [310] Watts and Strogatz showed that small-world networks are not exclusively found in social systems, but are a universal property of complex networks as they also occur in biological and technological systems.

2.4.3 Scale-free networks

Scale-free networks are the exemplar network topology of complex systems, and for Network Science they have a foundational status similar to the one of d-dimensional lattices for Euclidean geometry, crystallography and solid-state physics. In Ref. [15], Barabási and Albert showed that *scale-free* networks are ubiquitous as they describe the underlying topology of the majority of social, technological and biological systems. *Scale-free* networks are those networks whose degree distribution $P(k)$ follows a power-law tail, i.e. the degree distribution $P(k)$ follows for $k \gg 1$

$$P(k) \simeq Ck^{-\gamma}, \tag{2.36}$$

where C is a normalization constant and the power-law exponent γ is in the range $\gamma \in (2, 3]$. This property implies that scale-free networks are characterized by an extreme heterogeneity in the degrees of the nodes and by the absence of a typical scale of their degree, hence the term 'scale-free'. In order to understand the properties of scale-free networks, it is opportune to consider the stylized class of power-law networks whose degree distribution is exactly given by

$$P(k) = Ck^{-\gamma}, \tag{2.37}$$

where C is the normalization constant and the power-law exponent γ is greater than one, i.e. $\gamma > 1$. The condition $\gamma > 1$ is requested to ensure the normalization condition

$$\sum_k P(k) = 1. \tag{2.38}$$

Power-law networks are characterized by an average degree $\langle k \rangle$ and second moment $\langle k^2 \rangle$ defined as

$$\langle k \rangle = \sum_k k P(k),$$

$$\langle k^2 \rangle = \sum_k k^2 P(k). \tag{2.39}$$

Depending on the value of the power-law exponent γ, different regimes can be observed:

(a) For $\gamma \in (3, \infty)$ both the average degree $\langle k \rangle$ and the second moment $\langle k^2 \rangle$ are finite in the limit of infinite network size $N \to \infty$.

(b) For $\gamma \in (2, 3]$ the average degree $\langle k \rangle$ is finite but the second moment $\langle k^2 \rangle$ diverges in the limit of infinite network size $N \to \infty$.

(c) For $\gamma \in (1, 2]$ both the average degree $\langle k \rangle$ and the second moment $\langle k^2 \rangle$ diverge in the limit of infinite network size $N \to \infty$.

Most real networks have a finite average degree $\langle k \rangle$, implying that the average degree does not change if we consider a sample of the network with different numbers of nodes N. In this case, we say that the network is sparse. As a consequence, people often consider exclusively power-law exponents $\gamma > 2$. Nevertheless, even if we restrict our interest to power-law exponents $\gamma > 2$ the properties of power-law networks with exponent $\gamma \in (2, 3]$ strongly differ from the properties of power-law networks with exponent $\gamma > 3$.

Scale free networks are networks with a degree distribution $P(k)$ that for $k \gg 1$ can be approximated by a power-law distribution with power-law exponent $\gamma \in (2,3]$. Power-law networks with $\gamma \in (2, 3]$ and more in general scale-free networks are characterized by having a finite average degree $\langle k \rangle$ and a diverging second moment $\langle k^2 \rangle$. This implies that the average degree exists but is not typical, since in the networks there are large fluctuations of the degrees of the nodes. For instance, the standard deviation $\sigma(k)$ of the degree distribution, given by

$$\sigma(k) = \sqrt{\langle k^2 \rangle - \langle k \rangle^2}, \tag{2.40}$$

diverges with the network size, i.e. $\sigma(k) \to \infty$ as $N \to \infty$, indicating that there is no typical scale of their degree. As a consequence of this major characteristic, scale-free networks have nodes with very heterogeneous degrees and are dominated by some nodes with very high degree, i.e. a degree order of magnitude larger than the average, called *hub nodes*.

The large heterogeneities in the degrees of scale-free networks and the hubs present in these structures have very important consequences for the robustness properties of the networks and the interplay between network topology and dynamics (see next chapter for a discussion of the consequences of the scale-free property for dynamical processes).

Given the fundamental role of hubs, scale-free networks are usually not distinguished from pure power-law networks with exponent $\gamma \in (2, 3]$. An important characteristic of power-law networks is its highest degree K_N. When there are no other constraints limiting the maximum degree of the network, it is possible to show that the highest-degree K_N in a power-law network with power-law exponent $\gamma > 2$ scales as

$$K_N \sim N^{1/(\gamma-1)}. \tag{2.41}$$

Therefore, the maximum degree in power-law networks grows rapidly with N. Notice that for $\gamma \to 2$ we have $K_N \to N$, indicating that the maximum degree reaches the maximum possible value since the degree of a node cannot be larger than the number of nodes N of the network.

Interestingly, random power-law networks with exponent $\gamma > 2$ also have the small-world-distance property, as they have an average shortest distance $\bar{\ell}$ given by [80]

$$\bar{\ell} \sim \begin{cases} \ln N & \text{for} \quad \gamma \in (3, \infty,), \\ \frac{\ln N}{\ln \ln N} & \text{for} \quad \gamma = 3, \\ \ln \ln N & \text{for} \quad \gamma \in (2, 3]. \end{cases} \tag{2.42}$$

2.5 Structural correlations

2.5.1 General remarks

In Network Science, a complex system is characterized by the network of its interactions. The main underlying idea of the field is that the structure of the network encodes information and can reveal important properties of the dynamics and function of the complex system.

In this framework, looking for structural correlations in the network structure, i.e. structural properties that are not expected in a random hypothesis, is rather natural. In single networks, structural correlations include most notably the degree correlations and the correlation between weights of the links and network topology.

2.5.2 Degree correlations

In a large variety of real networks we observe that nodes are not connected randomly, but instead there is a significant bias for nodes of high degree to connect either to nodes of high degree (assortative networks) or to nodes of low degree (disassortative networks). In these cases we say that the network displays degree correlations. To establish whether the network is assortative, disassortative or without degree correlations it is necessary to have a valid model of uncorrelated network to refer to. An uncorrelated network is a network in which every link reaches a node of degree k with probability q_k proportional to k and independently of the degree of the node at the other end of the link, i.e.

$$q_k = \frac{k}{\langle k \rangle N}.$$ (2.43)

Under this hypothesis the probability p_{ij} that a node i connects to a node j is given by

$$p_{ij} = \frac{k_i k_j}{\langle k \rangle N}.$$ (2.44)

Given a real dataset, the degree correlations can be measured [242] by considering the average degree knn_i of the neighbours of node i

$$knn_i = \frac{1}{k_i} \sum_{j=1}^{N} a_{ij} k_j.$$ (2.45)

This quantity can be averaged over nodes of degree k, obtaining

$$knn(k) = \frac{1}{NP(k)} \sum_{i=1}^{N} knn_i \delta(k_i, k).$$ (2.46)

If we follow a random link of an uncorrelated network we reach a node of degree k with probability q_k given by Eq. (2.43) independently of the degree of the node at the other end of the link. Therefore, for uncorrelated networks it can be shown that $knn(k)$ does not depend on k and it is given by

$$knn(k) = \frac{\langle k^2 \rangle}{\langle k \rangle}.$$ (2.47)

If the function $knn(k)$, calculated for a real dataset, results in an increase in k, the network is assortative, having high-degree nodes connected preferentially to high-degree nodes, while if the function $knn(k)$ is decreasing the network is disassortative.

However, there are real networks where the function $knn(k)$ is not monotonic. In those cases it is opportune to consider a more coarse-grained measure of correlation, such as the Pearson assortativity coefficient.

The Pearson assortativity coefficient r [222] is a number with values between -1 and 1 defined as

$$r = \sum_{k,k'} \frac{kk' \left(e_{k,k'} - q_k q_{k'} \right)}{\sigma^2},$$ (2.48)

where $e_{k,k'}$ indicates the number of links connecting nodes of degrees k and k' respectively and where

$$\sigma^2 = \sum_k k^2 q_k - \left[\sum_k k q_k\right]^2.$$
(2.49)

Negative values of the Pearson assortativity coefficient r indicate that the network is overall disassortative, while positive values of r indicate that the network is overall assortative.

2.5.3   Strength and inverse participation ratio

In real weighted networks weights are not randomly assigned to links. For instance, the airport network formed by flight connections between different airports, with weights representing the passenger traffic, is highly correlated with the network topology. In fact, airports with many flight connections typically also tend to sustain high traffic on their connections. Let us define the strength s_i of a node i as the sum of the weights of its links, i.e.

$$s_i = \sum_{j=1}^{N} w_{ij}.$$
(2.50)

In order to capture the weight-topology correlations it is possible to correlate the strength s_i with the degree k_i of the same node i [23]. A way of assessing the relevance of these correlations consists in considering the average strength $s(k)$ of nodes of degree k

$$s(k) = \frac{1}{NP(k)} \sum_{i=1}^{N} s_i \delta(k_i, k)$$
(2.51)

as a function of the degree k. If the weights are distributed homogeneously in the network, we expect that the average strength $s(k)$ increases linearly with k, i.e. $s(k) \propto k$. If, instead, the weights are not distributed homogeneously in the network and high-degree nodes typically have links of higher weight, the average strength $s(k)$ increases non-linearly as a function of k, i.e $s(k) \propto k^\theta$ with $\theta > 1$.

Weighted networks can also display another type of weight–topology correlation. This is revealed by considering the heterogeneity between the weights of the links incident to a given node i. This heterogeneity is measured by the inverse participation ratio [4] given by

$$Y_i = \sum_{j=1}^{N} \left(\frac{w_{ij}}{s_i}\right)^2.$$
(2.52)

The inverse participation ratio is one ($Y_i = 1$) if there is a single link with a weight significantly larger than the other links, while if all the links have the same weight we have $Y_i = 1/k_i$. For this reason the inverse of Y_i is also used as a measure of the typical number of links with relevant weight. The inverse participation ratio can be averaged over nodes of degree k, yielding

$$Y(k) = \frac{1}{NP(k)} \sum_{i=1}^{N} Y_i \delta(k_i, k). \qquad (2.53)$$

The function $Y(k) = 1/k$ indicates that the weights of the links ending at each node of the network are the same, while $Y(k) = 1/k^{\tilde{\theta}}$ with $\tilde{\theta} \in [0, 1)$ indicates that the weights incident to any given node of the network are not the same.

2.6 Communities

2.6.1 Communities in complex networks

Most complex networks have communities. A community is formed by a set of nodes that are more densely connected to each other than to the rest of the network. In biological networks, including both molecular networks in the cell and brain networks, communities are generally considered to be related to the function of the network. In social networks they define the mesoscale organization of the network and often reveal the homophily patterns of the system under study.

Determining the community structure of a network from its adjacency matrix is a very challenging task. On the one side, network scientists are faced with the fact that although the concept of community is rather intuitive, up to now there has been no formal definition of community on which there is general consensus. On the other side, already-related computational problems such as the one of finding the best bipartition of a network are computationally very demanding.

Despite these very challenging problems, there is a rich literature on community detection algorithms (see, for instance, the rich review articles [121, 85]). A number of algorithms have been proposed that, although not free from limitations, can be very useful in practice to describe the mesoscale structure of the networks.

2.6.2 Hierarchical clustering

A series of works recast the problem of community detection in a complex network to a hierarchical clustering problem. Hierarchical clustering indicates the statistical and data-mining technique that determines the hierarchy of clusters that can be established to describe mesoscopically a set of points related to each other by a similarity measure (or equivalently a distance). In agglomerative clustering algorithms initially each point belongs to a different cluster. At each agglomerative step the two most similar clusters are merged and the similarity of the new cluster with all the other clusters is evaluated according to a predefined rule. For instance, in single linkage the similarity between cluster A and cluster B is given by the maximum similarity between any element of A and any element of B; in average linkage clustering, instead, the similarity between cluster A and cluster B is given by the mean of the similarity between every element of A and every element of B. The algorithm stops when there is just one cluster. The hierarchy of clusters is visualized with a dendrogram.

In order to extend this technique to complex networks we will assume that the nodes of the network play the role of the points in the hierarchical clustering. Additionally, we will need to define a measure of similarity between the nodes of the network. For instance, in Ref. [260] Ravasz et al. proposed to take the normalized number of common neighbours between two nodes

$$x_{ij} = \frac{\mathcal{J}(i,j)}{\min(k_i, k_j) + 1 - \theta(a_{ij})},$$
(2.54)

where $\mathcal{J}(i,j)$ is the number of common neighbours between node i and j plus one if there is a directed link between them and $\theta(x)$ is the Heaviside step function. Given this similarity measure between the nodes, the hierarchy between the clusters can be directly found by applying the hierarchical clustering technique.

2.6.3   Modularity

A community assignment $\{g_i\}$ indicates for each node i to which community $g_i \in \{1, 2, \ldots, P\}$ it belongs. Given a real network and a candidate community structure of the network, a very pressing question is to determine whether the detected communities are significant. The modularity defines a way of comparing the density of the links within each community to the expectation in the hypothesis that the link between a generic node i and a generic node j occurs with probability p_{ij}. Typically, the null hypothesis is that the network is random and uncorrelated while preserving the same degree sequence as the real dataset. Under this hypothesis, p_{ij} is given by Eq. (2.44). The modularity [226, 224] is defined as

$$\mathcal{Q} = \frac{1}{\langle k \rangle N} \sum_{ij} (a_{ij} - p_{ij}) \delta_{g_i, g_j}$$
(2.55)

and takes values smaller or equal to one, with a large value of $\mathcal{Q}$ indicating very significant communities.

The modularity can be used as a measure to evaluate the significance of a given community structure, but it can also be used for formulating community detection algorithms that try to find the community assignment that maximizes the modularity.

However, using modularity optimization for finding the community structure of the network has some limitations. For instance, there might be a large number of different community assignments which achieve large and comparable values of the modularity.

In these cases there is no well-defined maximum of the modularity function, and greedy algorithms can be trapped in local maxima.

2.6.4   Fast community detection algorithms

The possible use of community detection algorithms is typically limited by their computational complexity. Currently two greedy algorithms are frequently used to determine the communities of large complex networks: the Louvain algorithm [53] and the Infomap algorithm [267].

The Louvain algorithm [53] defines an efficient way of performing a greedy optimization of the modularity. Its algorithmic complexity scales linearly with the number of links L, i.e. it is $\mathcal{O}(L)$. Therefore, the algorithm can be applied to networks up to millions of nodes.

The Infomap algorithm [267] exploits the properties of the random walk on a network with community structure. In fact, the random walk tends to have a transient dynamics that is trapped within the communities of the network. By encoding the random walk dynamics in a bit of string, information theory tools are used to find a way to efficiently codify the string by assigning a codeword to each community. The computational complexity of the algorithm is $\mathcal{O}(L \ln L)$ so it can be used to detect communities in very large, sparse networks.

2.6.5   Overlapping communities

Until now, we have assumed that each node belongs to a single community. However, in a variety of networks it happens that a node can belong to several communities. For instance, a scientist might collaborate in different scientific fields and this can be reflected in his network of collaborators being effectively formed by several scientific communities. In this scenario we say that the communities of the network overlap.

Two main approaches have been proposed to describe these types of networks: k-clique community detection using the CFinder algorithm [237] and the clusterization of link communities [119, 3]. The CFinder algorithm [237] is based on the assumption that overlapping communities are formed by several adjacent k-cliques with $k > 2$. A k-clique is a graph formed by k nodes each connected to all the others. Two k-cliques are adjacent if they share $k - 1$ nodes. Starting from a given k-clique of the network, CFinder finds the set of all the k-cliques that can be found by hopping from a k-clique to an adjacent one and identify this set with a k-community. Different k-communities might overlap on nodes and/or links. For instance, in the word association network where words are connected to each other if they have a related meaning, the word 'bright' belongs to several communities associated with the concepts of intelligence, light, colours or astronomical terms.

Link community algorithms [119, 3] clusterize links instead of nodes. For instance, they assume that in a social network a link connecting a given individual to a work colleague might belong to a different cluster from a link connecting him to an old classmate. Two different types of link community algorithms have been proposed so far.

The first one [119] is based on the optimization of the link modularity that generalizes the modularity function. The second [3] is a hierarchical clustering algorithm that clusterizes links instead of nodes starting from a similarity matrix between the links.

2.7 Centralities

2.7.1 General remarks

In a variety of cases from social science and information technology to biology networks, scientists aim at ranking the nodes in order of their 'importance'. The proposed algorithms assign to each node a centrality measure according to which the nodes are sorted in descending order of their importance. The most successful centrality measure is undoubtedly PageRank, the algorithm that was originally constituting the Google search engine and has proven to be extremely efficient for searching online information. However, the most appropriate measure of centrality depends on the given network under consideration and on the network properties one wants to emphasize. For instance, the eigenvector, the Katz and the PageRank centrality all assume that a node is more central if central nodes already point to it, while the closeness centrality and the efficiency attribute more importance to nodes at short distance from the other nodes in the network. Finally, the betweenness centrality assumes that a node is more central if many shortest paths pass through the node, making the node essential to keeping the network together.

2.7.2 Eigenvector centrality

The eigenvector centrality, also called the Bonacich centrality [57], is a measure of centrality traditionally used in social science. It assumes that the centrality of a node is higher if already central nodes are connected to it. Therefore, it attributes to each node i a centrality measure x_i satisfying the eigenvalue problem

$$\lambda_1 x_i = \sum_{j=1}^{N} a_{ji} x_j, \qquad (2.56)$$

where λ_1 is the largest eigenvalue of the adjacency matrix $\mathbf{a}$. This measure is suitable for undirected networks, but for directed networks it has the shortcoming that all the nodes in the in-component have zero eigenvector centrality, independently of their role in the network.

2.7.3 Katz centrality

The Katz centrality [170] builds on the eigenvector centrality, as it also assumes that nodes increase their centrality if they are connected to central nodes. However, by assigning a minimum centrality to each node the Katz centrality overcomes the problem that the eigenvector centrality has on directed networks. Specifically, on directed networks the Katz centrality is non-zero also for nodes in the in-component. The *Katz centrality* x satisfies the equation

$$x_i = \mu \sum_{j=1}^{N} a_{ji} x_j + \omega, \tag{2.57}$$

where $\omega > 0$ and $\mu \in (0, 1/\lambda_1)$ where λ_1 is the largest eigenvalue of the adjacency matrix a. By explicitly solving Eq. (2.57), the Katz centrality vector x is given by

$$\mathbf{x} = \omega \left(\mathbf{I} - \mu \mathbf{a}^T\right)^{-1} \mathbf{1}, \tag{2.58}$$

where I indicates the $N \times N$ identity matrix, $\mathbf{a}^T$ indicates the transpose of the adjacency matrix and 1 indicates the N-dimensional column vector of elements $1_i = 1$. Equivalently, the Katz centrality x_i of node i can be written as

$$x_i = \omega \sum_{j=1}^{N} \left(\mathbf{I} - \mu \mathbf{a}^T\right)^{-1}_{ij}. \tag{2.59}$$

2.7.4 PageRank centrality

While in a number of situations it is reasonable to assume that if a very central node points to a given node this node acquires high centrality, sometime this effect is somewhat buffered by the fact that the important node might have many links. For instance, if a very important webpage contains a very large number of URL links, the prestige for the pointed webpages is not the same as if the webpage was containing only a few selected URL links. The PageRank centrality [60] builds on the Katz centrality by taking into account this buffering mechanism. Therefore, the *PageRank centrality* x_i of node i satisfies

$$x_i = \mu \sum_{j=1}^{N} a_{ji} \frac{1}{\kappa_j^{out}} x_j + \omega, \tag{2.60}$$

where $\kappa_j^{out} = \max(k_{j,out}, 1)$, $\mu \in (0, 1)$ and ω is given by

$$\omega = \frac{1}{N} \sum_{j=1}^{N} \left[(1 - \mu) + \mu \delta(k_{j,out}, 0)\right] x_j, \tag{2.61}$$

with $\delta(x, y)$ indicating the Kronecker delta. The PageRank centrality x_i of a node can also be interpreted as the probability that asymptotically in time a random walker is at node i. If located on a node with non-zero out-degree, this random walker hops to a random neighbour node with probability μ and jumps to a random node of the network (teleportation) with probability $1 - \mu$. If, instead, the random walker is on a node with zero out-degree, it jumps with probability one to a random node of the network. The coefficient μ is also called the damping parameter, while ω is usually called the teleportation coefficient. The typical value used for μ that constitutes a trade-off between local exploration and teleportation is $\mu = 0.85$.

2.7.5   Closeness and efficiency

The closeness and the efficiency are two centrality measures that attribute more relevance to nodes that are at short distance to the other nodes of the network.

The *Closeness Centrality* [310] Cl_i of a node i in an undirected network is given by

$$Cl_i = \frac{1}{\frac{1}{N-1} \sum_j d_{ij}} = \frac{1}{\bar{\ell}_i} \qquad (2.62)$$

where $\bar{\ell}_i$ is the average distance of the node i to the other nodes of the network. The closeness centrality has the limitation that it cannot be used for unconnected networks where it gives zero centrality for each node i of the network.

The efficiency is a measure that overcomes this problem and can also be defined in unconnected networks. The *efficiency* [183] E_i of a node i in an undirected network is given by

$$E_i = \frac{1}{N-1} \sum_{j \neq i} \frac{1}{d_{ij}}. \qquad (2.63)$$

The *global efficiency* E of a network is the average of the efficiencies E_i of its nodes, i.e.

$$E = \frac{1}{N} \sum_{i=1}^{N} E_i = \frac{1}{N(N-1)} \sum_{i,j|i \neq j} \frac{1}{d_{ij}}. \qquad (2.64)$$

Networks with high efficiency are networks that are particularly easy to navigate, notably a desirable property for transportation networks.

2.7.6   Betweenness centrality

The betweenness centrality attributes large relevance to nodes that act as bridges between different communities in the network. These are nodes that are traversed by a large number of shortest paths connecting different pairs of nodes in the network. The *betweenness centrality* [122, 132] b_i of node i in a network is given by

$$b_i = \sum_{r,s} \frac{n_{rs}^i}{g_{rs}}, \qquad (2.65)$$

where n_{rs}^i is the number of shortest paths between node r and node s that pass through node i, and g_{rs} is the total number of shortest paths between node r and node s. The betweenness centrality is in general not correlated with the degree of the node as also low-degree nodes might connect two large and otherwise unconnected regions of the network and therefore acquire high betweenness centrality. However, in random uncorrelated networks the nodes of higher betweenness are typically the nodes with higher degree.

2.7.7 Communicability

The communicability [116, 117] is a centrality measure that quantifies the importance of a node depending on the number of paths that connect it to the rest of the nodes in the network. Unlike the preceding centrality measures, the communicability does not just focus exclusively on shortest paths but instead captures the multiplicity and redundancy of paths of any length. Let us start from the consideration that the number (or total weight in the case of weighted networks) $\mathcal{N}_{ij}^L$ of the paths going from node i to node j in L steps is given by the element i,j of the L-th power of the adjacency matrix, i.e.

$$\mathcal{N}_{ij}^L = \left(\mathbf{a}^L\right)_{ij}. \tag{2.66}$$

Suppressing the relevance of paths of very long length, the communicability matrix $\hat{\mathbf{c}}$ is defined as

$$\hat{\mathbf{c}} = \sum_L \frac{\mathbf{a}^L}{L!} = e^{\mathbf{a}}. \tag{2.67}$$

Therefore, the matrix element $\hat{c}_{ij}$ indicates the weighted multiplicity of paths going from node i to node j. The centrality of the nodes can then be taken to be

$$x_i^{receive} = \sum_{j=1}^N \hat{c}_{ji},$$

$$x_i^{broadcast} = \sum_{j=1}^N \hat{c}_{ij}, \tag{2.68}$$

where $x_i^{receive} = x_i^{broadcast}$ on undirected networks. On directed networks, $x_i^{receive}$ ranks the nodes depending on the weighted number of paths that reach them, while $x_i^{broadcast}$ ranks them according to the weighted number of paths starting from node i and reaching any other node. Alternatively, the *Estrada index* $\hat{E}_i$ [118] ranks the nodes depending on the multiplicity of paths that start from node i and return to node i, i.e.

$$\hat{E}_i = \hat{c}_{ii} = \left[e^{\mathbf{a}}\right]_{ii}. \tag{2.69}$$

2.8   Models for complex networks

2.8.1   The relevance of modelling complex networks

Modelling complex networks is a major challenge of Network Science. The relevance of network modelling for Network Science is due to several factors.

On the one side, there is a need to identify the basic mechanisms determining the evolution of complex networks. In this respect, non-equilibrium-growing network models, including most notably the Barabási–Albert model, provide a fundamental explanation for the emergent universal properties of complex networks such as the scale-free degree distribution.

On the other side, reliable null models of networks in which a given set of structural constraints is enforced can be compared to real datasets, revealing which network properties follow the expectations and which ones deviate from it. Null models for networks include random graphs and the more general network ensembles such as the configuration model, the exponential random graphs and block models, among others. These equilibrium models define static networks obeying a maximum entropy principle and are therefore the least biased ensembles satisfying a given set of constraints.

Additionally, both non-equilibrium and equilibrium network models can be extensively used to generate artificial networks with controlled topological structures on which it is possible to study dynamical processes and evaluate the interplay between network topology and dynamics.

2.8.2   Random graphs

Random graphs were the first network models to include stochasticity when they were proposed by Erdös and Rényi in the 1960s [114]. Random graphs [56] are formed by N nodes connected by a fixed number of links or a fixed expected number of links and are otherwise completely random. Their relevance for the study of real complex networks is very significant. In fact, real networks, from the Internet to the brain, are determined by a stochastic dynamics and are not the result of a blue print. Therefore, understanding random graphs can shed light on the stochastic properties of real networks. However, real networks are not completely random and they obey organization principles that make them very interesting to analyse for extracting information about the underlying complex system. Therefore, to a large extent network scientists are interested in comparing real networks to random networks in order to measure which features of the real networks are not expected in a completely random hypothesis.

Random graphs are the exemplary network ensembles: the $\mathbb{G}(N, L)$ and $\mathbb{G}(N, p)$ network ensembles. In the $\mathbb{G}(N, L)$ ensemble one considers with equal probability all the simple networks of N nodes that have exactly L links and gives zero probability to every other network. The probability $P(G)$ of a simple network $G = (V, E)$ in this ensemble is given by

$$P(G) = \begin{cases} \frac{1}{Z} & \text{for } |V| = N \text{ and } |E| = L, \\ 0 & \text{otherwise,} \end{cases}$$

where

$$Z = \binom{N(N-1)/2}{L}$$

is the total number of networks with N nodes and L links.

The $\mathbb{G}(N,p)$ ensemble is instead formed by all the simple networks $G(V,E)$ with $N = |V|$ labelled nodes where each pair of nodes is linked with probability p.

In the $\mathbb{G}(N,p)$ ensemble the probability of a simple network $G = (V,E)$ with total number of nodes $N = |V|$ is given by

$$P(G) = p^L (1-p)^{N(N-1)/2-L} \text{ with } |E| = L. \tag{2.70}$$

Any network in this ensemble can be seen as a result of $N(N-1)/2$ independent coin tosses, one for each link, with a probability of success (i.e. drawing a link) equal to p.

The $\mathbb{G}(N,L)$ and $\mathbb{G}(N,p)$ ensembles are asymptotically equivalent, meaning that most of their statistical properties coincide in the large network limit $N \to \infty$ as long as one considers the following relation between L and p: $L = pN(N-1)/2$.

Since every link is drawn with probability p, it is immediate to show that the degree distribution $P(k)$ of the $\mathbb{G}(N,p)$ ensemble is a binomial distribution given by

$$P(k) = \binom{N-1}{k} p^k (1-p)^{N-1-k}. \tag{2.71}$$

Therefore, for this network ensemble the average degree is $\langle k \rangle = p(N-1)$. When the average degree of the nodes is fixed to a constant (i.e. $\langle k \rangle = c$) and we explore the limit of large network sizes $N \to \infty$ we have

$$p = \frac{c}{N-1}. \tag{2.72}$$

In this limit the binomial degree distribution given by Eq. (2.71) can be approximated by the Poisson distribution with average c,

$$P(k) = \frac{1}{k!} c^k e^{-c}. \tag{2.73}$$

Therefore, these networks are also called *Poisson networks*.

The expected clustering coefficient C_i of every node i in a random network is

$$C_i = C = p. \tag{2.74}$$

In fact, p is the probability that any two nodes of the network are linked, and therefore it is also the probability that any two neighbour nodes of node i are linked. Relevantly for sparse networks with constant average degree $\langle k \rangle = c$, since the probability of a link p is given by Eq. (2.72) it follows that the expected clustering coefficient C is given by

$$C = \frac{c}{N-1}. \tag{2.75}$$

This clustering coefficient is rapidly decaying with the network size N, indicating that these networks have very few triangles and are locally tree-like.

2.8.3   Small-world network models

The *Watts–Strogatz small-world model* [310] is the original model proposed to explain the small-world universality observed in real networks. The Watts–Strogatz small-world model is obtained starting from a regular one-dimensional lattice with nodes of degree k. In other words, we start from a lattice of N nodes placed on a ring and such that every node is linked to the k nearest neighbours on the ring. Each link of the network is removed from the lattice with probability p and its two ends are attached to randomly chosen distinct nodes of the network. For $p = 0$ the network is a regular one-dimensional ring, for $p = 1$ it is a random graph. For a wide range of intermediate values of p the network displays both the small-world distance property and a high clustering coefficient. Important variations of this original model include models in which instead of a one-dimensional ring the initial lattice has a finite dimension $d > 1$. Moreover, the rewiring of the links can be not completely random and might be performed instead by respecting the embedding space. In this case, when links are rewired nodes can be connected according to a probability that takes into account the distance between the nodes in the underlying lattice.

2.8.4   Growing network models

The emergence of the scale-free property

Many networks do expand and grow by increasing their number of nodes and links over time. Examples include the Internet and social online networks, but also biological networks such as brain networks and molecular networks of the cell which have been growing during the course of biological evolution. This is an evidence that the evolution of many networks is described by a non-equilibrium dynamics. Therefore, important information can be gained by studying non-equilibrium models of growing networks. In particular here we will show that the non-equilibrium framework not only models real networks but also provides explanatory arguments of their emergent properties and their universalities.

The major achievement of growing network models is to show that the scale-free properties can emerge from simple dynamical rules of network growth. In particular, the Barabási–Albert model [15] shows that when a network is evolved by the addition of new links, growth and preferential attachment are the fundamental mechanisms that yield scale-free network topologies.

Interestingly, growing network models enforcing a high density of triangles can generate scale-free networks that also have a high clustering coefficient and non-trivial community structure, showing that the mesoscale structure of networks can emerge from imposing a clustered local topology.

The Barabási–Albert model

The *Barabási–Albert model* [15] is the most fundamental growing network model and explains the emergence of the scale-free degree distribution starting from the preferential attachment mechanism. The preferential attachment mechanism, also called the 'rich gets richer' mechanism, describes the phenomenon observed in real complex networks (such as the World Wide Web, Wikipedia, actor networks) according to which nodes with larger degree have a higher probability of acquiring new links than nodes with smaller degree. Specifically, the Barabási–Albert model assumes that a node of degree k has a probability of acquiring new links that increases linearly with k. The model includes only two main ingredients, growth and preferential attachment, that are able to generate scale-free networks with power-law exponent $\gamma = 3$, demonstrating its powerful ability to reproduce the scale-free topology.

The Barabási–Albert model, also called the BA model, is simply defined. At time $t = 1$ the network is formed by $n_0 \geq m$ nodes connected by m_0 links. At each time $t > 1$ two processes define the network evolution.

- *Growth:* A node is added to the network. The node establishes m new connections with nodes of the rest of the network.

- *Linear preferential attachment:* Every new link of the new node is attached to an existing node i of the network not already linked to the new node with probability

$$\Pi_i = \frac{k_i}{\sum_j k_j},$$ (2.76)

where k_i is the degree of node i and where the sum over the nodes j extends over all the nodes of the network not already linked to the new node.

This model can be studied both in the mean-field approximation and using the master-equation approach providing exact asymptotic results.

In the mean-field approximation [15], the degree $k_i(t)$ that a node i arrived in the network at time t_i is taken to be a continuous deterministic variable depending on time t equal to the expected value of the degree of node i over different realizations of the stochastic network growth. In the mean-field approximation, the degree $k_i(t)$ of node i arrived in the network at time t_i measured at time t satisfies the differential equation

$$\frac{dk_i}{dt} = m\frac{k_i}{\sum_j k_j},$$ (2.77)

with initial condition $k_i(t_i) = m$. For large times $t \gg 1$ we can approximate the sum in the denominator as $\sum_j k_j \simeq 2mt$, obtaining that the degree k_i of node i arrived in the network at time t_i increases with time as a power law,

$$k_i = m \left(\frac{t}{t_i} \right)^{1/2},$$
(2.78)

for $t \geq t_i$. This implies that older nodes have higher degree. From this expression, given the network at time t, the probability $P(k_i(t) > k)$ that a node has degree $k_i(t)$ greater than k is given by

$$P(k_i(t) > k) = P\left(m \left(\frac{t}{t_i} \right)^{1/2} > k \right) = P\left(t_i < t \left(\frac{m}{k} \right)^2 \right).$$
(2.79)

Since at each time step we add a new node to the network, the probability that a random node of the network is arrived at time $t_i < \tau$, in the mean-field approximation it is given by

$$P(t_i < \tau) = \frac{\tau}{t},$$
(2.80)

as long as for $t \gg 1$. Using this expression in Eq. (2.79) follows that

$$P(k_i(t) > k) = \left(\frac{m}{k} \right)^2,$$
(2.81)

and consequently the degree distribution $P(k)$ is given by

$$P(k) = -\frac{dP(k_i > k)}{dk} = \frac{2m^2}{k^3}.$$
(2.82)

From this simple calculation it emerges that this model can generate scale-free networks with power-law exponent $\gamma = 3$ using exclusively two simple dynamical rules: growth and preferential attachment. Using the master equation approach [108, 178] it is possible to derive the exact degree distribution given by (see Appendix A for details)

$$P(k) = \frac{2m(m+1)}{k(k+1)(k+2)},$$
(2.83)

where this result is valid in the large network limit, i.e. asymptotically in time. This exact asymptotic expression for the degree distribution confirms the main conclusions derived from the mean-field approach and indicates that for large values of the degree $k \gg 1$ the degree distribution decays as a power law with exponent $\gamma = 3$.

Interestingly, the Bianconi–Barabási model [45, 44] which assigns to each node a *fitness* describing its ability of nodes to acquire new links and includes growth and generalized preferential attachment, yields scale-free networks with tunable power-law exponent $\gamma \in (2, 3]$. Additionally, this model can explain how latecomers might acquire high degree, a phenomenon also called fit-gets-rich.

Non-linear preferential attachment

The Barabási–Albert model proposes preferential attachment as a basic mechanism to generate scale-free networks. In particular, it assumes that the probability of attaching a new link to a node is linearly proportional to the number of links of the target node. One of the first questions that has been addressed in Network Science is whether also a non-linear preferential attachment could generate scale-free networks.

The modified network model with non-linear preferential attachment [178] considers a network evolution dictated by growth and non-linear preferential attachment, where the non-linearity can be tuned by adjusting an external parameter η.

The growing network model including the non-linear preferential attachment mechanism is defined as follows:

At time $t = 1$ from a network with $n_0 \geq m$ nodes connected by m_0 links.
At each time $t > 1$ two processes define the network evolution.

- *Growth:* A node is added to the network. The newly added node is connected to the other nodes of the network by m links.

- *Generalized non-linear preferential attachment:* The new links are attached to node i with probability Π_i with

$$\Pi_i \propto k_i^{\eta}, \tag{2.84}$$

where k_i is the degree of node i and $\eta > 0$.

For $\eta = 1$ this model reduces to the BA model. For $\eta < 1$ a sublinear preferential attachment in which hub nodes have a lower probability of attracting new links than in the BA model takes place. On the contrary, for $\eta > 1$ a superlinear preferential attachment in which hub nodes are more likely to acquire new links than in the BA model occurs.

In the context of this model, it is found that only linear preferential attachment yields scale-free networks. In fact, for $\eta < 1$ the degree distribution is homogeneous and given by a stretched exponential, while for $\eta > 1$ a gelation event is observed in which the oldest node acquires a finite fraction of all the links and all the other nodes have very small degree.

This result shows the special role of linear preferential attachment in generating scale-free networks as any growing network with $\eta \neq 1$ deviates from the scale-free network topology.

Growing network models enforcing triadic closure

Real networks often have a large clustering coefficient. Therefore, triangles can be considered a fundamental network structure. In the context of growing network models it is possible to consider the growth of networks by the addition of subsequent triangles glued along their links. Interestingly, when the building blocks of a growing network are not links but triangles or even 4-cliques (tetrahedra), preferential attachment is not a necessary input element of the growing network model for generating scale-free networks [109, 51, 52]. In fact, preferential attachment can spontaneously emerge as a property of the network growth. For instance, if we consider a model in which we start from a single triangle and at each time we add a new triangle having a single new node and one side glued to a random existing link of the network, we automatically generate a scale-free network. In this model, the preferential attachment is not imposed explicitly by the growth rule, but it is an emergent property of the dynamics because if we choose the link to which we attach the new triangle randomly, each node has a probability proportional to its number of links of acquiring new connections [109, 51, 52]. Interestingly, these growing network models not only have scale-free degree distribution and display the small-world distance property, but also have a high clustering coefficient and significant community structure [52]. This latter property shows that actually communities can emerge spontaneously from local dynamical rules that enforce high density of triangles [47].

In sociology, the tendency of closing triangles in social networks is called triadic closure. Triadic closure therefore refers to the high probability that two friends of a common person become friends with each other. Therefore, this explains the need to model growing networks that are not exclusively formed by triangles, but instead include both triangles and wedges (triplets of nodes connected by two links) and that mimic the social-network phenomenon of triadic closure.

Several models have been proposed; among others here we discuss the following one [47]:

Initially (at time $t = 1$), the network is formed by a clique of $n_0 \geq m$ nodes, i.e. a set of n_0 nodes in which every node is linked to all the others. At each time $t > 1$, a node is added to the network. Each newly added node is connected to the other nodes of the network by m links according to the following rules:

(a) *Random initial attachment.* The first link connects the new node i to a random node j of the network.

(b) *Triadic closure.* Each of the remaining $m - 1$ links are attached with probability p to a random neighbour of node j and with probability $1 - p$ to a random node of the network.

This model generates networks with broad degree distribution that become better approximated by a power law as the number of initial links m of every node increases. The network displays interesting degree correlations and, as long as p is large and m not too high, displays a clear community structure. The emergence of a non-trivial community structure is not an exclusive property of this model and has been observed numerically on a wide variety of network models, enforcing a high density of triangles and mimicking triadic closure [47].

2.8.5 Ensemble of networks

Maximum entropy ensembles

An ensemble of networks is the set of all the possible networks $G = (V, E)$ having N nodes in which each network G is associated with a probability $P(G)$. Therefore, the probability $P(G)$ uniquely determines the network ensemble. In principle, this probability can be chosen arbitrarily. However, in most practical applications the use of network ensembles is suitable because we do not know exactly the networks we want to model and only partial information about them is accessible. In these types of situations it is desirable not just to have a network ensemble to model the system under study, but also to have a *least-biased* network ensemble that uses the information available about the network but does not assume anything else a priori.

One of the most important results of information theory is that the least-biased probability $P(G)$ given a set of constraints should be the one that maximizes the entropy under the imposed conditions [81]. The entropy S of a network ensemble is given by

$$S = -\sum_G P(G) \ln P(G). \tag{2.85}$$

The entropy indicates the logarithm of the typical number of networks in the ensemble. Maximum entropy network ensembles can be obtained by maximizing S under a given set of constraints such as the total number of links, the degree sequence, the number of links between sets of nodes, etc. Therefore, the smaller the entropy, the less numerous the typical networks in the ensemble, and the larger it is, the more numerous they are.

The constraints that can be imposed can be classified into *hard constraints* and *soft constraints* [6]. The hard constraints are imposed on each network of the ensemble that has non-zero probability. These include the total number of links or the degree sequence. The soft constraints are imposed in average on the ensemble and include the expected number of links in the network and the expected degree sequence. The maximum entropy ensembles satisfying the hard constraints are also called *microcanonical network ensembles*, while the ones satisfying the soft constraints are called *canonical network ensembles*. Microcanonical and canonical network ensembles enforcing the same types of constraints as hard and soft constraints respectively are called conjugated ensembles. The exemplar conjugated microcanonical and canonical network ensembles are the random graph ensembles $\mathbb{G}(N, L)$ and $\mathbb{G}(N, p)$ with $L = pN(N-1)/2$. The terminology used here (microcanonical and canonical network ensembles) is borrowed from classical statistical mechanics when, for instance, the dynamical configurations of the particles in a gas are studied when the energy of the system is fixed (microcanonical ensemble) or when the gas is in a thermal bath and only the average energy is kept fixed (canonical ensemble). It is to be noted that the canonical network ensembles are also widely known in the statistical network community as exponential random graphs [265].

The entropy of network ensembles is an information measure that can be used to assess how much information is encoded in the network constraints. In fact, a network ensemble satisfying strict constraints that make the networks highly optimized for some specific task will be characterized by having few typical network realizations (small entropy), while more generic constraints will yield a network ensemble with many more typical networks (larger entropy). Ideally, we could think to relate the network constraints to the function of the networks. For instance, we could consider all the networks that perform a given function equally well. In practice until now, most often network ensembles typically consider few specific types of constraints such as the node degrees and the number of links among different classes of nodes. Going beyond these constraints and considering ensembles of networks with a given number of triangles or a given sequence of local clustering coefficient already constitutes a significant challenge for network theorists.

In the following we will consider specifically the microcanonical and canonical ensembles that preserve respectively the degree sequence and the expected degree sequence, and the block models that preserve the number of links between different classes of nodes (for details of the derivations see Appendix B).

Canonical network ensembles with expected degree sequence

The canonical network ensembles having an expected degree sequence $\{k_i\}$ assign to each network G a probability $P_C(G)$ that satisfies

$$\sum_G P_C(G) \left[\sum_{j=1}^N a_{ij} \right] = k_i. \tag{2.86}$$

Therefore, in this ensemble not all the network realizations have the same degree sequence, but on average each node i has an expected degree k_i. The probability of a network in this ensemble takes the exponential form

$$P_C(G) = \frac{1}{Z_C} \exp \left[-\sum_{i<j} (\lambda_i + \lambda_j) a_{ij} \right] \tag{2.87}$$

where λ_i is the Lagrangian multiplier enforcing the constraint expressed by Eq. (2.86) and Z_C is a normalization constant. Since $P_C(G)$ takes a characteristic exponential expression, this ensemble is also called an exponential random network with given expected degree sequence. The probability $P_C(G)$ factorizes into contributions coming from each individual link and can equivalently be written as

$$P_C(G) = \prod_{i<j} \left[p_{ij} a_{ij} + (1 - p_{ij})(1 - a_{ij}) \right], \tag{2.88}$$

where p_{ij} is the probability that a network of the ensemble contains the link between node i and node j, i.e. has $a_{ij} = 1$. The entropy of these networks can be expressed simply in terms of the marginal probabilities p_{ij} as

$$S = -\sum_{i<j} \left[p_{ij} \ln p_{ij} + (1 - p_{ij}) \ln(1 - p_{ij}) \right]. \qquad (2.89)$$

The probabilities p_{ij} are given in terms of the Lagrangian multipliers by

$$p_{ij} = \frac{e^{-\lambda_i - \lambda_j}}{1 + e^{-\lambda_i - \lambda_j}}, \qquad (2.90)$$

and the contraint in Eq. (2.86) can also be written as

$$k_i = \sum_{j=1}^{N} p_{ij}. \qquad (2.91)$$

Eq. (2.90) for the marginal probabilities p_{ij} in this network ensemble is remarkable. In fact, it implies that in general in this ensemble the probability p_{ij} does not factorize into contributions coming respectively from nodes i and j alone. This is a sign that in general these networks will be correlated and actually display disassortative correlations. However, if the degree distribution is not very broad and the maximum degree is bounded by the structural cutoff, these networks are in good approximation uncorrelated networks, as we will show in the following.

Microcanonical network ensemble with given degree sequence

The microcanonical network ensemble preserving a given degree sequence $\{k_i\}$ is also widely known as the configuration model. In this ensemble we associate an equal probability $P_M(G)$ with every network $G = (V, E)$ having a given degree sequence, i.e.

$$P_M(G) = \frac{1}{Z_M} \prod_{i=1}^{N} \delta\left(k_i, \sum_{j=1}^{N} a_{ij} \right), \qquad (2.92)$$

where Z_M is the normalization constant. Interestingly, this network ensemble is not statistically equivalent to its conjugated canonical ensemble discussed in the previous paragraph [6, 7]. This fact is revealed most notably by observing that the entropy of this ensemble that we indicate here with Σ is related to the entropy of the canonical ensemble S by [46, 7]

$$\Sigma = S - \Omega \qquad (2.93)$$

where Ω is a quantity that grows linearly with the number of nodes N and is therefore not negligible. Therefore, the entropy of the microcanonical ensemble Σ is not equal to the entropy of the canonical ensemble in the large-network limit. In this case we say that there is no ensemble equivalence. This is a consequence of the fact that here we are enforcing an extensive number of constraints while ensemble equivalence to hold requires a finite number of constraints.

The expression for Ω is in general given by

$$\Omega = -\ln\left[\sum_G P_C(G) \prod_{i=1}^N \delta\left(k_i, \sum_{j=1}^N a_{ij}\right)\right],\tag{2.94}$$

i.e. Ω represents the logarithm of the probability that in the canonical network ensemble we observe that the degree sequence is exactly given by the expected degree sequence. In the limit in which each node i of the network satisfies

$$k_i \ll \sqrt{\langle k \rangle N},\tag{2.95}$$

Ω takes the simple expression [46, 7]

$$\Omega = \sum_{i=1}^N \ln\left[\frac{1}{k_i!} k_i^{k_i} e^{-k_i}\right].\tag{2.96}$$

Uncorrelated networks

In the configuration model and in the exponential random graph the marginal probabilities p_{ij} take the non-factorizable expression given by Eq. (2.90). However, if in the network there is the structural cutoff

$$K_S = \sqrt{\langle k \rangle N}$$

indicating the maximum allowed degree in the network and every node i satisfies

$$k_i \ll K_s = \sqrt{\langle k \rangle N},$$

the probabilities p_{ij} can be approximated as

$$p_{ij} = \frac{k_i k_j}{\langle k \rangle N},\tag{2.97}$$

and the network is effectively uncorrelated.

This limit of the configuration model is widely used when studying dynamical models of networks such as percolation or epidemic spreading. Typically, these networks are also called random uncorrelated networks with degree distribution $P(k)$.

For uncorrelated random networks the clustering coefficient is given by

$$C = \frac{1}{\langle k \rangle N} \left(\frac{\langle k(k-1) \rangle}{\langle k \rangle} \right)^2.$$
(2.98)

It follows that for scale-free networks with structural cutoff, even if $\langle k(k-1) \rangle$ diverges with N as $\langle k(k-1) \rangle \simeq N^{(3-\gamma)/2}$ the clustering coefficient C vanishes for $N \to \infty$. This implies that these networks do not have a finite density of triangles and can be to a large extent treated as locally tree-like.

Block models

Block models are maximum-entropy ensembles in which nodes are classified into different types or blocks and the expected or exact number of links within each block and among each pair of different blocks is fixed. Block models can be divided into canonical network models enforcing the soft constraints and microcanonical models enforcing the hard constraints.

Let us consider a subdivision of nodes into Q blocks $q = 1, 2, \ldots, Q$, and let us indicate with q_i the block to which node i belongs. In the canonical block model we consider the maximum entropy ensemble in which the expected number of links among a block q and a block q' is given by $e_{q,q'}$ and the expected number of links within block q is $e_{q,q}$. The probability $P_C(G)$ of the network G on such canonical network ensembles takes the exponential form

$$P_C(G) = \frac{1}{Z_C} \exp\left[-\sum_{i<j} \lambda_{q_i,q_j} a_{ij} \right],$$
(2.99)

where Z_C is the normalization constant and $\lambda_{q,q'}$ is the Lagrangian multiplier enforcing the expected number of links $e_{q,q'}$. The probability p_{ij} of a link between node i and node j is given by

$$p_{ij} = \langle a_{ij} \rangle = \frac{e^{-\lambda_{q_i,q_j}}}{1 + e^{-\lambda_{q_i,q_j}}}$$
(2.100)

where the Lagrangian multipliers are fixed by the conditions

$$\sum_{i,j} p_{ij} \delta(q, q_i) \delta(q', q_j) = e_{q,q'} \quad \text{for } q \neq q',$$

$$\sum_{i<j} p_{ij} \delta(q, q_i) \delta(q, q_j) = e_{q,q},$$
(2.101)

where $\delta(x, y)$ is the Kronecker delta. The probabilities p_{ij} depend only on q_i, q_j. Therefore, we have $p_{ij} = p(q_i, q_j)$ with

$$p(q, q') = \frac{e_{q,q'}}{n_q n_{q'}} \text{ for } q \neq q'$$

$$p(q, q) = \frac{e_{q,q}}{n_q(n_q - 1)/2} \qquad (2.102)$$

where n_q indicates the total number of nodes in community q. The entropy S of these ensembles is given by

$$S = -\sum_{i<j} \left[p_{ij} \ln p_{ij} + (1 - p_{ij}) \ln(1 - p_{ij}) \right]. \qquad (2.103)$$

If the number of constraints is non-extensive $Q(Q+1)/2 \ll N$, in the large N limit this expression is given by

$$S = \ln \left[\prod_{q<q'} \binom{n_q n_{q'}}{e_{q,q'}} \prod_q \binom{n_q(n_q - 1)/2}{e_{q,q}} \right].$$

In the microcanonical block model the quantities $e_{q,q'}$ and $e_{q,q}$ indicate the exact number of links among blocks q and q' and within block q respectively. The number of network in this ensemble is simply given by

$$Z_M = \prod_{\alpha=1}^{M} \left[\prod_{q<q'} \binom{n_q n_{q'}}{e_{q,q'}} \prod_q \binom{n_q(n_q - 1)/2}{e_{q,q}} \right],$$

where n_q indicates the number of nodes in community q.

It follows that as long as the number of constraints is not extensive, i.e. $Q(Q+1)/2 \ll N$, the canonical and microcanonical network ensembles are asymptotically equivalent and

$$S \simeq \Sigma.$$

2.9   Spatial network ensembles

In this section we consider canonical maximum entropy network ensembles for spatially embedded networks [25] where each pair of nodes (i, j) is at distance d_{ij} in the embedding space. The ensemble of networks we consider here can be divided into two major classes. In the first class, we consider a given binning of the distances between the nodes and we fix on average the number of links connecting nodes whose distance falls in a given bin. Moreover, we also fix the expected degree k_i of each node i. Therefore, we first define N_B bins of distance intervals and then we consider the constraints

$$\sum_G P_C(G) \left[\sum_{i<j} a_{ij} \chi_\nu(d_{ij}) \right] = C_\nu \qquad (2.104)$$

and

$$\sum_G P_C(G) \left[\sum_{j=1}^N a_{ij} \right] = k_i, \qquad (2.105)$$

where $\chi_\nu(d_{ij})$ indicates whether the distance d_{ij} does $(\chi_\nu(d_{ij})=1)$ or does not $(\chi_\nu(d_{ij})=0)$ fall in the bin $\nu = 1,2,\ldots,N_B$ of distance intervals. The maximal entropy network ensemble satisfying this set of constraints has distribution $P_C(G)$ given by Eq. (2.88) and entropy S given by Eq. (2.89) where the marginal probability p_{ij} of the link (i,j) is given by

$$p_{ij} = \frac{e^{-\lambda_i-\lambda_j-\sum_\nu \lambda_\nu \chi_\nu(d_{ij})}}{1 + e^{-\lambda_i-\lambda_j-\sum_\nu \lambda_\nu \chi_\nu(d_{ij})}}. \qquad (2.106)$$

Here the λ_ν symbols indicate the Lagrangian multipliers of the constraints in Eq. (2.104) and the λ_i symbols indicate the Lagrangian multipliers that enforce the constraints in Eq. (2.105). This ensemble can be used to generate randomized network ensembles starting from a given network embedded into a geometrical space for which we have calculated the number C_ν of links whose distance falls in the bin ν of distance intervals and the degree k_i of each node i.

In the second case, we fix on average the total cost associated with the connections existing in the network and the expected degree of each node. Therefore, we consider the maximum entropy ensemble satisfying the constraints

$$\sum_G P_C(G) \left[\sum_{i<j} a_{ij} f(d_{ij}) \right] = \hat{C} \qquad (2.107)$$

$$\sum_G P_C(G) \left[\sum_{j=1}^N a_{ij} \right] = k_i, \qquad (2.108)$$

where $f(d)$ indicates the cost associated with a link (i,j) of distance $d_{ij} = d$. Typical choices for the function $d(d)$ are

$$f(d) = d,$$
$$f(d) = \ln(d). \qquad (2.109)$$

The first option corresponds to a cost which is a linear function of the distance of the link, the second option corresponds to a cost that scales linearly with the order of magnitude of the distance. In this case, the maximum entropy ensemble assigns to each network a probability $P_C(G)$, given by Eq. (2.88), and has an entropy S given by Eq. (2.89) where the marginal probability p_{ij} of each link (i,j) is given by

$$p_{ij} = \frac{e^{-\lambda_i - \lambda_j - \lambda_C f(d_{ij})}}{1 + e^{-\lambda_i - \lambda_j - \lambda_C f(d_{ij})}}. \tag{2.110}$$

Here the Lagrangian multiplier λ_C enforces the constraint in Eq. (2.107) on the expected total cost of the links and the Lagrangian multipliers λ_i enforce the constraints given by Eq. (2.108) on the expected degree of node i. This ensemble can be used to generate spatial networks with the desired expected cost of the link starting exclusively from a set of N nodes embedded in a geometrical space and an expected degree sequence.

3

The Dynamics on Single Networks

3.1 Interplay between structure and function

Whenever we aim at characterizing network robustness, the spreading of an epidemic, the properties of a random walk, or the onset of the synchronized state, we observe that complex network topologies dramatically affect the behaviour of the dynamical processes.

Traditionally, dynamical processes have been studied on d-dimensional lattices and great interest has been given to investigating the dependence of dynamical behaviour with the dimensionality d. However, with the rise of the interest in complex networks new surprising results have revealed a more significant interplay between structure and dynamics in complex networks.

It has been shown that the behaviour of many dynamical processes changes very significantly if the network is scale-free. For instance scale-free networks are much more robust to random damage than networks with a converging second moment $\langle k^2 \rangle$ of the degree distribution [78]. Moreover, scale-free topologies also significantly favour the spread of viruses, ideas and information within a network with relevant consequences for containing pandemics or devising viral advertising strategies [243].

Diffusion and synchronization are instead strongly dependent on the spectral properties of the network, characterizing the relaxation time to the steady state of the diffusion processes and the stability of the fully synchronized state in complex networks.

In this chapter we will give a brief summary of the main theoretical frameworks that have been proposed to characterize the interplay between structure and dynamics on single networks and of the main results achieved so far. These results constitute the starting point for exploring the specific signatures that multiplexity has on the dynamical processes defined in multilayer networks.

To the reader desiring to have a wider overview of the field we suggest referring to the monographies [24, 253] and the following review articles: [106] (on general critical phenomena), [241] (on epidemic spreading), [9] (on synchronization), [196] (on network control).

Multilayer Networks. Ginestra Bianconi, Oxford University Press (2018).
© Ginestra Bianconi. DOI: 10.1093/oso/9780198753919.001.0001

3.2   Phase transitions and emergent phenomena

In the vast majority of real scenarios, in Network Science we aim at characterizing dynamical processes that give rise to emergent phenomena. These are macroscopic changes in the dynamical state of the system that cannot be predicted by considering the dynamical system defined on a single node. The processes giving rise to these macroscopic changes are also called *critical phenomena* [24, 106]. For instance, when we want to formulate ways to contain an influenza epidemic or when our goal is to determine the robustness of a network, we are especially interested in dynamical states that affect the network as a whole, such as an endemic or the complete dismantling of a network. In these contexts the statistical mechanics framework, originally proposed to characterize the different states of matter (such as water/ice/vapour for water molecules and diamond/graphene/graphite for carbon atoms) turns out to be very useful.

In statistical mechanics, a phase transition from one phase of matter to another one is characterized by studying the changes of an *order parameter*, determining the macroscopic properties of a large (actually infinite) system as a function of an external *control parameter*. In the context of network theory we will use exactly the same procedure. For instance, in percolation theory we will study the fraction of nodes S that are in the giant component of a network as a function of the probability p that any given node is not initially damaged. Similarly, in epidemic spreading we will determine for which infectivity rates λ the network is in an endemic phase characterized by having a finite fraction of all the nodes of the network infected. Let us call the generic order parameter M and the generic control parameter x. The system undergoes a phase transition when there is a critical value of x, called x_c, such that for $x > x_c$ the order parameter is non-vanishing, i.e. $M > 0$, and for $x \leq x_c$ the order parameter is vanishing, i.e. $M = 0$.

Several phase transitions that we will encounter in this chapter can be characterized as continuous, second-order phase transitions in which the order parameter M close to the transition, i.e. for $x \simeq x_c$, takes the values

$$M = \begin{cases} D(x - x_c)^{\hat{\beta}} & \text{for } x \geq x_c \\ 0 & \text{for } x > x_c, \end{cases} \tag{3.1}$$

where $\hat{\beta} > 0$ is called the *dynamical critical exponent* and D is a constant. In this case as $x \to x_c^+$ the order parameter approaches zero, i.e. $M \to 0$. A major example of continuous second-order phase transitions is percolation of single networks that will be characterized in the next section.

However there are also discontinuous phase transitions in which as we approach the critical value of the control parameter, $x \to x_c^+$, the order parameter approaches a non-zero value, i.e. $M \to M_c > 0$, and only for $x = x_c$ the order parameter abruptly takes the value $M = 0$.

Among the discontinuous phase transitions, of particular interest are the hybrid transitions which have the discontinuity of the order parameter M as a function of the control parameter x, but, like second-order phase transitions, they are characterized by a singular behaviour for $x \to x_c^+$. This behaviour can be described as

$$M = \begin{cases} M_c + D(x - x_c)^{\hat{\beta}} & \text{for } x \geq x_c \\ 0 & \text{for } x > x_c, \end{cases} \tag{3.2}$$

where $\hat{\beta} > 0$ is called the *dynamical critical exponent* and $D > 0$ is a constant. A major example of a hybrid-phase transition is the percolation transition in interdependent networks that will be characterized in chapter 11.

3.3   Robustness and percolation

3.3.1   The percolation transition

In the theory of percolation, it is assumed that a fundamental proxy for the proper function of a given network is the existence of the giant component. In fact, as discussed in chapter 2 (see Sec. 2.3.6), the existence of a giant component allows the propagation of ideas, information and signals along the links of the giant component, reaching a finite fraction of all the nodes of the network.

In node percolation the robustness of a given network is tested by assuming that some nodes are randomly removed with probability f and monitoring the fraction of nodes S that remains in the giant component of the network after the inflicted damage. The parameter $p = 1 - f$, indicating the probability that a node is not initially damaged, is usually used to characterize the percolation process. The percolation transition is characterized by the behaviour of S (the order parameter of the percolation transition) as a function of p. The percolation transition occurs for a given value of p, called the *percolation threshold* p_c. It is found that for $p \leq p_c$ there is no giant component in the network and $S = 0$, while for $p > p_c$ there is a giant component and $S > 0$. For values of $p \simeq p_c$ the order parameter S follows

$$S = \begin{cases} D(p - p_c)^{\hat{\beta}} & \text{for } p \geq p_c \\ 0 & \text{for } p < p_c, \end{cases} \tag{3.3}$$

where $\hat{\beta} > 0$ is called the *dynamical critical exponent* of percolation and $D > 0$ is a constant.

The percolation transition is a beautiful example of a continuous second-order phase transition; in fact, in the limit for $p \rightarrow p_c^+$ $S \rightarrow 0$, meaning that if the nodes of the network are damaged with increasing probability the size of the giant component is continuously reduced until reaching size zero for $p = p_c$.

Similarly it is possible to define the bond percolation in which instead of removing the nodes randomly we remove the links of the network randomly. Also in this case the order parameter S undergoes a continuous second-order phase transition as a function of p, indicating in this case the probability that a link is not randomly removed. The critical behaviour of S also follows Eq. (3.3) in this case, but the values of the percolation threshold p_c and the dynamical critical exponent $\hat{\beta}$ in general might be different from the ones observed in node percolation. In sections 3.3.2 and 3.3.3 we will actually see that the epidemic threshold for node percolation and bond percolation is actually the same on random networks, while the dynamical critical exponent for some network topologies is different.

On locally tree-like networks both node percolation and bond percolation can be studied on single instances of networks by message-passing algorithms. Message-passing algorithms constitute a very efficient way to study a number of dynamical processes on complex networks and they have been used not only for percolation [168] but also for characterizing the behaviour of epidemic-spreading processes [5] and for predicting network controllability [197]. Percolation is one of the major examples of the possible application of message-passing techniques to study dynamical processes on networks. In the following we have decided to give a synthetic but complete account of their use in this context.

3.3.2   Node percolation

Message-passing approach

Let us consider a network where each node i is either damaged ($s_i = 0$) or not damaged ($s_i = 1$). As long as the network is locally tree-like, it is possible to determine whether a node belongs to the giant component of the network or not by running a message-passing algorithm. The message-passing algorithm consists of a set of messages, $\sigma_{i \to j}$, sent from each node i to each neighbour node j that follows a recursive set of equations. In the stationary state the messages have performed a kind of 'distributed computation' revealing the large-scale properties of the network. Specifically in this case they indicate which nodes belong or do not belong to the giant component.

For each link between node i and node j there are two distinct messages, $\sigma_{i \to j}$ and $\sigma_{j \to i}$. The message $\sigma_{i \to j} = 1$ sent from node i to node j has the meaning 'through me you are connected to other nodes in the giant component'. The message $\sigma_{i \to j} = 0$ instead has the meaning 'through me you are not connected to other nodes in the giant component'.

The messages $\sigma_{i \to j} = 0, 1$ are updated locally according to the value of the messages received by node i from neighbour nodes $\ell \neq j$. Specifically $\sigma_{i \to j} = 1$ if:

(a) node i is not initially damaged, i.e. $s_i = 1$;

(b) node i belongs to the giant component even if the link between node j and node i is removed from the network, i.e. node i receives at least one positive message $\sigma_{\ell \to i} = 1$ from nodes $\ell \neq j$ that are neighbours of node i.

If these conditions are not met, then $\sigma_{i \to j} = 0$. These messages determine whether a node i does ($\sigma_i = 1$) or does not ($\sigma_i = 0$) belong to the giant component. In fact node i belongs to the giant component ($\sigma_i = 1$) if and only if:

(a) node i is not initially damaged, i.e. $s_i = 1$;

(b) node i receives at least one positive message $\sigma_{\ell \to i} = 1$ from one of its neighbours ℓ.

This algorithm can be explicitly encoded in the following recursive equations for the messages $\sigma_{i \to j}$ and for the probabilities σ_i:

$$\sigma_{i \to j} = s_i \left[1 - \prod_{\ell \in N(i) \backslash j} (1 - \sigma_{\ell \to i}) \right]$$

$$\sigma_i = s_i \left[1 - \prod_{\ell \in N(i)} (1 - \sigma_{\ell \to i}) \right] \qquad (3.4)$$

where $N(i)$ indicates the set of neighbours of node i and $N(i) \backslash j$ indicates the set of all nodes $\ell \neq j$ which are neighbours of node i.

In a variety of cases the exact configuration of the initial damage $\{s_i\}$ is not known. Instead the probability $f = 1 - p$ that a node is damaged is known. Under these conditions the above message-passing equations can be averaged over the distribution $\hat{P}(\{s_i\})$ of the initial damage configurations $\{s_i\}$ given by

$$\hat{P}(\{s_i\}) = \prod_{i=1}^{N} p^{s_i} (1 - p)^{1 - s_i}, \qquad (3.5)$$

where p is the probability that a node is not initially damaged ($s_i = 1$) and $1 - p$ is the probability that a node is initially damaged ($s_i = 0$). By indicating with $\hat{\sigma}_{i \to j}$ and $\hat{\sigma}_i$ the average of the messages $\sigma_{i \to j}$ and the probabilities σ_i over the distribution $\hat{P}(\{s_i\})$ we obtain the following message-passing equations:

$$\hat{\sigma}_{i \to j} = p \left[1 - \prod_{\ell \in N(i) \backslash j} (1 - \hat{\sigma}_{\ell \to i}) \right]$$

$$\hat{\sigma}_i = p \left[1 - \prod_{\ell \in N(i)} (1 - \hat{\sigma}_{\ell \to i}) \right], \qquad (3.6)$$

where now $\hat{\sigma}_{i \to j}$ and $\hat{\sigma}_i$ take real values in the interval $[0, 1]$. Here $\hat{\sigma}_i$ indicates the probability that node i belongs to the giant component if the distribution $\hat{P}(\{s_i\})$ of the initial damage follows Eq. (3.5). From this equation it follows that the percolation threshold p_c of a single network is given $p_c = 1/\Lambda$ where Λ maximum eigenvalue of the non-backtracking matrix. The non-backtracking matrix is a $L \times L$ matrix of elements $B_{(i,j),(r,s)} = \delta(j, r)[1 - \delta(s, i)]$, where (i, j) and (r, s) are links of the network.

Finally, to study the percolation properties of random uncorrelated networks with given degree distribution $P(k)$ we average the messages $\hat{\sigma}_{i \to j}$ over the network ensemble, getting the probability S' that by following a link we reach a node in the giant component. Similarly, by averaging the probability $\hat{\sigma}_i$ over the network ensemble we get the probability S that a random node is in the giant component. From Eqs (3.6) averaging over the random network ensemble it is possible to deduce that S' and S follow

$$S' = p\left[1 - \sum_k \frac{k}{\langle k \rangle}P(k)(1 - S')^{k-1}\right],$$

$$S = p\left[1 - \sum_k P(k)(1 - S')^k\right]. \tag{3.7}$$

If we introduce the generating functions

$$G_0(x) = \sum_k P(k)x^k,$$

$$G_1(x) = \sum_k \frac{k}{\langle k \rangle}P(k)x^{k-1}, \tag{3.8}$$

Eqs (3.7) can equivalently be written as

$$S' = p\left[1 - G_1(1 - S')\right],$$

$$S = p\left[1 - G_0(1 - S')\right]. \tag{3.9}$$

These equations are always satisfied for $S' = S = 0$, but, depending on the properties of the degree distribution $P(k)$ and on the probability p, this solution might not be stable. When this happens another non-trivial solution $S' > 0, S > 0$ emerge, implying that the network displays a giant component. Let us note that $S = 0$ if and only if $S' = 0$. Therefore, in order to study the stability of the solution $S' = S = 0$ it is sufficient to study the stability of the non-linear equation determining S'. If we linearize the equation for $S' \ll 1$ we get

$$S' = p\frac{\langle k(k-1) \rangle}{\langle k \rangle}S'. \tag{3.10}$$

Therefore, the solution $S' = 0$ becomes unstable for

$$p\frac{\langle k(k-1) \rangle}{\langle k \rangle} > 1. \tag{3.11}$$

This result implies that a random network with degree distribution $P(k)$ has a giant component if and only if [78, 227]

$$p > p_c = \frac{\langle k \rangle}{\langle k(k-1) \rangle}. \tag{3.12}$$

Percolation in Poisson networks

For Poisson networks with average degree $\langle k \rangle = c$ and degree distribution

$$P(k) = \frac{c^k}{k!}e^{-c}, \tag{3.13}$$

the generating functions are given by

$$G_0(x) = G_1(x) = e^{c(x-1)}. \tag{3.14}$$

Therefore, Eqs (3.9) imply that $S = S'$ and that S satisfies

$$S = p(1 - e^{-cS}). \tag{3.15}$$

Finally, since for Poisson networks with average degree $\langle k \rangle = c$

$$\frac{\langle k(k-1) \rangle}{\langle k \rangle} = c, \tag{3.16}$$

the condition for having a giant component reads

$$p > p_c = \frac{1}{c}. \tag{3.17}$$

Therefore, for Poisson networks the percolation threshold is determined exclusively by the average degree of the network. In order to derive the critical exponent $\hat{\beta}$ defined by Eq. (3.3.3), let us expand Eq. (3.15) for $p \simeq p_c$ implying $S \ll 1$,

$$S = pcS - \frac{1}{2}c^2 S^2 + \dots \tag{3.18}$$

By truncating this expansion at the second order, since $S \geq 0$, we get for $p \simeq p_c$

$$S = \begin{cases} D(p - p_c)^{\hat{\beta}} & \text{if } p \geq p_c \\ 0 & \text{if } p < p_c \end{cases} \tag{3.19}$$

where $D = 2/c$ and $\hat{\beta} = 1$. Therefore the critical exponent is $\hat{\beta} = 1$. It can be shown that this result extends to every random network with degree distribution $P(k)$ having a convergent first, second and third moment ($\langle k \rangle$, $\langle k(k-1) \rangle$ and $\langle k(k-1)(k-2) \rangle$). In fact all these networks have critical exponent $\hat{\beta} = 1$.

Percolation in scale-free networks

Uncorrelated sparse power-law networks with degree distribution $P(k) = Ck^{-\gamma}$ and $\gamma > 2$ have a percolation threshold given by [78, 227]

$$p_c = \frac{\langle k \rangle}{\langle k(k-1) \rangle}. \tag{3.20}$$

Therefore, as a function of γ we observe a remarkable phenomenon:

(a) For $\gamma > 3$ we have a finite moment average degree $\langle k \rangle$ and finite $\langle k(k-1) \rangle = \langle k^2 \rangle - \langle k \rangle$ even in the limit of large network sizes $N \to \infty$. Therefore the epidemic threshold p_c is finite and the network is dismantled when $0 < p \le p_c$.

(b) For $\gamma \in (2,3]$ the network is *scale-free*, implying that although the first moment $\langle k \rangle$ of the degree distribution remains finite in the large network limit $N \to \infty$, this average degree is not typical for a random node of the network. In fact the network has nodes with very different degrees and the second moment of the degree distribution diverges in the large network limit, i.e.

$$\langle k(k-1) \rangle \to \infty \text{ as } N \to \infty.$$

Using this result in Eq. (3.20) we see that for scale-free networks the percolation threshold vanishes in the infinite network limit [78], i.e.

$$p_c \to 0 \text{ as } N \to \infty.$$

The result obtained for $\gamma \in (2, 3]$ implies that scale-free networks are much more robust than homogeneous networks with finite second moment of the degree distribution. In fact, a vanishing percolation threshold implies that almost all the nodes of the network can be damaged and still the network contains a giant component. Intuitively this phenomenon is related to the fact that scale-free networks have hub nodes that, having a large degree, are able to keep the network together even when the system is strongly damaged. This is one of the pivotal results of Network Science showing that actually the scale-free network universality observed in technological as well as social and biological networks can be rooted in the fact that scale-free network topologies are particularly robust to random damage.

The dynamical critical exponent $\hat{\beta}$ for percolation on power-law networks depends on the power-law exponent γ (see Table 3.1) and acquires the mean-field value $\hat{\beta} = 1$ only for $\gamma > 4$.

3.3.3 Bond percolation

Bond percolation (also called link percolation) refers to the case in which the initial damage inflicted on the network removes some of its links. As in node percolation, the robustness of the network can be monitored by determining the size of the giant component resulting after the initial damage. In bond percolation the initial configuration of the damage can be characterized by the variables $s_{ij} = 0, 1$, indicating for every link (i, j) of the network whether it has been initially damaged ($s_{ij} = 0$) or not ($s_{ij} = 1$). As long as the network is locally tree-like, it is possible to determine if a node belongs ($\sigma_i = 1$) or does not belong ($\sigma_i = 0$) to the giant component of the network using a message-passing algorithm. Let us indicate with $\sigma_{i \to j} = 0, 1$ the messages sent from node

Table 3.1 *Dynamical critical exponent $\hat{\beta}$ for node percolation on power-law networks with degree distribution $P(k) = Ck^{-\gamma}$ [106]. For values of the power-law exponent $\gamma = 3, 4$, logarithmic corrections to the scaling behaviour defined by Eq. (3.4) are observed.*

γ	$\hat{\beta}$
$\gamma > 4$	1
$\gamma \in (3,4)$	$1/(\gamma - 3)$
$\gamma \in (2,3)$	$1/(3 - \gamma)$

i to node j. The message-passing algorithm for bond percolation is a simple modification of the one used for node percolation. Specifically, $\sigma_{i\to j} = 1$ if node i belongs to the giant component even if the link between node j and node i is removed from the network. This occurs if node i receives at least one positive message $\sigma_{\ell\to i} = 1$ from nodes $\ell \neq j$ that are neighbours of node i and are connected to node i by a non-damaged link, i.e. $s_{\ell i} = 1$. If this condition is not met, then $\sigma_{i\to j} = 0$.

Node i belongs to the giant component ($\sigma_i = 1$) if and only if node i receives at least one positive message $\sigma_{\ell\to i} = 1$ from a neighbour ℓ of node i connected to node i by a non-damaged link, i.e. $s_{\ell i} = 1$.

This algorithm directly translates into the message-passing equations

$$\sigma_{i\to j} = \left[1 - \prod_{\ell\in N(i)\backslash j} (1 - s_{\ell i}\sigma_{\ell\to i}) \right],$$

$$\sigma_i = \left[1 - \prod_{\ell\in N(i)} (1 - s_{\ell i}\sigma_{\ell\to i}) \right]. \tag{3.21}$$

As for node percolation, it can also be the case for bond percolation that only the distribution of the initial damage is known. Specifically, we might assume that each link is damaged with probability $f = 1 - p$ and that therefore the distribution $\hat{P}(\{s_{ij}\})$ over all the possible initial damage configurations is given by

$$\hat{P}(\{s_{ij}\}) = \prod_{<i,j>} p^{s_{ij}}(1 - p)^{1-s_{ij}}, \tag{3.22}$$

where $< i,j >$ indicate the set of all the links of the network.

Let us indicate with $\hat{\sigma}_{i\to j}$ the average of the messages $\sigma_{i\to j}$ over the distribution $\hat{P}(\{s_{ij}\})$ and let us indicate with $\hat{\sigma}_i$ the average of σ_i, where $\sigma_{i\to j}$ and σ_i satisfy the message-passing equations (3.21) for bond percolation. It can be shown easily that $\hat{\sigma}_{i\to j}$ and $\hat{\sigma}_i$ satisfy the recursive equations

$$\hat{\sigma}_{i \to j} = \left[1 - \prod_{\ell \in N(i) \backslash j} (1 - p\hat{\sigma}_{\ell \to i}), \right]$$

$$\hat{\sigma}_i = \left[1 - \prod_{\ell \in N(i)} (1 - p\hat{\sigma}_{\ell \to i}) \right]. \tag{3.23}$$

From this equation it follows that the percolation threshold p_c of a single network is given $p_c = 1/\Lambda$ where Λ is the maximum eigenvalue of the non-backtracking matrix. The non-backtracking matrix is a $L \times L$ matrix of elements $B_{(i,j),(r,s)} = \delta(j, r)[1 - \delta(s, i)]$, where (i, j) and (r, s) are links of the network. Finally, let us assume that the specific network topology is unknown and the only accessible information is that the network is generated by a random uncorrelated network ensemble with degree distribution $P(k)$. In this case the average message S indicating the probability that by following a non-damaged link we reach a node in the giant component and the probability S that a random node is in the giant component satisfy

$$S' = 1 - G_1(1 - pS'),$$
$$S = 1 - G_0(1 - pS'), \tag{3.24}$$

where the generating functions $G_0(x)$ and $G_1(x)$ are given by Eqs (3.8).

Eqs (3.24) determine whether $(S > 0)$ or not $(S = 0)$ the random network is expected to have a giant component. As in the case of node percolation for values of $p < p_c$, where p_c indicates the percolation threshold, there is no giant component in the network, i.e. $S = 0$, while for $p > p_c$ the giant component is expected to be $S > 0$. The emergence of the giant component is determined by the onset of the instability of the solution $S = S' = 0$ of Eqs (3.24). Therefore, in order to find the percolation threshold we can linearize the equation for S' (given by (3.24)) for $S' \ll 1$, getting

$$S' = p \frac{\langle k(k-1) \rangle}{\langle k \rangle} S'. \tag{3.25}$$

This equation indicates that the solution $S' = 0$ is unstable for

$$p \frac{\langle k(k-1) \rangle}{\langle k \rangle} > 1. \tag{3.26}$$

It follows that the percolation threshold for bond percolation is the same as the percolation threshold for node percolation and that the network has a giant component if and only if

$$p > p_c = \frac{\langle k \rangle}{\langle k(k-1) \rangle}. \tag{3.27}$$

This implies that scale-free networks have a vanishing percolation threshold also in bond percolation and are therefore very robust to random damage of the links. For Poisson networks and for networks with finite first, second and third moment of the degree distribution the dynamical exponent $\hat{\beta}$ for bond percolation takes the mean-field value

Table 3.2 *Dynamical critical exponent $\hat{\beta}$ for bond per-colation on power-law networks with degree distribution $P(k) = Ck^{-\gamma}$ [106]. For values of the power-law expo-nent $\gamma = 3,4$, logarithmic corrections to the scaling behaviour defined by Eq. (3.3.3) are observed.*

γ	$\hat{\beta}$
$\gamma > 4$	1
$\gamma \in (3,4)$	$1/(\gamma - 3)$
$\gamma \in (2,3)$	$(\gamma - 2)/(3 - \gamma)$

$\hat{\beta} = 1$. However the dynamical critical exponent $\hat{\beta}$ for bond percolation on scale-free networks is different from the one of node percolation (see Table 3.2 for the dynamical exponents of bond percolation on power-law networks).

3.3.4 Targeted attack

In percolation it is also possible to consider the case in which nodes or links are not randomly damaged but instead damaged according to a non-random strategy (targeted attack) [79]. For instance, one strategy could be to target the fraction f of nodes with highest degree or in general with the highest value of a centrality measure of interest. In particular, let us assume that nodes of degree k are initially damaged with probability $\phi(k)$. Then, following the same steps as described in the previous paragraphs, it can be shown that the network has a giant component if and only if [79]

$$\frac{\langle k(k-1)\phi(k)\rangle}{\langle k\rangle} > 1. \tag{3.28}$$

When a fraction f of nodes of highest degree is targeted by the attack strategy we have

$$\phi(k) = \theta(k - k_c), \tag{3.29}$$

where $\theta(x)$ is the step function and k_c is determined by the condition

$$\sum_{k \geq k_c} P(k) = f. \tag{3.30}$$

Interestingly, it has been observed that scale-free networks become fragile when the nodes of highest degree are targeted by the attack strategy. In fact, by targeting the high-degree nodes first, scale-free networks rapidly acquire a finite cutoff and behave like homogeneous networks displaying a finite percolation threshold [79].

3.4   Epidemic spreading

3.4.1   The relevance of epidemic spreading

Epidemic spreading processes are among the most studied dynamical processes in Network Science [24, 241]. In fact, they have many applications in biological, social and technological networks. Epidemic spreading models can be used to model diffusion of infectious diseases, diffusion of computer viruses across the Internet, diffusion of adoption of behaviour, rumour spreading and so on. A general problem in the context of infectious diseases relates to the prediction of the occurrence of an epidemic outbreak and its prevalence and relates to the design of efficient methods of immunization of the population. In the context of social networks and diffusion of rumours or behaviour, the prediction of the virality of news, or a product, is attracting large interest. Moreover, the identification of the influential spreaders can be central for marketing strategies of products.

Despite the large interest in the topic, epidemic spreading models can be very challenging, and predicting the size of an outbreak is a problem that requires large computer simulations and detailed information about the microstructure of the problem.

However, at the theoretical level it is possible to gain important insights on the mechanism, determining the properties of the epidemic spreading by considering the Susceptible-Infected-Susceptible (SIS) model and the Susceptible-Infected-Removed (SIR) model and characterizing their non-equilibrium behaviour with simulations and analytical approaches.

3.4.2   SIS and SIR models

The two major models of epidemic spreading are the Susceptible-Infected-Susceptible (SIS) model and the Susceptible-Infected-Removed (SIR) model. These models can be used to study the spreading process within a population in which the interaction pattern is dictated by an underlying network.

In the SIS model on a network we consider N individuals, each one associated to a different node $i = 1, 2, \ldots N$ of the network. Each individual i can be either susceptible (indicated with S_i) or infected (indicated with I_i). When an infected individual and a susceptible individual are on nearest neighbour nodes, the susceptible individual becomes infected at rate ξ. We indicate this process with

$$S_i + I_j \xrightarrow{\xi} I_i + I_j. \tag{3.31}$$

An infected individual becomes a susceptible individual at rate μ. We indicate this process with

$$I_i \xrightarrow{\mu} S_i. \tag{3.32}$$

This model is adopted in cases where the infected individuals are not removed from the population and can become eventually susceptible again. For instance, this model can be adopted as a stylized model of the spread of influenza where individuals can get influenza multiple times in the course of several years. Other situations in which the adoption of the SIS model could be appropriate are the infection of computers by electronic viruses, or the case of the shopping behaviour of some consumer who might decide to buy a product from a given brand repeatedly over time.

A different scenario leads to the definition of the SIR model where instead nodes can be infected only once, and after a certain period they are effectively removed from the population. This modelling framework is adopted in the case in which biological viruses either induce an immunization or are lethal. In social sciences the SIR model should be considered when the adoption of a given behaviour can occur only once. In the SIR model on a network we consider N individuals $i = 1, 2, \ldots, N$. Each individual is associated with a node $i = 1, 2, \ldots N$ of the network and can be in three possible states: susceptible, infected or removed. A susceptible individual (indicated by S_i) is an individual that is not infected but can get the infection. An infected individual (indicated by I_i) is an individual that has the infection and can spread it to neighbour nodes that are susceptible. Finally, a removed individual (indicated by R_i) is an individual that has had the infection but cannot spread it any longer.

The SIR dynamics on a complex network is defined as follows. When an infected individual and a susceptible individual are nearest neighbour nodes, the susceptible individual becomes infected at rate ξ. We indicate this process with

$$S_i + I_j \xrightarrow{\xi} I_i + I_j. \tag{3.33}$$

An infected individual becomes removed at rate μ. We indicate this process with

$$I_i \xrightarrow{\mu} R_i. \tag{3.34}$$

Both the SIS and the SIR models display a phase transition between an absorbing state in which the epidemic affects only an infinitesimal fraction of the nodes of the network and a phase consistent with epidemic outbreaks involving a finite fraction of nodes of the network.

3.4.3   Phase transitions in epidemic spreading

All epidemic spreading processes are non-equilibrium dynamical models in which there is an *absorbing state*. The absorbing state is the state in which the number of infected individuals eventually reaches the value zero and the epidemics cannot spread any longer. The SIS and SIR epidemic models have a phase transition as a function of the *infectivity* λ. The *infectivity* λ is the control parameter of the phase transition in both the SIS and the SIR and is given by

$$\lambda = \frac{\xi}{\mu}. \tag{3.35}$$

For low infection rates

$$\lambda \leq \lambda_c$$

the epidemic quickly reaches the absorbing phase involving a negligible fraction of the network nodes. However, for infection rates

$$\lambda > \lambda_c,$$

there is an epidemic outbreak in the network and a finite fraction of nodes are infected by the epidemics.

Epidemic spreading has different characteristics depending on the type of dynamics that takes place. For example, a significant difference can be observed in the typical profile of the number of infected individuals as a function of time in the case of the SIS model or in the case of the SIR model.

In the case of the SIS model, for $\lambda > \lambda_c$ we observe for large network sizes $N \to \infty$ an endemic state in which the fraction of infected individuals is greater than zero at all times, and therefore the epidemic never dies. In the case of the SIR model, for any values $\lambda > \lambda_c$ we observe that the epidemic will affect a finite fraction of the nodes, but will always eventually die out asymptotically in time.

The order parameter of the epidemic transition in both the SIS and SIR models is given by the fraction of nodes ρ that are not susceptible in the limit $t \to \infty$, and in the limit of infinite population $N \to \infty$, i.e.

$$\tilde{\rho} = \lim_{t \to \infty} \lim_{N \to \infty} \frac{N^I(t) + N^R(t)}{N}, \qquad (3.36)$$

where $N^I(t)$ is the number of infected nodes at time t, $N^R(t)$ is the number of removed nodes at time t and N is the total number of nodes. The order parameter has the following behaviour for $\lambda \simeq \lambda_c$:

$$\tilde{\rho} \propto \begin{cases} D(\lambda - \lambda_c)^{\hat{\beta}} & \text{for} \quad \lambda > \lambda_c \\ 0 & \text{for} \quad \lambda \leq \lambda_c \end{cases},$$

where λ_c is called the *epidemic threshold*, $\hat{\beta}$ is the *dynamical critical exponent* and $D > 0$ is a constant.

3.4.4 SIS model

The behaviour of the SIS epidemic spreading can be studied using a number of approximations showing the rich interplay between the network structure and the dynamics for the SIS epidemic spreading.

Let us consider a discrete time version of the SIS model and let us indicate with $X_i(t) = 1$ that node i is infected and with $X_i(t) = 0$ that node i is susceptible at time t.

A node i susceptible at time t becomes infected at time $t+1$ if at least one of its neighbours has infected it at time t. A node i, infected at time t, becomes susceptible at time $t+1$ with constant probability μ. Therefore, SIS dynamics can explicitly be written as

$$X_i(t) = 0 \rightarrow X_i(t+1) = 1 \text{ with probability } \left[1 - \prod_{j \in N(i)} \left(1 - \xi X_j(t) \right) \right]$$

$$X_i(t) = 1 \rightarrow X_i(t+1) = 0 \text{ with probability } \mu. \tag{3.37}$$

Therefore, the probability $p_i = \langle X_i \rangle$ that node i is infected satisfies

$$p_i(t+1) = \langle X_i(t+1) \rangle = \left\langle (1 - X_i(t)) \left[1 - \prod_{j \in N(i)} \left(1 - \xi X_j(t) \right) \right] \right\rangle$$

$$+ (1 - \mu) \langle X_i(t) \rangle . \tag{3.38}$$

The above equations are not closed over the variables p_i. In fact, the value of $p_i = \langle X_i \rangle$ depends on the average of the products of more than one variable X_j. To solve Eqs (3.38) exactly one would need to couple them with other equations for the average of the products between several stochastic variables X_j. Unfortunately, this approach yields a hierarchy of equations that do not close, so inevitably it is necessary to make an approximation to close a finite set of such equations. The most drastic approximation is called the individual mean-field approximation. Essentially the individual mean-field approximation neglects fluctuations and assumes that

$$\langle X_{n(1)} X_{n(2)} \dots X_{n(r)} \rangle \simeq \langle X_{n(1)} \rangle \langle X_{n(2)} \rangle \dots \langle X_{n(r)} \rangle \tag{3.39}$$

for any sequence of indices $n(1), n(2), \dots n(r)$. In this approximation Eqs (3.38) close over the variables $p_i(t)$ and read

$$p_i(t+1) = (1 - p_i(t)) \left[1 - \prod_{j \in N(i)} \left(1 - \xi p_j(t) \right) \right] + (1 - \mu) p_i(t). \tag{3.40}$$

The steady-state solution of this equation

$$p_i(t) = p_i(t+1) = p_i^\star, \tag{3.41}$$

achieved by the system for $t \gg 1$ satisfies

$$0 = -\mu p_i^\star + (1 - p_i^\star) \left[1 - \prod_{j \in N(i)} \left(1 - \xi p_j^\star \right) \right]. \tag{3.42}$$

This equation always admits an epidemic-free solution $p_i^\star = 0$, but this solution can become unstable for infectivities

$$\lambda = \frac{\xi}{\mu} > \lambda_c \tag{3.43}$$

where λ_c is called the epidemic threshold. In order to study the onset of this instability we linearize Eq. (3.42) for $p_i^\star \ll 1$, obtaining

$$\mu p_i^\star = \xi \sum_{j=1}^{N} a_{ij} p_j^\star. \tag{3.44}$$

This equation reveals that small perturbations of the epidemic-free state $p_i^\star$ are enhanced, leading to an instability of the epidemic-free state for

$$\frac{\xi}{\mu} \Lambda > 1, \tag{3.45}$$

where Λ is the maximum eigenvalue of the adjacency matrix **a**. Therefore, the epidemics become endemic for

$$\lambda > \lambda_c = \frac{1}{\Lambda}. \tag{3.46}$$

A closer look at this derivation reveals that this result obtained within the mean-field approximation is restricted to networks in which the eigenvector associated with the maximum eigenvalue Λ is delocalized on the network.

Assuming that the SIS dynamics takes place over infinitesimal intervals of time $\Delta t \ll 1$, we rescale the parameter of the dynamics as

$$\xi \to \xi \Delta t$$
$$\mu \to \mu \Delta t \tag{3.47}$$

obtaining

$$\left[1 - \prod_{j \in N(i)} \left(1 - \xi p_j(t) \right) \right] \to \left[1 - \prod_{j \in N(i)} \left(1 - \xi p_j(t) \Delta t \right) \right]. \tag{3.48}$$

For $\Delta t \ll 1$ we can approximate

$$\left[1 - \prod_{j \in N(i)} \left(1 - \xi p_j(t) \Delta t \right) \right] \sim \xi \sum_{j=1}^{N} a_{ij} p_j(t) \Delta t. \tag{3.49}$$

Inserting this expression into Eqs (3.40), in the limit $\Delta t \to 0$ we obtain the continuous time equation

$$\frac{dp_i}{dt} = -p_i + \lambda(1 - p_i) \sum_{j=1}^{N} a_{ij} p_j, \tag{3.50}$$

where, without loss of generality, we have rescaled the time according to $t \to t/\mu$. The stability of the epidemic-free stationary state can be studied directly from this equation, yielding exactly the same epidemic threshold as derived from the discrete time mean-field dynamics. Additionally, this equation can be studied in the framework of the so-called annealed approximation. In the *annealed approximation* one assumes that each node rewires its connections on the same temporal scale at which the epidemic spreads, while keeping the number of its connections constant. In this scenario it is allowed to substitute the adjacency matrix elements a_{ij} with their average value over an ensemble of graphs, $a_{ij} \to p_{ij}$. For uncorrelated networks with a given degree sequence, one makes the approximation

$$a_{ij} \to \frac{k_i k_j}{\langle k \rangle N}. \tag{3.51}$$

Therefore, in the annealed network approximation p_i only depends on the degree of the node i and we have

$$p_i|_{k_i=k} = \rho_k. \tag{3.52}$$

By using the annealed approximation in Eq. (3.50) it can be derived that the probability ρ_k that a node of degree k is infected follows the equation [243]

$$\frac{d\rho_k}{dt} = -\rho_k + \lambda k(1 - \rho_k) \sum_{k'} \frac{k'}{\langle k \rangle} P(k') \rho_{k'}. \tag{3.53}$$

At stationarity we have

$$\rho_k = \lambda k \frac{\Theta(\lambda)}{1 + \lambda k \Theta(\lambda)}, \tag{3.54}$$

where

$$\Theta(\lambda) = \sum_{k} \frac{k}{\langle k \rangle} P(k) \rho_k. \tag{3.55}$$

The epidemic-free solution $\rho_k = 0$ can become unstable for infection rate λ greater than the epidemic threshold λ_c. In order to derive λ_c in this annealed approximation framework we linearize Eq. (3.54) for $\rho_k \ll 1$, obtaining

$$\rho_k = \lambda k \sum_k \frac{k'}{\langle k \rangle} P(k') \rho'_k. \tag{3.56}$$

This eigenvalue problem determines the condition for the instability of the epidemic-free stationary solution $\rho_k = 0$. Specifically, the system will display an epidemic outbreak for

$$\lambda \frac{\langle k^2 \rangle}{\langle k \rangle} > 1. \tag{3.57}$$

Therefore the epidemic threshold λ_c is given by [243]

$$\lambda_c = \frac{\langle k \rangle}{\langle k^2 \rangle}. \tag{3.58}$$

This relevant result is signalling that also in this case scale-free network topologies have dramatic effects on the network dynamics. In fact, while for homogenous network with finite first and second moment of the degree distribution the epidemic threshold λ_c is finite, for scale-free networks the epidemic threshold vanishes, i.e.

$$\lambda_c \to 0 \text{ as } N \to \infty.$$

This implies that every epidemic, regardless of its infection rate, becomes endemic in infinite scale-free networks.

3.4.5   SIR model

The SIR dynamics admits a mapping to the bond-percolation problem and an exact solution on locally tree-like networks.

Let us define the *transmissibility* T as the probability that an infected node transmits the infection to a neighbour susceptible node while it is infected. The mapping of the SIR dynamics to bond percolation can be performed as follows. The cluster of removed nodes generated by an SIR epidemic outbreak starting from a single infected node is mapped to the percolation cluster of the network connected to the initial seed of the infection, in which the probability p that each link is initially retained (not removed from the network) is given by $p = T$. Therefore, an SIR epidemic outbreak affecting a finite fraction of the nodes of the network corresponds to the percolating phase of bond percolation. Given the mapping of the SIR to bond percolation with an epidemic threshold determined by Eq. (3.27), putting $p = T$ we have that the epidemic outbreak is predicted to occur in a network if [223]

$$T > T_c = \frac{\langle k \rangle}{\langle k(k-1) \rangle}. \tag{3.59}$$

Let us now relate the transmissibility T with the infection rate λ following the steps outlined in Ref. [223]. To this end, let us first notice that since each infected node is removed at constant rate μ, the distribution $P(\tau)$ of the time τ that an infected node remains infected is a Poisson distribution with mean $\langle \tau \rangle = \frac{1}{\mu}$, i.e.

$$P(\tau) = \mu e^{-\mu \tau}. \tag{3.60}$$

Secondly, let us define the τ-transmissibility T_τ as the probability that an infected node spreads the infection to a neighbour susceptible node over time τ. The τ-transmissibility T_τ is given by

$$T_\tau = 1 - \lim_{\delta t \to 0} (1 - \xi \delta t)^{\tau/\delta t} = 1 - e^{-\xi \tau}. \tag{3.61}$$

The transmissibility T that any infected node transmits the infection is given by the average of T_τ over the distribution $P(\tau)$ describing the distribution of times that a random infected node remains infected, i.e.

$$T = 1 - \int_0^\infty d\tau P(\tau) e^{-\xi \tau} = \frac{\lambda}{\lambda + 1}. \tag{3.62}$$

Using Eq. (3.59) together with Eq. (3.62), we get that a network will sustain an epidemic outbreak when the infection rate λ satisfies

$$\lambda > \lambda_c = \frac{\langle k \rangle}{\langle k^2 \rangle - 2\langle k \rangle}. \tag{3.63}$$

It follows that if the SIR dynamics takes place on a network having a finite second moment of the degree distribution, the epidemic threshold is finite. If, however, the network is scale free, the epidemic threshold goes to zero, i.e. $\lambda_c \to 0$ as $N \to \infty$. Therefore, independently of the value of λ, every SIR epidemic process generates an outbreak.

In some cases, the SIR dynamics is modified by fixing τ the duration of an infection (the time required by an infected individual to be removed) to a constant. In this case the transmissibility T is simply given by

$$T = 1 - e^{-\xi \tau} \tag{3.64}$$

and the epidemic threshold is determined by Eq. (3.59).

3.4.6 Immunization strategies

Both in the framework of the SIS and the SIR model we have shown that scale-free topology appears to be very advantageous for epidemic spreading. In fact, the vanishing epidemic threshold implies that even the smaller infectivity gives rise to an epidemic outbreak. This phenomenon is particularly relevant for the worldwide spread of infectious diseases, because in the modern world long-distance travel is dominated by the scale-free topology of the airport networks. In the context of electronic viruses this phenomenon is also very important, and can explain the long lifetime of computer viruses spreading on the Internet despite the prompt availability of anti-virus software. A major question in this context is how can we tame epidemic spreading? Are there more specific immunization strategies that are more efficient? In percolation theory we have seen that targeting nodes of high degree is a strategy that dismantles scale-free graphs quickly. This phenomenon has positive consequences in the SIR dynamics when we consider targeted immunization of high-degree nodes. In fact, under targeted immunization scale-free networks also acquire a finite epidemic threshold and the spread of the epidemics can actually be contained [244].

In the following, we give the expression for the epidemic thresholds of the SIS and SIR models when each node i is immunized with probability r_i. To this end, let us assume both in the SIS model and in the SIR model that a susceptible node i in contact with a neighbour infected node gets the infection with probability $\xi(1 - r_i)$. By following derivations similar to the ones of the precedent paragraphs for the SIS model treated with the individual mean-field approximation, the epidemic threshold λ_c becomes

$$\lambda_c = \frac{1}{\Lambda_g},\tag{3.65}$$

where Λ_g is the largest eigenvalue of the matrix $\mathbf{g}$ of elements $g_{ij} = (1-r_i)a_{ij}$. If we assume that the probability r_i that node i is immunized depends only on its degree $k_i = k$, and therefore $r_i = r_{k_i}$, we can perform the annealed approximation that gives for the SIS model the epidemic threshold

$$\lambda_c = \frac{\langle k \rangle}{\langle k^2(1 - r_k) \rangle}.\tag{3.66}$$

Finally for the SIR model, assuming that r_i is only a function of the degree of node i, i.e. $r_i = r_{k_i}$, by performing the mapping with bond percolation the critical value T_c for the transmissibility T is given by

$$T_c = \frac{\langle k \rangle}{\langle k(k-1)(1 - r_k) \rangle}.\tag{3.67}$$

From these equations it emerges that if a fraction of about 10% of nodes with highest degree is immunized, typically scale-free networks with minimum degree equal to one develop a finite epidemic threshold.

3.5   Diffusion and random walks

3.5.1   Diffusion

One of the most fundamental dynamical processes on networks is diffusion described by the transport of a continuous quantity along the links of a network. Diffusion on random networks is described by the equation

$$\frac{dx_i}{dt} = \sum_{j=1}^{N} a_{ij}(x_j - x_i) \qquad (3.68)$$

where x_i is a continuous variable assigned to the generic node $i = 1, 2, \ldots N$ of the network. This equation extends the famous heat equation defined for a continuous medium to discrete networks. It can be written in matrix form as

$$\frac{d\mathbf{x}}{dt} = -\mathbf{L}\mathbf{x} \qquad (3.69)$$

where $\mathbf{x}$ is the vector describing the dynamical state of each node, i.e. $\mathbf{x} = (x_1, x_2, \ldots x_N)^T$, and $\mathbf{L}$ is the Laplacian matrix of the network having elements

$$L_{ij} = k_i \delta_{ij} - a_{ij}. \qquad (3.70)$$

The dynamical Eq. (3.69) given the initial condition $\mathbf{x}(0) = \mathbf{x}_0$ has the solution

$$\mathbf{x(t)} = e^{-\mathbf{L}t}\mathbf{x}_0. \qquad (3.71)$$

This implies that the Laplacian matrix fully determines the properties of diffusion on a network. Since the spectral properties of a given Laplacian codify important structural aspects of the corresponding network, the Laplacian plays a key role in establishing the interplay between the structure of the network and the dynamics of the diffusion process. Let us briefly summarize a few major spectral characteristics of the Laplacian for a simple network.

(i) The Laplacian is semidefinite positive, implying that all the eigenvalues satisfy $\lambda_n \geq 0$. Here and in the following we order the eigenvalues in non-decreasing order: $\lambda_1 \leq \lambda_2 \leq \ldots \lambda_N$.

(ii) The Laplacian always admits a zero eigenvalue $\lambda_1 = 0$ whose corresponding eigenvector, as long as the network is connected, is uniform over all the nodes of the network.

(iii) The degeneracy of the zero eigenvalue is exactly equal to the number of connected components in the network. Therefore, any connected network has a single zero eigenvalue.

Large connected networks (with $N \gg 1$) can be distinguished according to their spectral properties, as in the following:

(a) If the smallest non-zero eigenvalue λ_2 is well separated from the zero eigenvalue λ_1, we say that the network has a spectral gap.

(b) If the smallest non-zero eigenvalue is very close to zero, $\lambda_2 \ll 1$, we say that the network does not have a spectral gap. In this scenario, if the density of eigenvalues for small values of λ scales like

$$\rho(\lambda) \simeq \lambda^{d_s/2-1}, \tag{3.72}$$

we say that the network has spectral dimension d_s. Notably lattices in d dimensions have spectral dimension $d_s = d$.

Diffusion on a complex network is strongly affected by the presence or absence of a spectral gap. In fact, if we decompose the initial condition $\mathbf{x}_0$ into the basis of eigenvectors $\mathbf{u}^{(n)}$ of the Laplacian as

$$\mathbf{x}_0 = \sum_n c_n \mathbf{u}^{(n)}, \tag{3.73}$$

Eq. (3.71) reads

$$\mathbf{x}(t) = \sum_n e^{-\lambda_n t} c_n \mathbf{u}^{(n)}. \tag{3.74}$$

Now in the presence of a spectral gap we can approximate this expression as

$$\mathbf{x}(t) \simeq \mathbf{u}^{(1)} + e^{-\lambda_2 t} \mathbf{u}^{(2)}, \tag{3.75}$$

where $\mathbf{u}^{(1)}$ is the homogeneous eigenvector of the zero eigenvalue $\lambda_1 = 0$ representing the steady state of the diffusion dynamics, and $\mathbf{u}^{(2)}$ is the eigenvector associated with the first non-zero eigenvalue λ_2. It follows that in the presence of a spectral gap, the typical relaxation timescale τ of the diffusion dynamics is

$$\tau = \frac{1}{\lambda_2}. \tag{3.76}$$

On the contrary, if there is no spectral gap, the relaxation dynamics might not be exponential anymore but can instead be power-law.

3.5.2   Random walk

The random walk describes the diffusion of a particle on a complex network. According to an unbiased random walk, the particle located at node i moves with equal probability to every neighbour node j of node i. In an undirected, unweighted and connected network, assuming that each node of the network has non-zero degree, the rate P_{ij} at which the particle hops from node i to node j is given by

$$P_{ij} = \frac{a_{ij}}{k_i}. \tag{3.77}$$

Assuming that the random walker starts at time $t = 0$ from node i_0, the probability $\pi_i(t)$ that the random walker is on node i at time t follows the master equation

$$\frac{d\pi_i(t)}{dt} = \sum_{j=1}^{N} P_{ji}\pi_j(t) - \left(\sum_{j=1}^{N} P_{ij} \right) \pi_i(t), \tag{3.78}$$

with initial condition $\pi_i(0) = \delta\,(i, i_0)$, where $\delta(x, y)$ is the Kronecker delta. Given that $\sum_{j=1}^{N} P_{ij} = 1$, the above equation can be equivalently written as

$$\frac{d\pi}{dt} = -\tilde{L}\pi\,(t) \tag{3.79}$$

where the vector π is given by $\pi = (\pi_1, \pi_2, \ldots \pi_N)^T$ and $\tilde{L}$ is the normalized Laplacian with elements

$$\tilde{L}_{ij} = \delta_{ij} - \frac{a_{ij}}{k_j}. \tag{3.80}$$

Therefore Eq. (3.79) has the solution

$$\pi\,(t) = e^{-\tilde{L}t}\pi\,(0). \tag{3.81}$$

The stationary state of the unbiased random walk is given by $\pi_i(t) = \mu_i$ satisfying

$$0 = \sum_{j=1}^{N} \tilde{L}_{ij}\mu_j = \mu_i - \sum_{j=1}^{N} \mu_j P_{ji}, \tag{3.82}$$

i.e. μ_i is the right eigenvector associated with the zero eigenvalue of the normalized Laplacian. Since μ_i is the probability that asymptotically in time the random walk is on node i, the set of all μ_i is normalized, i.e.

$$\sum_{i=1}^{N} \mu_i = 1, \tag{3.83}$$

which yields together with Eq. (3.82)

$$\mu_i = \frac{k_i}{\langle k \rangle N}. \tag{3.84}$$

This implies that asymptotically in time the probability that the random walker is at node i is proportional to its degree k_i independently of the network structure and the initial condition of the random walk as long as the network is connected.

The relaxation dynamics that describes the transient regime on the contrary depends significantly on the spectral properties of the normalized Laplacian. It is to be noted that the spectrum of the normalized Laplacian reduces to the spectrum of the Laplacian only in the case of networks with constant degree of the nodes. Therefore the two spectra are typically distinct. Let us briefly summarize a few major spectral characteristics of the normalized Laplacian.

(i) The normalized Laplacian is semidefinite positive, having all the eigenvalues real and positive satisfying $\tilde{\lambda}_n \geq 0$. Here and in the following, we order the eigenvalues in non-decreasing order $\tilde{\lambda}_1 \leq \tilde{\lambda}_2 \leq \ldots \leq \tilde{\lambda}_N$.

(ii) The matrix is asymmetric, therefore for every eigenvalue $\tilde{\lambda}_n$ there are corresponding right $(\tilde{\mathbf{u}}_R^{(n)})$ and left $(\tilde{\mathbf{u}}_L^{(n)})$ eigenvectors normalized according to

$$\sum_i \tilde{u}_{i,R}^{(n)} \tilde{u}_{i,L}^{(n')} = \delta(n, n'). \tag{3.85}$$

The right and left eigenvectors related to the same eigenvalue have elements that satisfy

$$\tilde{u}_{i,R}^{(n)} = k_i \tilde{u}_{i,L}^{(n)}. \tag{3.86}$$

(iii) The Laplacian always admits a zero eigenvalue $\tilde{\lambda}_1 = 0$ whose corresponding left eigenvector is uniform over all the nodes of the network, and whose corresponding right eigenvector is given by

$$\tilde{\mathbf{u}}_R^{(n)} = \sqrt{\langle k \rangle N} (\mu_1, \mu_2, \ldots \mu_N)^T, \tag{3.87}$$

with μ_i given by Eq. (3.84) and the left eigenvector

$$\tilde{\mathbf{u}}_L^{(n)} = \frac{1}{\sqrt{\langle k \rangle N}} (1, 1, \ldots, 1)^T. \tag{3.88}$$

(iv) The degeneracy of the zero eigenvalue is exactly equal to the number of connected components of the network. Therefore, any connected network has a single zero eigenvalue.

Large connected networks (with $N \gg 1$) can be distinguished according to their spectral properties as in the following:

(a) If the first non-zero eigenvalue $\tilde{\lambda}_2$ is well separated from the zero eigenvalue $\tilde{\lambda}_1 = 0$ we say that the network has a spectral gap.

(b) If the first non-zero eigenvalue is very close to zero, $\tilde{\lambda}_2 \ll 1$ we say that the network does not have a spectral gap. In this scenario, it is said that the network has spectral dimension $\tilde{d}_S$ if the density of eigenvalues $\tilde{\rho}(\tilde{\lambda})$ follows

$$\tilde{\rho}(\tilde{\lambda}) \simeq \tilde{\lambda}^{\tilde{d}_s/2 - 1} \tag{3.89}$$

for $\tilde{\lambda} \ll 1$.

In the presence of the spectral gap, Eq. (3.81) reveals that $\tilde{\lambda}_2$ characterizes the typical scale of the relaxation dynamics. In fact, by projecting the equation on the eigenvector of the normalized Laplacian we get

$$\pi_i(t) = \sum_{n=1}^{N} e^{-\tilde{\lambda}_n t} \tilde{u}_{i,R}^{(n)} \tilde{u}_{i_0,L}^{(n)}. \tag{3.90}$$

By observing that the right eigenvector and the left eigenvector associated with the eigenvalue $\tilde{\lambda}_1 = 0$ have elements given by Eq. (3.87) and by Eq. (3.88) respectively, the above expression for $\pi_i(t)$ indicates that the typical timescale τ of the random walk relaxation dynamics to the stationary state $\pi_i(t) = \mu_i$ is given by

$$\tau = \frac{1}{\tilde{\lambda}_2}. \tag{3.91}$$

The relaxation in the presence of a finite spectral dimension can be much slower. Let us consider for instance the probability $p_0(t)$ that a random walker initially at a random node of the network returns to this node at time t. This probability is given by

$$p_0(t) = \frac{1}{N} \sum_{i_0=1}^{N} \sum_{n=1}^{N} e^{-\tilde{\lambda}_n t} \tilde{u}_{i_0,R}^{(n)} \tilde{u}_{i_0,L}^{(n)},$$

$$= \frac{1}{N} \sum_{n=1}^{N} e^{-\tilde{\lambda}_n t}. \tag{3.92}$$

For networks without a spectral gap we have

$$p_0(t) = \int d\tilde{\lambda} \tilde{\rho}(\tilde{\lambda}) e^{-\tilde{\lambda} t}, \tag{3.93}$$

yielding in the presence of a finite spectral dimension $\tilde{d}_S$ slow dynamics of the random walk characterized by having

$$p_0(t) \simeq t^{-\tilde{d}_s/2}. \tag{3.94}$$

This result implies that for $\tilde{d}_S > 2$ the random walk is transient and there is a non-zero probability that the random walk never returns to its origin.

Until now, we have characterized the properties of the unbiased random walk exclusively. However, sometimes it is important to study the dynamics of a particle that hops from a node to any of its neighbours according to some predefined preference or bias. For instance, a random walk could prefer to hop to nodes of high degree or low degree or to nodes having a high centrality measure. In these cases we will call the random walk a biased random walk.

Let us indicate with f_i the property determining the bias of the random walk, then the probability that the particle hops from node i to node j is taken to be

$$P_{ij} = \frac{a_{ij}f_j}{\sum_r a_{ir}f_r}. \tag{3.95}$$

If the random walk is biased toward nodes of high degree we could for instance take $f_i = k_i^\beta$ with $\beta > 0$. If it is biased toward nodes of high centrality x_i we could take $f_i = x_i$. The analysis of the biased random walk can be conducted along the same lines as for characterizing the unbiased random walk. It is worth mentioning that the stationary state of the unbiased random walk is given by μ_i satisfying

$$\mu_i = \sum_{j=1}^{N} \mu_j \frac{a_{ji}f_i}{\sum_r a_{jr}f_r}. \tag{3.96}$$

3.6 Synchronization

3.6.1 Master Stability Function

Given a generic dynamical system of coupled differential equations defined on a network structure, the Master Stability Function [246, 17] establishes under which conditions the fully synchronized state is stable. The success of this approach is rooted in the very nature of its results that provide reliable stability conditions independently of the details of the dynamical process under consideration.

Here the structural properties of a network will be shown to have profound implications on the dynamics taking place on it. In particular, the Master Stability Function determining the stability of the fully synchronized state will be directly dependent on the spectral properties of the Laplacian of the network.

Let us assume that each node i of the network is assigned a dynamical variable $x_i \in \mathbb{R}$, the dynamics of which is dictated by the differential equation

$$\frac{dx_i}{dt} = f(x_i) + \sigma \sum_{j=1}^{N} a_{ij} \left[H(x_j) - H(x_i) \right], \tag{3.97}$$

where $f(x)$ and $H(x)$ are continuous and differentiable functions and $\sigma > 0$ is a tunable parameter modulating the intensity of the coupling between the dynamical states of neighbour nodes. This dynamical system always admits a fully synchronized state $x_i(t) = s(t)$, where $s(t)$ is the solution of the differential equation

$$\frac{ds}{dt} = f(s). \tag{3.98}$$

In order to study the stability of this solution we assume that $x_i = s(t) + \xi_i(t)$ with $\xi_i(t) \ll 1$ and we linearize the dynamical system obtaining for $\boldsymbol{\xi} = (\xi_1, \xi_2, \ldots, \xi_N)^T$

$$\frac{d\boldsymbol{\xi}}{dt} = \left[f'(s)\delta_{ij} + \sigma \sum_{j=1}^{N} L_{ij} H'(s) \right] \boldsymbol{\xi}, \tag{3.99}$$

where L_{ij} indicates the elements of the Laplacian matrix given by Eq. (3.70). By projecting the vector $\boldsymbol{\xi}$ into the basis of eigenvectors of the Laplacian we get the system of equations

$$\frac{d\eta_n}{dt} = \left[f'(s) + \sigma \lambda_n H'(s) \right] \eta_n, \tag{3.100}$$

where λ_n with $n = 1, 2, \ldots, N$ indicates the eigenvalues of the Laplacian in non-increasing order and η_n indicates the component of the vector $\boldsymbol{\xi}$ along the direction of the nth eigenvector of the Laplacian. For $n = 1$, Eq. (3.100) characterizes the dynamics of the fully synchronized state. In fact, the eigenvector associated with the eigenvalue $\lambda_1 = 0$ is uniform over all the nodes of the network. For $n > 1$, Eq. (3.100) instead characterizes the evolution of small perturbations from the fully synchronized state.

Requiring that the fully synchronized state is stable implies that for every value of s the perturbation is not exponentially enhanced by the dynamics. By defining $\Lambda(\sigma \lambda_n)$ given by

$$\Lambda(\sigma \lambda_n) = \max_s \left[f'(s) + \sigma \lambda_n H'(s) \right] \tag{3.101}$$

where the maximum is taken over the trajectory $\dot{s}(t) = f(s(t))$, we derive that the condition for the stability of the synchronized state reads

$$\Lambda(\sigma \lambda_n) \leq 0 \quad \forall n = 2, 3, \ldots, N. \tag{3.102}$$

3.6.2   The Kuramoto model

The Kuramoto model [9, 266] is a stylized model which describes the onset of a synchronized state when a set of oscillators, each having a different internal frequency, are coupled to each other. The coupling can be assumed to be weak but all-to-all, or to take place between oscillators that are on neighbour nodes of a given network. Here we consider the latter scenario and we assume that each node i of the network is associated with the dynamical variable θ_i of a given oscillator. The internal frequency ω_i of each node i is assumed to be drawn from a distribution $g(\omega)$ that is unimodal and symmetric around the mean frequency $\Omega = 0$. Nearby nodes on the network are coupled dynamically according to the equations

$$\frac{d\theta_i}{dt} = \omega_i + \sigma \sum_{j=1}^{N} a_{ij} \sin(\theta_j - \theta_i), \qquad (3.103)$$

where σ indicates the coupling constant. The dynamical state of the network can be monitored by considering the global variable

$$re^{i\psi} = \frac{1}{N} \sum_{j=1}^{N} e^{i\theta_j(t)}. \qquad (3.104)$$

In fact, if the system is in a synchronized state, all the oscillators have the same phase $\theta_i(t) = \theta^\star(t)$, resulting in $r = 1$. On the contrary, if the dynamical system is not in the synchronized state the phases of the oscillators will be incoherent, resulting in a value of $r \simeq 0$.

As a function of the coupling constant σ the system displays a continuous phase transition from an incoherent phase (for $\sigma < \sigma_c$) to a synchronized phase (for $\sigma \geq \sigma_c$). Specifically, in the limit of infinite network sizes $N \to \infty$ the order parameter r for $\sigma \simeq \sigma_c$ takes values

$$r = \begin{cases} A(\sigma - \sigma_c)^{\hat{\beta}} & \text{for } \sigma \geq \sigma_c, \\ 0 & \text{for } \sigma < \sigma_c, \end{cases} \qquad (3.105)$$

with $\hat{\beta} > 0$ indicating the critical exponent of the transition and $A > 0$ being a constant.

In the annealed network approximation the critical value σ_c of the coupling constant can be evaluated to be dependent on the degree distribution of the network and is given by

$$\sigma_c = \frac{2}{\pi g(0)} \frac{\langle k \rangle}{\langle k^2 \rangle}. \qquad (3.106)$$

In this approximation scale-free networks have $\sigma_c \to 0$ as $N \to \infty$ and therefore are easier to synchronize than networks with homogeneous degree distributions. However the validity of this result is only guaranteed if the network continuously rewires its links by keeping the same degree sequence. Interestingly, we observe additional signs of the rich interplay between structure and dynamics. In fact, significant differences are observed between random Erdös and Rényi (ER) networks and the BA scale-free network when one focuses on the pattern to synchronization [137]. Here by pattern to synchronization we mean the organization process that occurs in the network as the coupling constant σ is increased, ultimately giving rise to the synchronized state. For ER networks we observe that the clusters of synchronized nodes are disconnected for low values of the coupling constant σ, and as the coupling constant is increased they aggregate according to a percolation-like process. For scale-free networks, instead hubs play the role of catalysts for the formation of a large synchronization cluster and therefore a single synchronization cluster including the hubs forms the core of the synchronized region of the network and aggregates progressively other small synchronized clusters.

Interestingly, it has recently been shown that the synchronization transition can become discontinuous when the internal frequencies of the nodes are dependent on their degrees. In particular, in [136] it has been shown both numerically and analytically that when the internal frequency ω_i of node i depends on the node degree k_i as

$$\omega_i \propto k_i^\theta \tag{3.107}$$

with $\theta > 0$, the synchronization is abrupt and discontinuous. As a consequence of this phenomenon, this interesting phase transition has been called 'explosive synchronization'.

3.7   Network control

Network control [196] determines the conditions under which, by applying a set of external signals to some of the nodes of a network (the *driver nodes*), it is possible to drive the network to any desired dynamical state in finite time [274, 197, 207]. This problem is of fundamental importance for biological networks, where a major goal is, for instance, to reprogram cancer cells or understand brain function, but also has wider applications in social, technological and financial networks.

We will indicate with the P-dimensional vector $\mathbf{u}$ the number of external signals. Assuming that the internal dynamics of the network is linear, the dynamical state of the network is described by a linear differential equation for the N-dimensional vector $\mathbf{x}$ whose element x_i indicates the state of the node i. This differential equation reads

$$\frac{d\mathbf{x}}{dt} = \mathbf{ax} + \mathbf{bu}, \tag{3.108}$$

where $\mathbf{a}$ is the $N \times N$-weighted and -directed adjacency matrix of the network describing the coupling between different dynamical variables linked by the network, and $\mathbf{b}$ is the $N \times P$ matrix determining the action of the external signals $\mathbf{u}$ on the dynamical state of the network. In particular, the matrix $\mathbf{b}$ indicates which nodes are the driver nodes, i.e. directly receive the external signals.

A network is controllable when its dynamical state can be determined by suitably choosing the external signals. According to the Kalman's controllability rank condition theorem [166], a network is controllable if the matrix

$$\mathbf{c} = (\mathbf{b}, \mathbf{ab}, \mathbf{a}^2\mathbf{b}, \dots, \mathbf{a}^{N-1}\mathbf{b}) \tag{3.109}$$

has full rank.

This condition depends non-trivially on the entries of the weighted adjacency matrix $\mathbf{a}$. However, in a number of cases the values of the internal couplings between the dynamical variables of the network are not known and the accessible information is only topological, i.e. the available information only indicates which pairs of nodes interact with each other.

Under these conditions it is appropriate to consider the framework of structural controllability, which guarantees the controllability of a network for a random choice of the non-zero and real-valued entries of the matrices **a** and **b** with probability one. Specifically, in a number of applications the minimum number of driver nodes that ensures the network's structural controllability is the major parameter to evaluate how challenging it is to control a given network.

Using the framework of structural controllability, Liu et al. [197] showed that the minimum set of driver nodes can be found by mapping the problem into a Maximum Matching Problem.

The Maximum Matching Problem is a combinatorial optimization problem that assumes that every link of the network is either matched or unmatched. Additionally, it imposes that every node must have at most one incoming and one outgoing matched link. Nodes are matched if they have one matched incoming link, otherwise they are unmatched. The Maximum Matching Problem finds the matching configuration of the links for which the set of unmatched nodes is minimal.

The link between the Maximum Matching Problem and network controllability is established by the Minimum Input Theorem [197]. This theorem states that the minimum set of driver nodes that guarantees the full structural controllability of a single network is the set of unmatched nodes in a maximum matching of the same directed network. Therefore, the network controllability can be recast into the Maximum Matching Problem which can be studied using the Hopcroft–Karp maximum matching optimization algorithm or using the statistical mechanics message-passing algorithm called Belief Propagation.

Interestingly, for network controllability we see that the key structural properties that determine the network's structural controllability are not the scale-free distribution and the presence of big hubs in the network like in percolation or epidemic spreading. Rather, it can be shown that in locally tree-like networks the factor that determines the number of driver nodes is the density of nodes with low in- and out-degree, specifically with in-degree or out-degree less or equal than two [207]. For these networks, the larger the fraction of nodes with low in- or out-degree, the larger the number of driver nodes. When all the nodes of the network have an in- and out-degree greater than two, it is possible to structurally control the network with just one external signal [207].

Part III

Multilayer Networks

4

Multilayer Networks in Nature, Society and Infrastructures

4.1 Multilayer networks: the general multidisciplinary framework

As the vast majority of complex systems, from the brain to the Internet, are formed by interacting elements, networks are ubiquitous. As a result, Network Science can be considered as one of the most prosperous scientific fields of current times. The multilayer network approach constitutes a very recent development of the field, where the focus is characterizing the interactions of several interconnected networks. In fact, complex systems are rarely formed by single, isolated networks with links of equivalent meaning and connotation. For instance, in social networks we can distinguish between several types of social ties (friends, colleagues, acquaintances, family ties, etc.), in transportation networks we can distinguish between different means of transportation (bus, metro, train, etc.). Also in molecular biology and brain networks the interactions can have different valences and it is important to study these systems with a comprehensive multilayer framework that allows us to treat the differences existing between different types of interactions.

Multilayer networks have been first introduced in the context of social science to characterize the different types of social ties existing between the nodes of a social network [120, 308, 245]. However, only recently the relevance of considering multilayer network structures has been recognized more widely. Currently multilayer networks are investigated in many fields, including neuroscience, molecular biology, ecology, economy, transportation networks, infrastructures and climate. In this chapter we provide an overview of the different contexts where the multilayer network has been proven to provide a much more comprehensive view of complex systems than the single network framework. The research in this field is growing at a very intense pace, therefore we would like to stress that our account of the possible multilayer network applications is only partial, and we apologize if we are unable to cover the entire growing literature in the field.

Multilayer Networks. Ginestra Bianconi, Oxford University Press (2018).
© Ginestra Bianconi. DOI: 10.1093/oso/9780198753919.001.0001

4.2   Multilayer networks: a gentle introduction

In this section, we give a first, informal definition of multilayer networks that will allow the reader to appreciate the differences between the large variety of possible applications of this framework better.

A multilayer network is formed by several interacting networks. A given multilayer formed by distinct M layers is formed by a set of M networks describing the interactions within each layer and $M(M-1)/2$ networks describing the interactions between nodes in every pair of different layers. This general scenario can be simplified in multiple ways. Here we mention three particular classes of multilayer networks: multiplex networks, multi-slice networks and networks of networks.

Multiplex networks constitute the simplest example of multilayer networks. Multiplex networks are often used in a case in which the same set of nodes is connected by links indicating the different types of interactions. By associating a different colour with each type of link, multiplex networks can be represented as a single coloured network. However, it is possible also to assume that links of the same type form the layers of the multiplex network. For example, a social network can be represented by a multiplex in which the same set of people might interact via email (first layer) or via mobile phone (second layer). Similarly, transportation networks within a city (let us say London) can be represented by a multiplex with, for instance, one layer indicating bus connections and another layer indicating tube/metro connections. The links within each layer represent different types of interactions (email/phone-call contacts; bus/tube connections). The links across different layers are placed exclusively between corresponding (replica) nodes in the different layers. For example, in the previous example a given link across the two layers might connect the node representing John Smith in the email network with the node representing John Smith in the mobile-phone network, or in the transportation network of London we can connect Oxford Circus tube station with Oxford Circus bus station. We note here that since multiplex networks have a well-defined structure, they can be used also to model multilayer networks where the nodes in different layers are distinct as long as there is a one-to-one mapping between the nodes of different layers and the corresponding nodes of each pair of layers are interacting. For example, a multiplex network can be formed by one layer indicating gene correlations and another indicating protein interaction networks as long as there is a one-to-one mapping between genes and proteins.

The second class of multilayer networks are multi-slice networks describing temporal networks, i.e. networks in which links are present only for a given amount of time. Given a set of nodes in which interactions are time-dependent, a multi-slice network is a multilayer network in which each layer is formed by the network of interactions occurring in a time window of duration δt. Therefore, every layer is formed by the same set of nodes as in multiplex networks. The only difference between multi-slice networks and multiplex networks is the natural ordering between layers in a consecutive temporal sequence. This ordering is reflected in the way the links across different layers are placed. In fact, in multi-slice networks the links across the different layers connect only nodes belonging to subsequent layers. Let us, for example, consider the temporal network describing the functional, time-varying correlations across different brain regions. In this

(a)

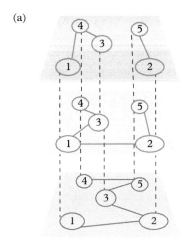

(b)

(c)

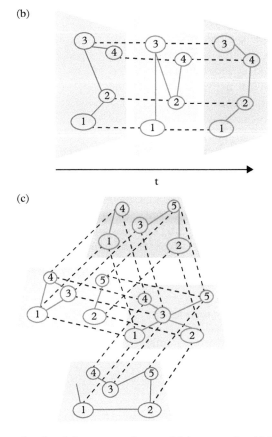

Fig. 4.1 *Different examples of multilayer networks: a multiplex network with* M $=$ *3 layers (panel (a)); a multi-slice network with* M $=$ *3 layers (panel (b)); a network of networks with* M $=$ *4 layers (panel (c)).*

case, the multi-slice network constructed from this data will have each layer indicating the correlations existing between different regions averaged over a given time window. Additionally, the links across the layers will connect each given brain region at time t with the same brain region at time $t + \delta t$.

The third class of multilayer network that we will discuss here is formed by networks of networks. In a network of networks each layer is formed by a different set of nodes. At the layer level there is a network of networks describing which layers interact with each other.

In Fig. 4.1 we show examples of multiplex networks, multi-slice networks and networks of networks. In the following chapter we will give a more formal definition of multilayer networks, multiplex and multi-slice networks, networks of networks and the most general multilayer networks, describing their most fundamental structural properties in full detail.

4.3 Social networks

Social networks are the original setting where the multiplexity of networks has been proposed. In fact, agents of a social network are related by different types of social ties including friendship, collaboration, family ties, etc. Additionally, different means of communication in a social network (email, mobile phone, chat, conference call, etc.) can provide another multilayer structure. In this scenario, the same set of agents are linked by different networks (layers) formed by interactions occurring with a different communication technology. Although the multiplex framework was proposed in the sociological literature several decades ago to unveil the fine-graining structure of social networks [120, 308, 245], only recently have new large-scale data on social multilayer networks become available.

An example of the historic datasets that have been used in social science is the network between Florentine families during the fifteenth century [236]. In this dataset, influential families are connected both by business and marriage alliances (see Fig. 4.2). Specifically in Ref. [59] a measure of *personal hierarchy* has been proposed to create meaningful partitions of the network. In this network the central role of the Medici family is particularly notable.

4.3.1 Online social networks

Recently, with the extensive diffusion of online social networks, a new generation of social datasets are becoming available, allowing us to extend the analysis of social networks to much larger datasets. The multiplex network approach has been proposed to study most of the online social networks including Facebook [190, 219, 152, 189], Twitter [235], Youtube [1], Netflix [35, 158, 159], Flickr [220, 171, 62]. These datasets can be studied as multilayer networks, given the variety of possible interactions between the users, and as temporal multi-slice networks, given the intrinsic temporal nature of the interactions. The general problems in this context are typical of the field of Data Science: finding

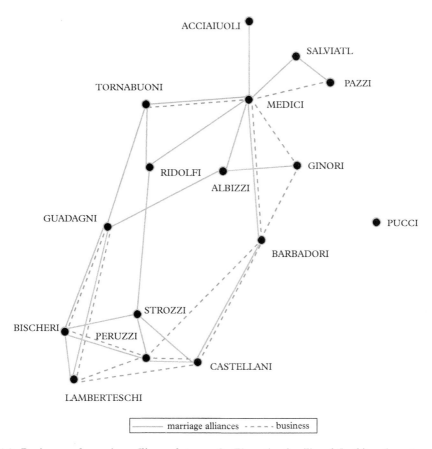

Fig. 4.2 *Business and marriage alliances between the Florentine families of the fifteenth century. Data from [236].*

the best clustering of the dataset, assessing the centrality of the nodes and developing recommendation algorithms. In all these problems the multilayer network approach is essential and provides the best framework for integrating at the same time the information about the network structure and the different connotations of the interactions existing between the nodes.

A very interesting social network of this novel generation is the one formed by the online social networks of the virtual society of the *Pardus* online game [290]. In this dataset, avatars can establish different types of interactions: friendship, collaboration, trade, enmity, attack, bounty. This *in silico* social network formed by 300 000 nodes reveals emergent properties, including a scale-free distribution of the different layers and a relevant link overlap across the layers. The link overlap among two layers is given by the total number of node pairs linked in both layers. Interestingly, a significant link overlap is observed both for pairs of layers corresponding to positive types of interactions and for pairs of layers corresponding to negative types of interactions.

Another relevant and open-access dataset that integrates information about online and offline social interactions is the so-called *Aarhus dataset* which consists of five layers of social interactions (Facebook, Lunch, Co-authors, Leisure, Work) between the employees of the Computer Science department of Aarhus University [199].

4.3.2   Collaboration and citation networks

A very rich social network dataset is constituted by the American Physical Society (APS) metadata that includes the complete bibliographic information about all the papers published in the APS journals since 1893 (see Fig. 4.3). This dataset includes for every paper the names of the Authors, the Physics and Astronomy Classification Scheme (PACS) numbers, the date and the journal; additionally it includes information about the citations obtained by each paper from other APS papers. This dataset can be used to extract different types of multilayer networks. In fact, for every journal it is possible to construct a collaboration/citation-weighted multiplex network of Authors collaborating with each other in one layer and citing each other in the other layer [209]. Moreover, it is possible to build a collaboration multiplex network [230, 165] where the different layers are collaborations on different scientific topics classified according to the PACS numbers of the resulting publications (see Fig. 4.3). The collaboration/citation multiplex networks provide important new insight into how scientific credit and reputation is established in scientific social networks starting from scientific collaborations and citation rate. The collaboration multiplex networks where the different layers are built from collaborations in different scientific topics can contribute to the understanding of the organization of scientific knowledge from a bottom-up perspective. In fact, the similarities existing

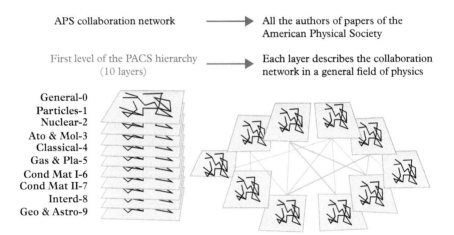

Fig. 4.3 *Schematic representation of the multiplex collaboration networks extracted from the APS dataset. Each layer corresponds to the network of collaboration between APS authors on one of the ten major scientific subjects. From the multiplex network it is possible to extract the network between layers where each node is a layer and links indicate a measure of their similarity.*

between the community structure of different layers provide information on how close in the scientific practice different scientific subjects are [230, 165, 29]. Additionally, this dataset can be studied as a temporal network since collaborations are established at a given time along the wide temporal frame of the dataset [182]. This allows us to characterize emergence of new scientific topics as it is reflected by the formation and evolution of communities of scientists working on related subjects.

Other networks include collaborations and citations between papers published in the arXiv and the network of citations between US patents. The first one can be studied as a multiplex or as a multi-slice network, the latter is only studied from the multi-slice temporal network perspective.

Collaboration networks do not only include scientific collaborations; they are ubiquitous. In particular, actor collaboration datasets have also been widely explored since the beginning of the field of Network Science. Recently, actor collaboration networks have been explored from the multilayer network perspective. In fact, by assigning to each layer a given film genre, it is possible to construct a multiplex network out of the Internet Movie Database (IMDb) [230]. Distinguishing different movie genres allows us to extract more information from the dataset exploring the versatility of actors, or, on the contrary, their specialization and the establishment of communities of actors across different genres.

4.3.3 Temporal social networks

Social networks are also intrinsically temporal, and are currently intensively studied using the temporal and multilayer framework. Specifically, recent technological developments have allowed us to record the temporal nature of social networks. These technologies include RFID (Radio-Frequency-Identification-Devices), sociopattern data [74] that are able to record temporal face-to-face interactions (see Fig. 4.4). With this technology it has been possible to collect data of face-to-face interactions in different settings (conferences, schools, hospitals). It has been found that face-to-face interactions are bursting as the duration of contacts and the inter-contact time interval are power-law distributed, indicating that the establishment of a social interaction is not a Poisson process but a process with memory. These datasets allow us also to investigate how the temporal nature of networks changes the properties of dynamical systems defined on them. By aggregating the interactions occurring in a given time window, the multi-slice description of the dataset allows us to reconstruct the multilayer nature of the community structure of the data. The RFID are not the only technology that is used to record face-to-face interactions. Other datasets include most notably the Reality Mining dataset [112], collecting human contact data recorded between 100 students of the MIT over a period of nine months in 2004.

Other notable social networks are political networks formed by parliamentary deputies that vote for the same laws. These networks are temporal and they can be described by multi-slice networks. The network between US Congress members is one of the first multilayer networks that has been used in the literature [219], and more recently a similar political network between the Brazil Congress deputies has been analysed [247]. Both

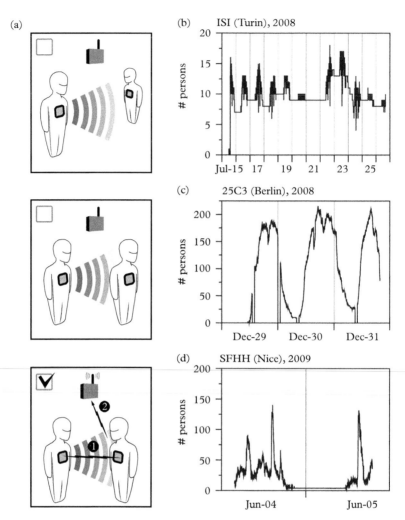

Fig. 4.4 *Schematic illustration of the RFID sensor system (panel (a)). RFID tags are worn as badges by the individuals participating in the deployments. A face-to-face contact is detected when two persons are close and facing each other. The interaction signal is then sent to the antenna. The panels (b)–(d) display the activity pattern measured in terms of the number of tagged individuals as a function of time in the three deployments: ISI refers to the deployment in the offices of the ISI foundation in Turin, Italy, with 25 participants (panel (b)); 25C3 to the 25th Chaos Communication Congress in Berlin, Germany, with 575 participants (panel (c)); and SFHH to the congress of the Societé Francaise d'Hygiéne Hospitaliére, Nice, France, with 405 participants (panel (d)). Dashed vertical lines indicate the beginning and end of each day. Typical daily rhythms are observed in the office and conference settings. Reprinted figure from Ref. [74]. ©2010 Cattuto et al.*

datasets have been used for detecting the relevance of multiplexity in determining the mesoscale organization of the multi-slice network.

4.4   Complex infrastructures

Complex infrastructure provide a beautiful example of multilayer networks where important questions related to the optimal design of the infrastructures and the interplay between structure and dynamics can be tested.

4.4.1   Transportation networks

Transportation networks have a natural multilayer structure and constitute a major example of multiplex networks. One of the first multiplex datasets that has been studied in this context is the European Multiplex Air Transportation Network [71]. This dataset is built by all flight connections joining European airports. The different layers are formed by flight connections of different airline companies (see Fig. 4.5). This dataset allows us to investigate the role of different airports in the European Airway Traffic, and can be used to design new flight connections and establish economic partnership between different airline companies. Additionally, it is a fully annotated dataset to test novel multilayer network algorithms such as strategies to aggregate the information of different layers while keeping relevant multilayer information. This network is characterized by important correlations including a significant overlap among the links of different layers and correlations between the degree of the same airport in different layers. It has to be noted that similar airport multiplex network datasets are currently freely available for all continents in multiplex network data repositories.

Multilayer transportation networks formed by layers characterizing transport over different means of transportation [93] are fundamental to understanding diffusion of people and diseases. If we travel to another city or another country it is the norm rather

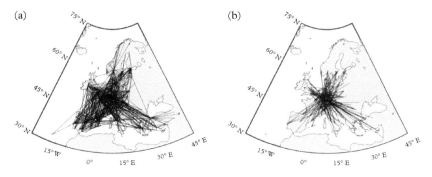

Fig. 4.5 *Visual representation of the European Multiplex Air Transportation Network [71]. Panel (a) represents the aggregated network of all the layers in which only links belonging to more than one layer are displayed, panel (b) represents the layer corresponding to a major airline company.*

than the exception that we take different means of transportation, including metro, train, taxi and airplane. Also, in large cities it is often the case that a combination of bus, metro and regional train routes are taken by commuters daily. This type of pattern describes what we call here diffusion in a multiplex network that reveals interesting novel characteristics with respect to diffusion in a single layer. In this context, the properties of the diffusion patterns can be significantly modulated by the cost associated with the different types of transportation and by the switching cost associated with changing means of transportation.

4.4.2   Interconnected infrastructures

If we consider the increasingly interconnected web of interactions between infrastructure networks, we observe that actually it is fundamental to embrace a multilayer approach and characterize global infrastructures as multilayer networks.

Multilayer global infrastructures include different services such as the power grid, the water supply networks the gas networks, and the banking system. These different infrastructures are related by interdependencies of different types (see Fig. 4.6). Therefore, the resilience, the robustness and the efficiency of these structures cannot be fully evaluated if the interdependencies between different infrastructures are not taken into account [66]. In the infrastructures interdependencies have a different implication from interactions occurring within each infrastructure layer. Taking into consideration the

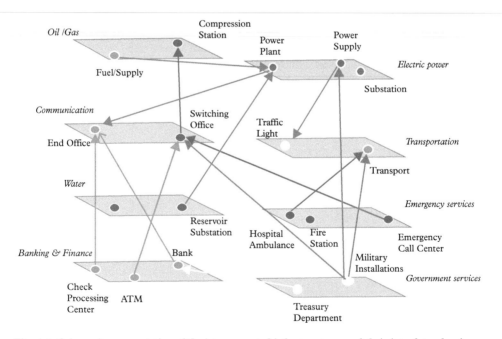

Fig. 4.6 *Schematic representation of the interconnected infrastructures and their interdependencies.*

different nature of the links and the different role of interdependencies can significantly enhance the understanding of the major factors determining the robustness and resilience properties of global infrastructures. Problems related to the robustness and resilience of multilayer networks have been central in the field. Specifically, it has been found that interdependencies might dramatically increase the fragility of the networks, yielding large avalanches of cascading failures.

In Ref. [63] Brummitt et al. analysed the interactions between three different power grids in the US network corresponding respectively to three US regions: Western, Eastern and Texas. By studying the propagation of avalanches of failures the Authors of Ref. [63] found that adding some connectivity between the different regions is beneficial, as it suppresses the largest cascade in each system by providing alternative paths. However, higher levels of interconnectivity have a negative impact, both because they open pathways for neighbouring networks to inflict large cascades and because new interconnections increase capacity and total possible load, which fuels even larger cascades.

4.5 Economical and financial networks

Economical and financial networks have recently been studied using the multilayer perspective. Specifically, the multilayer network perspective has been shown to provide an advantage in analysing the International Trade Network (ITN) also known as the World Trade Web (WTW). This is the network of trading relations (import–export) between different countries. The network has been extensively studied within the single network framework by treating each trade on the same footing, independently of the traded commodity. Nevertheless, disaggregating the information of the trade considering a multiplex network in which each layer corresponds to a trade of a given class of commodities reveals organization patterns of the network that remain covered when analysing the aggregated trade network. Another class of multilayer networks attracting large attention are the increasingly interconnected financial networks and interbank networks. In this context the major research question is to what extent the multilayer nature of these networks affects systemic risk. In the following we will discuss these two major classes of economical and financial networks separately.

4.5.1 Trade multiplex networks

The International Trade Network (ITN), also known as the World Trade Web (WTW), indicates the trade activity (export–import) between different countries of the world. A link is placed between two countries if there is a trade activity between them and potentially a weight can be associated with the link quantifying the amount of trading activity. Trade networks are fundamental for studies that aim at evaluating effects of the present globalized economy [127]. Recently, these networks have also been extensively used for assessing the current and predicting the future economic development of countries [153, 291]. Interestingly, trade networks have also been extensively used

[128, 129] as datasets to test randomization algorithms using hidden variables and randomized network ensembles (exponential random graphs).

While most of the works discussed so far treat the trading of different commodities on the same footing, the multilayer network provides a natural framework to include the information about which commodities are traded in the network description. In Refs [20, 21], a multiplex trade network composed of 162 countries (nodes) and 97 layers evolving yearly from 1992 to 2003, has been analysed. Specifically, in the first work the Authors have characterized the specific properties of each separate layer, while in the latter work they have shown that the community structure within each layer can be lost by considering the aggregated layer where all the trades of different commodities are not distinguished. Along similar lines, in Ref. [200] a multiplex network analysis has been conducted over the *UNCOMTRADE* dataset, curated by the United Nations, including 94 countries starting from the year 1962.

The Authors of Ref. [202] considered another version of the *UNCOMTRADE* dataset, forming a multiplex trade network with 162 countries and 97 commodities over the period 1992–2000 including weighted links. This database has been used by the Authors to test their reconstruction algorithm based on randomized network ensembles, largely reproducing the main statistical properties of the network.

A more specific dataset includes exclusively trades of commodities exchanged in maritime business: the Maritime Multiplex Trade Network. In Ref. [110] the Authors propose the study of a Maritime Multiplex Trade Network including five layers of main categories of commodities including liquid bulk (i.e. crude oil, oil products, chemicals, etc.), solid bulk (like aggregates, cement, or ores), containers, passengers/vehicles and general cargo (see Fig. 4.7). The data cover the months of October and November 2004 and the multiplex network has been extracted from *Lloyd's Voyage Records*. The multiplex network analysis of the dataset reveals that distinguishing between trade of different commodities allows the Authors to correlate the diversification of the trade of given ports with their traffic and the typical distance of their connections.

4.5.2   Financial multiplex networks

The rich multilayer structure of financial networks has recently been investigated as a cause of increased systemic risk. In particular, in Ref. [252] the Authors estimated that the single-layer approach might underestimate systemic risk by 90%. This study investigates the banking system of Mexico as a multiplex network with four layers including: exposure from derivatives, securities cross-holdings, foreign exchange exposure and deposits/loans during the years 2007–13. Additionally, a multilayer measure to evaluate the expected loss due to systemic risk in multiplex financial networks is proposed. This measure allows the Authors to compare the systemic risk over time on a daily scale, showing that the calculated expected loss follows several features of the market risk indicators.

In Refs [18, 19] the Italian Interbank Market during the period 2008–12 is studied from the multiplex network perspective. The layers of the interbank network are formed by distinguishing between three different types of transactions (overnight, short term, long term) and two different collaterizations (unsecured and secured loans). The network

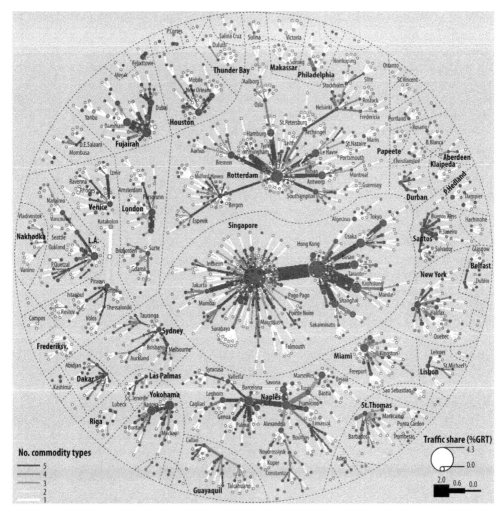

Fig. 4.7 *The Multiplex Maritime Trade Network. Nodes represent ports, pairwise connected by links when a commodity is transported between them. Different commodities are considered, thus making this graph a multiplex network. In order to represent this heterogeneity, links are coloured according to the number of commodity types travelling between ports. Furthermore, both node and link sizes represent the corresponding traffic share. Reprinted from Ref. [110] ©2013 with permission from Elsevier.*

is formed by 533 banks, almost all active in three unsecured layers. The layers of the multiplex network display strong correlations between the degrees and the strengths of the nodes and have a highly heterogeneous degree distribution. The entity of the link overlap across layers as measured by the Jaccard index is 17%, which is low if compared with the similarity of the same layers across consecutive years (70%). Therefore, banks tend to diversify their connections in different layers allowing for a faster propagation of contagion in the multiplex network than on single layers taken in isolation. In Ref. [19] the centrality measures of the banks across the different layers of the Italian Interbank Market are compared, and strong correlations of the centrality measures for large banks are found. Finally, the results of the interbank network are compared with null models of a randomized ensemble of multiplex networks with independent layers. Relevant deviations of the observed centrality measures with respect to the centralities obtained with the randomized ensemble are found.

A multilayer interbank model for assessing system risk has been proposed in Ref. [216] based on information on capital, short-term and long-term interbank borrowings, short-term and long-term interbank loans, aggregated securities holdings and the cash of 50 large EU banks. The multiplex network between the banks is not known, but randomly generated using a stochastic model based on the data. On these multiplex networks an agent-based contagion process is characterized. It is found that multiplex interbank networks include relevant non-linearities in the propagation of shocks that can have large effects in the contagion losses.

The stability of financial networks has also been characterized by considering the multiplex network formed by contracts of a different level of seniority [64, 144]. In the case of bankruptcy, debts are repaid according to the seniority of the contracts in order of decreasing seniority. Therefore, it is possible to distinguish different types of bankruptcy depending on the highest level of seniority at which a bank can repay all its debts. From a selfish perspective, a single bank having all the contracts at the highest seniority will be preferable. However, in Ref. [64] it is shown in the framework of a stylized multiplex cascade model that an optimal ratio of senior and junior debts can be predicted. This optimal combination of contracts with different seniorities is predicted to make the system less prone to large financial crises.

Finally, the financial network has been investigated using exclusively financial time series [221]. In fact, the information encoded in these datasets can be reduced to a multiplex network between the assets where the different layers correspond to significant statistical linear, non-linear, tail and partial correlations. The multilayer network is able to reveal important properties of the financial market which are not detectable using a single network approach, in particular signalling the different role of industrial sectors during crises.

4.6   Molecular networks and the interactome

The multilayer network is a very promising framework for describing the interactome, including all physical interactions between molecules in the cell. Network Science has been a very natural framework that has allowed the treatment of molecular interaction networks coming from high-throughput experiments and from large databases integrating many single-molecule experiments. In the beginning of the field, the different types of molecular interactions were studied separately, and biological networks such as the transcription network, the protein–protein interaction networks and the metabolic networks were studied in isolation. Network Science is now completely integrated into system biology, which aims to characterize biological functions, diseases and the response to drugs.

More recently, with the emergence of network medicine [16], it has become clear that in order to describe the network of interactions that are involved in a single disease it is essential to take into account the entire interactome [206] that is formed by all the known molecular interactions of the cell at the same time (see Fig. 4.8). The interactome integrates information coming from genomic studies such as genome-wide associations studies (GWAS) with information about different types of biological interactions such as regulations, signalling and protein–protein interactions coming from different experiments. The characterization of the interactome has already allowed us to identify the set of interactions involved in a given disease that appear to be organized in clusters of the interactome network. This line of research is very promising, since relevant information about the disease cluster can already be extracted from the current version of the interactome network, which contains only a fraction of the full set of interactions of the human cell. As new experiments are constantly expanding the set of known interactions, it is believed that the next conceptual step will be to treat the interactome as a multilayer network, finally considering all the interactions at the same time but distinguishing their different biological roles. At the moment, several groups

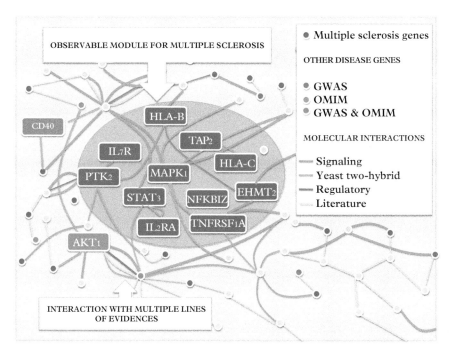

Fig. 4.8 *A schematic representation of the neighbourhood of the interactome reveals the multilayer nature of the entire network where each interaction has a different biological nature. Genes associated with multiple sclerosis are shown in the shaded area. Data from Ref. [206].*

are working around this line of research and there is no doubt that such a development of network medicine will provide significant new insight into the formulation of a new personalized medicine.

Along a different direction, a multiplex network analysis has been performed in Ref. [36] over multiplex networks combining yeast molecular interactions and gene co-expression. Since genes can be mapped one-to-one with their encoded proteins, the multiplex network is formed on the first layer by proteins that can bind to form larger protein complexes, on the second by the regulatory network indicating which are the activators and repressors of any given gene and in the third by the gene co-expression network. The Authors of Ref. [36] propose an original multi-community detection algorithm based on the optimization of a generalized modularity measure called SimMod (see chapter 8). The proposed algorithm is shown to identify functionally enriched modules and to provide enhanced results in comparison with other clustering methods without requiring training on known biological functions.

In Ref. [70] Cantini et al. propose an algorithm to find cancer drivers from multiplex networks constructed by integrating gene expression correlation networks for cancer tissues (gastric, lung, pancreas, colorectal) with transcription factor co-targeting, microRNA co-targeting and protein interaction networks. Therefore, for each type of cancer the Authors of Ref. [70] have considered a four-layer multiplex network in which only the first layer (gene coregulation) is dependent on the cancer type. The rationale for the choice of this four-layer multiplex network is that gene co-expression and protein interactions are known to require tight coregulation of the partners, involving fine-tuning of the translational and post-translational layers of the regulation. The gene multilayer communities in the resulting multiplex network are then studied by using the consensus clustering proposed in Ref. [182] (see chapter 8). It is found that the multilayer communities have small overlap with the communities of individual layers and that multilayer communities are more informative than the communities detected taking into account only the co-expression layer of cancer tissues. Finally, by comparing the multi-communities of cancer tissue with the multilayer communities of healthy cells, the Authors of Ref. [70] have been able to identify communities involved in the oncogenic process, finding several already known oncogenes and some new candidate cancer drivers.

A different research direction is currently being developed for applying multilayer network tools to provide a comprehensive view of given biological experiments across different biological conditions. For example, the multilayer network perspective can be adopted to study in parallel results coming from different biological tissues. This is particularly useful in the context of gene expression data where already a large dataset of gene expression has been analysed using a multiplex network approach. In particular, by considering the set of $M = 130$ gene expression correlation matrices as a tensor the Authors of Ref. [195, 194] have been able to simultaneously cluster a set of genes and a set of specific tissues, forming *recurrent heavy subgraphs* (see chapter 8). The detected RHS have been shown to correlate with clusters of genes with similar biological function (see Fig. 4.9).

Finally, the multilayer network perspective has been used to find conserved modules across the protein interaction networks of different species in Ref. [210]. The general

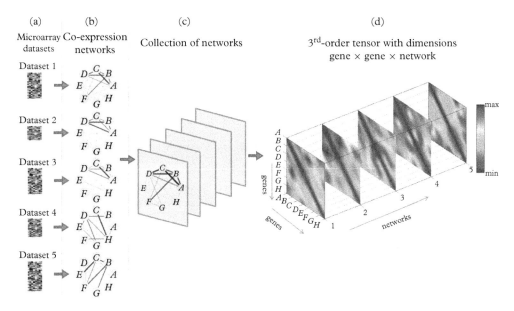

(a) Microarray datasets   (b) Co-expression networks   (c) Collection of networks   (d) 3rd-order tensor with dimensions gene × gene × network

Fig. 4.9 *Illustration of the multilayer (tensorial) representation for multiple co-expression networks. Microarray datasets (panel (a)) are modelled as a collection of co-expression networks (panel (b)); these co-expression networks form a multilayer network (panel (c)) that can be represented as a third-order tensor such that each slice represents the adjacency matrix of one network (panel (d)). Reprinted from Ref. [195]. ©2011 Li et al.*

theoretical challenge addressed by this paper is the network alignment problem that aims at finding the best matching between communities of two different networks by taking into account the many-to-many mapping existing between the nodes of two networks. In Ref. [210] the Authors propose a novel spectral-clustering algorithm for aligning communities across the network. This method is especially tailored to treat hypergraphs and is based on a generalization of the Perron–Frobenius theorem. This algorithm is then applied to a duplex multilayer network formed by the yeast and the human protein interaction networks. The results show that there is a high level of conservation on a global scale.

4.7   Brain networks

Recently the characterization of brain networks using multilayer networks has been gaining momentum, as multilayer networks allow a more detailed representation of brain networks than single networks. This important multilayer nature of brain networks has been recognized both at the neuronal level and at the macroscopic level of brain regions.

At the neuronal level, most of the interest has been focusing on the connectome of the nematode *C. elegans*. The brain of the worm *C. elegans* comprises 302 neurons connected

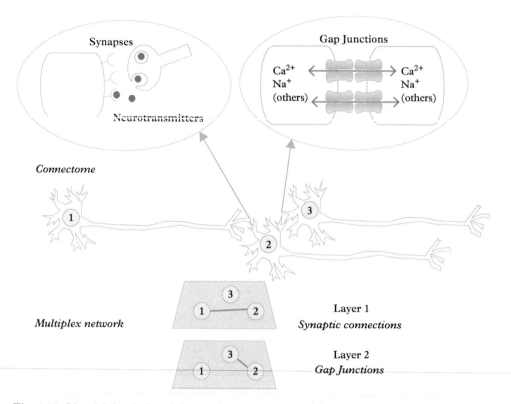

Fig. 4.10 *Pictorial description of the wired connectome comprising chemical (synaptic) and electrical (gap junction) connections.*

by a fully annotated set of synaptic (chemical) connections and gap junctions (electrically connected). The resulting structure of the connectome is therefore a multiplex network with two layers (see Fig. 4.10). Note that sometimes synaptic connections are distinguished between Chemical Monoadic and Chemical Polyadic connections, forming a multiplex network with three layers. This multiplex network reveals positive correlations between the degrees of the neurons in the different layers, indicating for instance that hub neurons in one layer tend to be hub neurons also in the other layer [230]. Recently the novel centrality measure called Functional Multiplex PageRank has been shown to characterize, starting from the multiplex network structure, the different functions (cell types) of the neurons [164]. Interestingly, a recent paper has revealed that the wiring diagram of the brain of *C. elegans* only partially describes the connectome of interactions between its neurons. In fact, the neurons communicate also through extra-synaptic volume transmission, occurring especially via monoamines (including serotonin, dopamine, octopamine, tyramine) and neuropeptides (including 16 layers) (see Fig. 4.11). These 'wireless' networks, due to the diffusion of molecules such as the monoamides and the neuropeptides, can be inferred from gene expression data indicating whether the

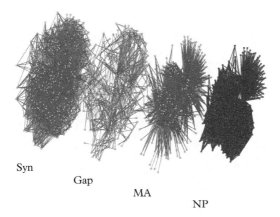

Syn

Gap

MA

NP

Fig. 4.11 *Multilayer connectome of* C. elegans *including the synaptic layer (Syn), the gap junction layer (Gap), the aggregated monoamine signalling layer (MA) and the aggregated neuropeptide signalling layer (NP). Node positions are the same in all layers. Reprinted figure from Ref. [37]. ©2016 Bentley et al.*

neurons express the corresponding receptors. It is found that the monoamide layers do not significantly overlap with the synaptic layers and that they are characterized by a hub-and-spoke structure with a few nodes forming a well-connected subgraph. The analysis of the multiplex network using multilinks (see Sec. 7.3) reveals locations in the network where aminergic and neuropeptide signalling modulate synaptic activity. The connectome of *C. elegans*, including also the novel addition of the extra-synaptic layers, is likely to become the gold standard in the study of brain multiplex networks.

At the other end of the spatial scale of brain networks are the brain functional networks investigated through fMRI experiments. These non-invasive experiments allow us to characterize brain function during resting states or during the performance of single tasks in healthy individuals and in individuals with mental disorders.

Recently, a large body of works emphasized the relevance of investigating the relation between functional brain networks and structural brain networks using tools of network theory [68]. These studies address a veritable multilayer problem, as the brain can be interpreted as a multiplex network in which different macroscopic regions of the brain interact structurally, thanks to brain fibres, and functionally, as detected by their correlated activity.

A recent paper [100] proposes a clustering of brain regions which corresponds to a graph alignment between the two networks. The proposed method allows us to detect a common skeleton shared by structure and function from which a new brain partition can be extracted which is rather distinct from the commonly employed anatomical or functional parcelations. These results underline the strong correspondence between brain structure and resting-state dynamics, as well as the emerging coherent organization of the human brain.

Functional brain networks have been studied quite extensively from the temporal multi-slice network perspective. Interestingly, a careful detection of multilayer communities allows for the evaluation of the network flexibility, a measure that evaluates the persistence of multilayer communities [219] across different time slices (see Sec. 8.2.1). It is found that a moderately high flexibility of brain functional networks recorded during a simple motor task correlates with learning [28].

Finally, from functional brain networks extracted from fMRI data multilayer networks have also been extracted [261] by distinguishing between layers formed by strongly correlated brain regions interacting through weaker links. Interestingly, the structure of these networks has been found to display a set of degree correlations that ensures improved robustness properties with respect to random multilayer networks.

4.8 Ecological networks

In these last two decades, food webs have emerged as a poweful tool to combine ecosystem ecology focusing on fluxes of matter and energy and community ecology focusing on the biodiversity of the ecological networks [111, 311]. In food webs in which the strength of the trophic interactions determines the weights of the links, network analysis provides an estimate of the energy flow in the web and reveals the rich interplay between the structure of food webs and their function. Since species of an ecological community can interact in different ways, ecological networks are not formed exclusively by food webs. Particular attention has been drawn to the study of mutualistic plant–pollinator interactions [26] which are bipartite networks characterized by a structural property called nestedness.

Ecological networks provide multiple opportunities for conducting an informative multilayer network analysis [250]. In the multilayer network framework it is possible to characterize several types of interactions occurring between the species of a given ecological community at the same time. In Ref. [205] mutualistic and antagonistic interactions are combined and the role of multiplexity in determining the stability of the multilayer ecological network is analysed. In Ref. [172] a multiplex network between trophic and different non-trophic interactions of a marine rocky intertidal community in Chile has been studied. The work has focused on a structural study of the typical interaction patterns between the species and on the characterization of the ecological stability of the community. In this context it is suggested that multiplexity can enhance the stability of the entire ecosystem.

Studies conducted over a given interval of time allow the modelling of ecological networks as multi-slice networks. In Ref. [234] the characterization of a plants–pollinator network over 12 years reveals that the variability in the ecological network depends on two different processes: (i) the ecological network can change over time because the species composition of the community changes; (ii) the ecological network can change because the interactions between existing species change in time. This study shows in particular that an ecological network structure is determined not exclusively by the presence of a given set of species in the ecosystem. Additionally, this study shows that different species can have a different behaviour in temporal networks; in fact, some species can be much more persistent than others in a time-varying ecosystem. This implies that typically the species turnover in an ecosystem might affect a subset of species only.

Finally, the multilayer network framework can be used to study the effect of biogeography on ecological networks. In fact, to each ecological patch it is possible to attribute a layer of the multilayer network structure. In this way it is possible [251] to compare different species interaction networks and analyse their diversity in terms of species composition and species interactions.

This field is rapidly expanding using methods and techniques of multilayer network analysis; for further details on the results achieved so far and future perspectives we refer the reader to Ref. [250].

4.9   Climate networks

In recent years, climate networks have been successfully used to analyse climate data [103, 102]. The nodes of these networks are isobaric locations identified by their latitude, longitude and pressure. Pairs of geographical locations constituting the nodes of the network can be linked according to the zero-lag linear correlations between different time series of climatological data, or according to their mutual information [103, 102]. Links corresponding to long-range interactions between different locations, called *teleconnections*, have attracted great interest among the climatologists. Additionally, the analysis of the topology of climate networks, using local measures (such as the degree and the clustering coefficient) and global measures (such as the betweenness centrality and the average distance) have been shown to be very useful to characterize the flow of matter and energy in the climate system.

The analysis of climatological data can be further enriched by using the multilayer network framework, opening new venues for understanding the function of this complex system. Relevant results have already been obtained by performing these multilayer network studies. On the one side, the temporal nature of the climatological data allows for a multi-slice network analysis [312]. Using the multi-slice network framework to analyse the temporal variation of climate networks, in Ref. [312] the Authors have shown evidence that El Niño–Southern Oscillation has a strong impact on the stability of the climate system. On the other hand, climate networks corresponding to different isobaric surfaces can be coupled together forming a large multilayer network [101] quantifying the effect of large-scale convection processes in the troposhere and in the stratosphere.

These results provide great evidence of the relevance of multilayer network analysis for characterizing the properties of climate networks.

5

The Mathematical Definition

5.1 General multilayer networks and more specific topologies

In this chapter we give the mathematical definitions of general multilayer networks, multiplex networks, multi-slice networks and networks of networks. We will highlight the differences between the different types of multilayer networks and we will present the different mathematical tools that are used for their representation, including adjacency matrices, supra-adjacency matrices and tensors.

As in any young field, in the field of multilayer networks there is still no general agreement on the correct terminology to use for different types of multilayer networks. This general tendency of fast-developing fields is even exacerbated in this case by the large degree of freedom in defining multilayer network structures. Therefore, several different terms have been used in the past to indicate the same type of structure. Here, we aim at defining the minimal number of terms that is sufficient to describe and classify the vast majority of different multilayer networks.

5.2 The most general multilayer network

A multilayer network $\mathcal{M}$ is given by the triple

$$\mathcal{M} = (Y, \vec{G}, \mathcal{G}).$$

Here Y indicates the set of layers

$$Y = \{\alpha | \alpha \in \{1, 2, \dots, M\}\}, \tag{5.1}$$

of the multilayer network and M indicates the total number of layers, i.e. the cardinality of Y

$$M = |Y|.$$

Multilayer Networks. Ginestra Bianconi, Oxford University Press (2018).
© Ginestra Bianconi. DOI: 10.1093/oso/9780198753919.001.0001

Additionally, $\vec{G}$ indicates the ordered list of networks characterizing the interactions within each layer $\alpha = 1, 2, \ldots, M$, i.e.

$$\vec{G} = (G_1, G_2, \ldots, G_\alpha, \ldots, G_M) \tag{5.2}$$

where

$$G_\alpha = (V_\alpha, E_\alpha)$$

is the network in layer α. The set of nodes of layer α is indicated by V_α and the set of links connecting nodes within layer α is indicated by E_α. The links belonging to the sets E_α with $\alpha = 1, 2, \ldots, M$ are also called *intralinks*. The total number of nodes in layer α will be indicated by N_α, i.e. N_α is the cardinality of the V_α node set,

$$N_\alpha = |V_\alpha|.$$

Finally, the $M \times M$ list $\mathcal{G}$ of bipartite networks $\mathcal{G}$ characterizes the interactions across pairs of different layers and has elements $\mathcal{G}_{\alpha,\beta}$ given by

$$\mathcal{G}_{\alpha,\beta} = (V_\alpha, V_\beta, E_{\alpha,\beta}) \tag{5.3}$$

for each $\alpha < \beta$ and $\alpha, \beta \in \{1, 2 \ldots M\}$. Here $\mathcal{G}_{\alpha,\beta}$ indicates the bipartite network with node sets V_α and V_β and link set $E_{\alpha,\beta}$. The links of the networks $\mathcal{G}_{\alpha,\beta}$ are called *interlinks* and connect the nodes of layer α to the nodes of a different layer β.

Please note that both $\vec{G}$ and $\mathcal{G}$ are defined as lists of sets and not as sets of sets. This is a necessary choice because the layers are typically labelled, and therefore the labels of the different layers are not exchangeable. For instance, in a duplex layer $\alpha = 1$ might indicate the mobile-phone contact network and layer $\alpha = 2$ might indicate the email contact network. It follows that the order of the graphs in the lists is relevant for the interpretation of the results. For example,

$$\vec{G} = (G^A, G^B) \tag{5.4}$$

indicates that G^A is the mobile-phone contact network and G^B is the email contact network where

$$\vec{G} = (G^B, G^A) \tag{5.5}$$

indicates that G^B is the mobile-phone contact network and G^A is the email contact network.

5.3 Multiplex networks

5.3.1 General properties

A multiplex network structure constitutes a simplified multilayer network with the following properties:

(a) *Multiplex networks are multilayer networks in which there is a one-to-one mapping of the nodes in different layers, also called replica nodes.*

(b) *Interlinks can exclusively connect corresponding replica nodes.*

Very often a multiplex network is used to describe the interactions between the same set of nodes, with each layer characterizing a different type of interaction. Examples of multiplex networks are notably social networks, where the same set of people can be related by different sets of interactions (friendship, common hobby, collaboration) or can interact through several means of communication (phone, email, chat, conference call, online social networks). Other examples range from transportation networks to brain networks. For example, airports are connected by flights of different airline companies, two stations of a city can be connected by a bus route or a train and metro route and in the brain neurons are connected both by synapses and gap junctions.

Sometimes, however, multiplex networks can be used also for representing interactions between different sets of nodes as long as the nodes are mapped one-to-one. For instance, in the study of robustness of multilayer networks such as coupled infrastructures it is sometimes assumed that the nodes in different networks are connected pairwise by interlinks. In the case of interdependent power grid and communication infrastructures, it is often assumed that each power plant is interdependent on a single node of the communication network that is monitoring its dynamics [66]. Therefore, in this context a multiplex network structure is adopted to characterize the entire interdependent system also if the replica nodes indicate two distinct entities (see chapter 11).

A multiplex network admits two different representations. In the first representation there is no distinction between the identity of corresponding replica nodes and there is no explicit use of interlinks. In the second representation corresponding replica nodes are treated as distinguishable entities and there is an explicit description of interlinks. In the following, we will discuss the two types of description separately.

5.3.2 Multiplex networks without interlinks

The first representation of multiplex networks refers to the case where there is no explicit treatment of the interlinks, and in which corresponding nodes in different layers indicate the same identity (see Fig. 5.1 panel (a)).

In this case a multiplex network can be viewed as a multilayer network $\mathcal{M}$ where there are no interlinks, i.e.

$$\mathcal{M} = (Y, \vec{G})$$

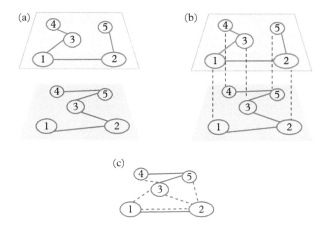

Fig. 5.1 *Different representations of a multiplex network formed by* M = 2 *networks. Every layer is formed by a network determined by a different type of interaction (panel (a)), the replica nodes are connected by interlinks (panel (b)), the multiplex network is represented as a single network with links of different types (panel (c)). In panel (c) the dot-dashed links and the solid line links represent the two layers of the multiplex network.*

with

$$\vec{G} = (G_1, G_2, \ldots, G_\alpha, \ldots, G_M).$$

Each network $G_\alpha = (V_\alpha, E_\alpha)$ is formed by the same set of nodes, i.e.

$$V_\alpha = V = \{i | i \in \{1, 2, \ldots, N\}\}, \tag{5.6}$$

and by the set of links E_α. In this case, the full information about the multiplex network is encoded in M distinct adjacency matrices $\mathbf{a}^{[\alpha]}$ indicating the network in layer α. The adjacency matrices $\mathbf{a}^{[\alpha]}$ of undirected, unweighted multiplex networks are $N \times N$ matrices of elements

$$a_{ij}^{[\alpha]} = \begin{cases} 1 & \text{if node } i \text{ is linked to node } j \text{ in layer } \alpha, \\ 0 & \text{otherwise.} \end{cases} \tag{5.7}$$

This definition can immediately be adapted to take into account the directionality or the weights of the links. Therefore, in the directed multiplex network case we have

$$a_{ij}^{[\alpha]} = \begin{cases} 1 & \text{if node } i \text{ points to node } j \text{ in layer } \alpha, \\ 0 & \text{otherwise.} \end{cases} \tag{5.8}$$

For weighted, undirected multiplex networks elements $a_{ij}^{[\alpha]}$ are given by

$$a_{ij}^{[\alpha]} = \begin{cases} w_{ij}^{[\alpha]} & \text{if node } i \text{ is linked to node } j \text{ in layer } \alpha \text{ with weight } w_{ij}^{[\alpha]}, \\ 0 & \text{otherwise.} \end{cases} \tag{5.9}$$

In the most general case of the weighted and directed multiplex network the matrix elements $a_{ij}^{[\alpha]}$ are given by

$$a_{ij}^{[\alpha]} = \begin{cases} w_{ij}^{[\alpha]} & \text{if node } i \text{ points to node } j \text{ in layer } \alpha \text{ with weight } w_{ij}^{[\alpha]}, \\ 0 & \text{otherwise.} \end{cases} \tag{5.10}$$

Given the simplicity of the multiplex network structures, when we neglect interlinks multiplex networks can also be represented as single networks in which links have a different 'colour' indicating the different types of interaction. In this description, the set of links of the same colour identifies each layer of the multiplex network (see Fig. 5.1 panel (c)).

5.3.3   Multiplex networks with interlinks

Although in a multiplex network the nodes of different layers are mapped one-to-one, usually indicating the same identity, in practice in some cases it might still be useful to distinguish the identity of nodes in each layer. For example, in air transportation networks two airline companies might use two different airport terminals, or in a public transportation system within a city like London it might be relevant to distinguish between the bus station and the tube station of the same location. Corresponding nodes belonging to different layers are called *replica nodes*. When distinguishing between different replica nodes, it is also possible to treat explicitly interlinks linking each pair of replica nodes (see Fig. 5.1 panel (b)). This allows us, for instance, to associate a cost for switching between different means of transportation such as transit between bus and tube transportation networks. Additionally, this representation is also suitable for treating the case in which the nodes of different layers are distinct entities as in Ref. [66]. In this second representation, a multiplex network is formed by N nodes $i = 1, 2, \ldots N$ and M layers, each node i admits M 'replica nodes' (i, α) with $\alpha = 1, 2, \ldots, M$ indicating the identity of node i in layer α. As a consequence of this notation, in a social network formed by the mobile phone, the email and the Facebook layers, a given person will be represented as three replica nodes representing his/her identity on each of the three layers.

Starting from the set of nodes V of the multiplex network

$$V = \{i | i \in \{1, 2, \ldots, N\}\} \tag{5.11}$$

we construct M sets of nodes V_α, each one representing the replicas of the nodes in V in layer α, i.e.

$$V_\alpha = \{(i, \alpha) | i \in \{1, 2, \ldots N\}\}. \tag{5.12}$$

Note here that the labelling of the nodes within each layer is not arbitrary, indicating the fact that all the nodes (i, α) with the same label i but belonging to different layers α are 'replica nodes'.

If one distinguishes between different replica nodes, the multiplex network can be represented as a multilayer network $\mathcal{M}$ comprising the pair

$$\mathcal{M} = (Y, \vec{G}, \mathcal{G})$$

where both $\vec{G}$ given by Eq. (5.2) and $\mathcal{G}$ (formed by bipartite networks $\mathcal{G}_{\alpha,\beta}$ given by Eq. (5.4)) are non-empty.

Specifically, each network G_α in layer α is formed by the interactions between the set V_α of the replica nodes in layer α, i.e. $G_\alpha = (V_\alpha, E_\alpha)$.

The network $G_\alpha = (V_\alpha, E_\alpha)$ is determined by the $N \times N$ adjacency matrix $\mathbf{a}^{[\alpha]}$ of elements given by Eqs (5.8) − (5.11) for undirected/directed and unweighted/weighted multiplex networks.

The networks $\mathcal{G}_{\alpha,\beta} = (V_\alpha, V_\beta, E_{\alpha,\beta})$ across different layers have a trivial structure. In fact, the interlinks in the set $E_{\alpha,\beta}$ connecting the replica nodes in layer α with the replica nodes in layer β are exclusively of the type $[(i,\alpha),(i,\beta)]$, i.e. they connect each replica node (i,α) on layer α with its replica node (i,β) on layer β,

$$E_{\alpha,\beta} = \{[(i,\alpha),(i,\beta)] | i \in \{1,2,\ldots,N\}\}. \tag{5.13}$$

In this multiplex network representation both the intralinks and the interlinks can be indicated by the $N \cdot M \times N \cdot M$ supra-adjacency matrix $\mathcal{A}$ [133] whose generic element $\mathcal{A}_{i\alpha,j\beta}$ indicates whether the replica node (i,α) is linked to the replica node (j,β) is given by

$$\mathcal{A}_{i\alpha,j\beta} = \begin{cases} a_{ij}^{[\alpha]} & \text{if } \alpha = \beta \\ \delta(i,j) & \text{if } \alpha \neq \beta \end{cases} \tag{5.14}$$

where $\delta(x,y)$ indicates the Kronecker delta. Therefore, the supra-adjacency matrix $\mathcal{A}$ takes a block structure form

$$\mathcal{A} = \begin{pmatrix} \mathbf{a}^{[1]} & \mathbf{I} & \cdots & \mathbf{I} \\ \mathbf{I} & \mathbf{a}^{[2]} & \cdots & \mathbf{I} \\ \vdots & \vdots & \ddots & \vdots \\ \mathbf{I} & \mathbf{I} & \cdots & \mathbf{a}^{[M]} \end{pmatrix}, \tag{5.15}$$

where $\mathbf{I}$ indicates the $N \times N$ identity matrix.

Note that the supra-adjacency matrix can be considered as an adjacency matrix between replica nodes (i,α), but all replica nodes (i,α) with $\alpha = 1,2,\ldots,M$ indicate the same node (for example, an individual in a social network, a station or an airport in a transportation network and so on). The procedure of describing a multiplex network using the supra-adjacency matrix is also called *flattening, unfolding* or *matricisation*.

Given the trivial structure of the non-diagonal blocks of the supra-adjacency matrix, using a supra-adjacency matrix to describe the structure of a multiplex network is a redundant operation that does not encode more information than the adjacency matrices

$\{\mathbf{a}^{[\alpha]}\}_{\alpha=1,2,\dots,M}$. Nevertheless, when considering dynamical processes on multiplex networks it is often convenient to use a supra-adjacency matrix in order to associate a different type of dynamics to the interlinks. For example, in multilayer transportation networks there might be an incentive to diffuse on the same layer (metro, bus, etc.) with respect to changing from one means of transportation to another. In this case it is often useful to describe the multiplex network with a supra-adjacency matrix and to attribute a different diffusion constant to the interlinks mimicking the cost of shifting from one transportation layer to another.

5.3.4   Multiplex edge list and multiplex data

Since most multiplex network data are sparse, a usual representation of the data consists of a multiplex edge list.

The multiplex edge list of an undirected, unweighted multiplex network is the list of triples (α, i, j) indicating that node i and node j are linked in layer α.

The multiplex edge list of an undirected and weighted multiplex network is the list of quadruples $(\alpha, i, j, w_{ij}^{[\alpha]})$ indicating that node i and node j are connected in layer α with weight $w_{ij}^{[\alpha]}$.

The multiplex edge list of an unweighted, directed multiplex network is a list of triples (α, i, j) indicating a link from node i to node j in layer α.

The multiplex edge list of a weighted, directed multiplex network is a list of quadruples $(\alpha, i, j, w_{ij}^{[\alpha]})$ indicating a link from i to node j in layer α with weight $w_{ij}^{[\alpha]}$.

5.3.5   Aggregated network

The *aggregated network* of a multiplex network is the network that results from all the interactions of the multiplex network when the differences between the different types of interaction are neglected. The *aggregated network* $\tilde{\mathcal{G}} = \left(V, \tilde{E}\right)$ is formed by the set of nodes V of the multiplex network and has a set of links $\tilde{E}$. The aggregated network can be assigned an (unweighted) adjacency matrix $\tilde{\mathbf{a}}$. The adjacency matrix $\tilde{\mathbf{a}}$ has elements $\tilde{a}_{ij}$ equal to one only if the link (i, j) exists at least in one layer, i.e.

$$\tilde{a}_{ij} = \begin{cases} 1 & \text{if } \sum_{\alpha=1}^{M} a_{ij}^{[\alpha]} > 0, \\ 0 & \text{otherwise.} \end{cases} \qquad (5.16)$$

5.4   Multi-slice networks

5.4.1   General properties

A multi-slice network is a special type of multilayer network with the following properties:

(a) *Multi-slice temporal networks are multilayer networks in which there is a one-to-one mapping of the nodes in different layers also called replica nodes. Every layer indicates a different snapshot of a temporal network.*

(b) *The interlinks can exclusively connect the replica nodes of pairs of layers in temporal succession.*

Temporal networks are networks in which the links can appear and disappear in time [156, 157, 204]. They describe a large variety of complex systems ranging from social contact networks to functional brain networks. Recently we have gained new insights into temporal networks thanks to the availability of time-resolved network data. It is now possible to have high-quality data on the dynamics of face-to-face contacts, phone calls or online social interactions that constitute the microscopic structure of social networks. Similarly, in brain research functional brain networks describing the correlated activity of different regions of the brain have an inherently temporal nature.

Temporal networks are represented by a sequence of events (interactions) occurring over a period of time T. The interactions can be either instantaneous or practically instantaneous, like sending a text message or an email, or can be characterized by a duration like for example a phone call, or a face-to-face interaction. In both cases the information encoded in a temporal network can be aggregated in *multi-slice networks* formed by a set of networks, each one indicating all the interactions occurring during a given time-window δt.

Given a temporal network formed by N nodes interacting over a period of time T and the time-window δt that is chosen to aggregate the data, the *multi-slice network* can be interpreted as (a variation of) a multiplex network formed by $M = T/\delta t$ layers in which every layer $\alpha = 1, 2 \ldots, M$ encodes all the interactions occurring during the interval of time $[(\alpha - 1)\delta t, \alpha \delta t]$. As a consequence of the similarities between multi-slice networks and multiplex networks, there are also two representations for multi-slice temporal networks: the first one does not make explicit use of interlinks, the second one does.

5.4.2 Multi-slice networks without interlinks

If we neglect the information about the interlinks, a multi-slice network can be reduced to a multiplex network

$$\mathcal{M} = (Y, \vec{G})$$

with

$$\vec{G} = (G_1, G_2, \ldots, G_\alpha, \ldots, G_M).$$

Each network $G_\alpha = (V_\alpha, E_\alpha)$ indicates the interactions occurring between the nodes in the same set V, i.e.

$$V_\alpha = V = \{i | i \in \{1, 2, \ldots, N\}\}$$

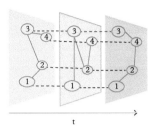

Fig. 5.2 *A multi-slice network. Every layer corresponds to the network of interactions occurring in a temporal window* t. *The replica nodes* (i, α) *characterize a node* i *in the time-slice* α. *Interlinks only connect subsequent replica nodes.*

during a given temporal window indicated by $\alpha \in \{1, 2, \ldots, M\}$. Each of these networks is fully characterized by the $N \times N$ adjacency matrix $\mathbf{a}^{[\alpha]}$ with $\alpha = 1, 2, \ldots, M$. In the case of an unweighted and undirected multi-slice network, the generic adjacency matrix $a_{ij}^{[\alpha]}$ has elements

$$a_{ij}^{[\alpha]} = \begin{cases} 1 & \text{if node } i \text{ is linked to node } j \text{ in the time interval } [(\alpha - 1)\delta t, \alpha \delta t), \\ 0 & \text{otherwise.} \end{cases} \quad (5.17)$$

In the case in which the interactions are directed and unweighted, instead we will have

$$a_{ij}^{[\alpha]} = \begin{cases} 1 & \text{if node } i \text{ points to node } j \text{ in the time interval } [(\alpha - 1)\delta t, \alpha \delta t), \\ 0 & \text{otherwise.} \end{cases} \quad (5.18)$$

The extension to a multi-slice network with weighted layers is straightforward, as it is obtained by allowing the non-zero entries of all the adjacency matrices $\mathbf{a}^{[\alpha]}$ to take arbitrary positive values $w_{ij}^{[\alpha]} > 0$ indicating the weight of the link.

5.4.3   Multi-slice networks with interlinks

Interestingly, in multi-slice networks different replica nodes (i, α) representing the same node i in a different temporal slice α can be distinguished and *interlinks* can be included in the description of the multi-slice network. Although interlinks only connect corresponding nodes (replica nodes) in the different layers, the way in which interlinks are assigned is not the same as the one used for assigning interlinks to multiplex networks. In fact, it reflects the temporal sequence of the layers. In particular, interlinks of multi-slice networks connect each node in one layer to the same node in the consecutive layer (see Fig. 5.2).

Starting from the set of nodes V of the temporal network

$$V = \{i | i \in \{1, 2, \ldots, N\}\} \quad (5.19)$$

we construct M sets of nodes V_α, each one representing the replicas of the nodes in V in the temporal slice α, i.e.

$$V_\alpha = \{(i, \alpha) | i \in \{1, 2, \ldots N\}\}. \tag{5.20}$$

In this representation a multi-slice network $\mathcal{M}$ comprises a triplet of sets

$$\mathcal{M} = (Y, \vec{G}, \mathcal{G})$$

where $\vec{G}$ and $\mathcal{G}$ are both non-empty. Each network G_α in layer α is formed by the set V_α of the replica nodes in layer α and by their interactions, i.e. $G_\alpha = (V_\alpha, E_\alpha)$. The network $G_\alpha = (V_\alpha, E_\alpha)$ is determined by the $N \times N$ adjacency matrix $\mathbf{a}^{[\alpha]}$ which has the same definition as in Eqs (5.17), (5.18).

The networks $\mathcal{G}$ across different layers include exclusively interactions between subsequent replica nodes, i.e. links $[(i, \alpha), (i, \alpha + 1)]$ connecting each replica node (i, α) with its replica node $(i, \alpha + 1)$ in the subsequent time-slice. Therefore, the set of links $E_{\alpha, \alpha+1}$ with $\alpha = 1, 2, \ldots (M - 1)$ is given by

$$E_{\alpha, \alpha+1} = \{[(i, \alpha), (i, \alpha + 1)] | i \in \{1, 2, \ldots, N\}\}, \tag{5.21}$$

while the set of links $E_{\alpha, \beta}$ with $\beta \neq \alpha + 1$ is empty, i.e.

$$E_{\alpha, \beta} = \emptyset. \tag{5.22}$$

In this case, the supra-adjacency matrix $\mathcal{A}$ of the multi-slice network is given by the $(N \cdot M) \times (N \cdot M)$ matrix $\mathcal{A}$ with the block structure form

$$\mathcal{A} = \begin{pmatrix} \mathbf{a}^{[1]} & \mathbf{I} & 0 & \cdots & 0 & 0 \\ 0 & \mathbf{a}^{[2]} & \mathbf{I} & \cdots & 0 & 0 \\ \vdots & \vdots & \vdots & \ddots & \vdots & \vdots \\ 0 & 0 & 0 & \cdots & \mathbf{a}^{[M-1]} & \mathbf{I} \\ 0 & 0 & 0 & \cdots & 0 & \mathbf{a}^{[M]} \end{pmatrix}. \tag{5.23}$$

5.4.4   Multi-slice edge list

Multi-slice datasets can be efficiently stored as multiplex networks using a multi-slice edge list consisting in the weighted and directed case of a list of quadruples $(\alpha, i, j, w_{ij}^{[\alpha]})$ indicating the existence of a link from node i to node j in the temporal slice α with weight $w_{ij}^{[\alpha]}$. The multi-slice edge list for undirected or unweighted multi-slice networks is similarly obtained by a straightforward extension of the definition of the multiplex edge list (see Sec. 5.3.4).

5.4.5 Aggregated network

The *aggregated network* of a multi-slice network is constructed by neglecting the temporal nature of the interactions. It is therefore defined, as in the case of the multiplex network, by $\tilde{\mathcal{G}} = \left(V, \tilde{E}\right)$ where V is the set of nodes of the multi-slice network and $\tilde{E}$ is the set of edges of the aggregated network. The aggregated network can be assigned an (unweighted) adjacency matrix $\tilde{a}$ with matrix elements $\tilde{a}_{ij}$ equal to one only if the link (i,j) exists at least in one layer, and is given by Eq. (5.16).

5.5 Other types of multilayer network

5.5.1 Formalism

Several different types of multilayer networks $\mathcal{M} = (Y, \vec{G}, \mathcal{G})$, with Y, $\vec{G}$ and $\mathcal{G}$ defined by Eqs (5.1), (5.2), and (5.4), can be treated using the same formalism.

In this formalism every node of layer α is indicated by two indices (i, α). The index α determines the layer of the node. The index i identifies the particular node under consideration among all the nodes of layer α. Therefore, the set V_α of nodes of layer α is defined as

$$V_\alpha = \{(i, \alpha) | i \in \{1, 2, \dots N_\alpha\}\}. \tag{5.24}$$

Each network G_α is fully determined by the $N_\alpha \times N_\alpha$ adjacency matrix $\mathbf{a}^{[\alpha,\alpha]}$. The bipartite network $G_{\alpha,\beta}$ of interactions between nodes in layer α and nodes in layer $\beta < \alpha$ is instead described by the $N_\alpha \times N_\beta$ incidence matrix $\mathbf{a}^{[\alpha,\beta]}$ and the $N_\beta \times N_\alpha$ incidence matrix $\mathbf{a}^{[\beta,\alpha]}$. For the case of an undirected multilayer network, the matrices $\mathbf{a}^{[\alpha,\beta]}$ with $\alpha, \beta \in \{1, 2 \dots, M\}$ have elements

$$a_{ij}^{[\alpha,\beta]} = \begin{cases} 1 & \text{if } (i, \alpha) \text{ is linked to } (j, \beta), \\ 0 & \text{otherwise.} \end{cases} \tag{5.25}$$

It follows that in this case the matrix $\mathbf{a}^{[\beta,\alpha]}$ is the transpose of the matrix $\mathbf{a}^{[\alpha,\beta]}$. The treatment of directed and weighted multilayer networks is straightforward. For the directed networks we will have

$$a_{ij}^{[\alpha,\beta]} = \begin{cases} 1 & \text{if } (i, \alpha) \text{ points to } (j, \beta) \\ 0 & \text{otherwise.} \end{cases} \tag{5.26}$$

The extension to weighted adjacency matrices is immediate. For example, in the general weighted and directed case we will have

$$a_{ij}^{[\alpha,\beta]} = \begin{cases} w_{ij}^{[\alpha,\beta]} & \text{if } (i, \alpha) \text{ points to } (j, \beta) \text{ with weight } w_{ij}^{[\alpha]}, \\ 0 & \text{otherwise.} \end{cases} \tag{5.27}$$

The global set of interactions of a multilayer network can be described by a single *supra-adjacency matrix* $\mathcal{A}$ [133] including all the information carried by the matrices $\{\mathbf{a}^{[\alpha,\beta]}\}_{\alpha=1,2\ldots,M;\beta=1,2\ldots,M}$. To this end we indicate each node of the multilayer network with a pair of indices (i,α) with $\alpha = 1, 2, \ldots, M$ indicating the layer to which the node belongs and $i = 1, 2, \ldots, N_\alpha$ indicating the label of the node within the layer α. The total number $\hat{N}$ of nodes across all the layers of the multilayer network is given by

$$\hat{N} = \sum_{\alpha=1}^{M} N_\alpha. \tag{5.28}$$

The supra-adjacency matrix $\mathcal{A}$ is an $\hat{N} \times \hat{N}$ matrix of elements

$$\mathcal{A}_{i\alpha,j\beta} = a_{ij}^{[\alpha,\beta]}. \tag{5.29}$$

Therefore, the supra-adjacency matrix $\mathcal{A}$ takes the block structure form

$$\mathcal{A} = \begin{pmatrix} \mathbf{a}^{[1,1]} & \mathbf{a}^{[1,2]} & \cdots & \mathbf{a}^{[1,M]} \\ \mathbf{a}^{[2,1]} & \mathbf{a}^{[2,2]} & \cdots & \mathbf{a}^{[2,M]} \\ \vdots & \vdots & \ddots & \vdots \\ \mathbf{a}^{[M,1]} & \mathbf{a}^{[M,2]} & \cdots & \mathbf{a}^{[M,M]} \end{pmatrix}, \tag{5.30}$$

when the nodes (i,α) are ordered according to their index $n = i + \sum_{\beta<\alpha} N_\beta$. The description of a multilayer network in terms of the supra-adjacency matrix somewhat hides the multiplicity of the systems because the full multilayer network is described by a single large (supra-)adjacency matrix. The reader should nevertheless be aware that a multilayer network is not just a larger network including all the layers. In fact, the networks within each layer and across the different layers describe different types of interactions. This additional information and its coupling with the dynamical processes allows for a very different dynamical behaviour with respect to the single network scenario.

We note that the supra-adjacency matrix $\mathcal{A}$ is a major example of the general class of $\hat{N} \times \hat{N}$ matrices $\mathcal{C}$ called supra-matrices that are characterized by having elements $\mathcal{C}_{i\alpha,j\beta}$ and a block structure of the type

$$\mathcal{C} = \begin{pmatrix} \mathbf{c}^{[1,1]} & \mathbf{c}^{[1,2]} & \cdots & \mathbf{c}^{[1,M]} \\ \mathbf{c}^{[2,1]} & \mathbf{c}^{[2,2]} & \cdots & \mathbf{c}^{[2,M]} \\ \vdots & \vdots & \ddots & \vdots \\ \mathbf{c}^{[M,1]} & \mathbf{c}^{[M,2]} & \cdots & \mathbf{c}^{[M,M]} \end{pmatrix}, \tag{5.31}$$

where $\mathbf{c}^{[\alpha,\beta]}$ are $N_\alpha \times N_\beta$ matrices. Beside the supra-adjacency matrix $\mathcal{A}$, another major example of such a supra-matrix is the supra-Laplacian matrix $\mathcal{L}$ [133] that will be defined in chapter 14.

In general, in the multilayer networks following this description there is no restriction on the way the interlinks are placed across different layers. Every node of a given layer can be connected to multiple nodes of another layer and nodes of a given layer might be connected to nodes of any other layer.

However, some of these multilayer networks can be classified depending on the presence or absence of the replica nodes and the presence or absence of the supernetwork, as discussed in the next paragraphs.

5.5.2   Replica nodes

In a multilayer network with replica nodes there is a one-to-one mapping of the nodes in different layers and corresponding nodes are called *replica nodes*. Since there is a one-to-one mapping between the nodes in different layers, every layer is formed by the same number of nodes

$$N_\alpha = N,$$

for every $\alpha = 1, 2, \ldots, M$. The nodes (i, α) with the same label i belonging to different layers are called the replica nodes. Note that this property implies that the labels of the nodes within the same layer α cannot be freely reshuffled.

On the contrary, in multilayer networks without replica nodes, since there is no one-to-one map between the nodes of different layers the number of nodes in each layer can be different. Therefore, in this case nodes in different layers usually indicate different node entities which are not in a one-to-one relation. Additionally, in the absence of replica nodes there is no preferred ordering for labelling nodes in any given layer.

5.5.3   Networks of networks

Multilayer networks with a supernetwork are also called *networks of networks*. The supernetwork $G_L = (V_L, E_L)$ is formed by the supernetwork node set

$$V_L = \{\alpha | \alpha \in \{1, 2, \ldots, M\}\}, \tag{5.32}$$

where the set V_L indicates the set of all the layers α of the network of networks, and by the set E_L indicates the connections between different layers. The links of the supernetwork are fully characterized by the $M \times M$ adjacency matrix $\mathbf{A}$ of elements

$$A_{\alpha\beta} = \begin{cases} 1 & \text{if layer } \alpha \text{ is connected to layer } \beta, \\ 0 & \text{otherwise.} \end{cases} \tag{5.33}$$

The supernetwork of a network of networks determines uniquely which set of layers can be connected by interlinks. In particular, *interlinks only join nodes belonging to layers that are connected in the supernetwork*. It is important to note, however, that in multilayer networks without a supernetwork interlinks can join nodes belonging to any pair of different layers.

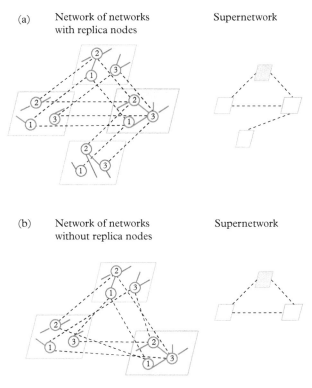

Fig. 5.3 *A network of networks with replica nodes (panel (a)) and a network of networks without replica nodes (panel (b)) are here schematically shown together with their corresponding supernetworks.*

Networks of networks can be classified according to the presence or absence of replica nodes. We will therefore consider the following two classes of networks of networks:

(a) **Networks of networks with replica nodes**

The interlinks of these networks of networks connect replica nodes of different layers only if the layers are connected in the supernetwork. It follows that the matrix $a^{[\alpha,\beta]}$ with $\alpha \neq \beta$ has elements

$$a_{ij}^{[\alpha,\beta]} = A_{\alpha\beta}\delta(i,j) \tag{5.34}$$

for unweighted networks of networks and

$$a_{ij}^{[\alpha,\beta]} = A_{\alpha\beta}\delta(i,j)w_{ij}^{[\alpha,\beta]} \tag{5.35}$$

for weighted networks of networks. An example of a network of networks with replica nodes is shown in Fig. 5.3 (panel (a)).

We note that multiplex networks with interlinks belong to this class of networks of networks where the supernetwork is formed by a fully connected network of M layers. Similarly, the multi-slice networks with interlinks belong to this class of networks of networks where the supernetwork is a chain of layers in temporal sequence.

(b) **Networks of networks without replica nodes**

In networks of networks without replica nodes the interlinks only connect nodes that belong to layers that are connected in the supernetwork. It follows that the adjacency matrix $a^{[\alpha,\beta]}$ with $\alpha \neq \beta$ has elements that satisfy the constraint

$$a_{ij}^{[\alpha,\beta]} = 0 \quad \text{if} \quad A_{\alpha\beta} = 0,$$
$$\sum_{i,j} a_{ij}^{[\alpha,\beta]} > 0 \quad \text{if} \quad A_{\alpha\beta} = 1.$$

An example of a network of networks without replica nodes is shown in Fig. 5.3 (panel (b)).

5.5.4   Multilayer edge list

The multilayer edge list of an undirected, unweighted multilayer network including both interlinks and intralinks is the list of quadruples (α, i, β, j) indicating a link between node (i, α) and node (j, β).

The multilayer edge list of an undirected and weighted multilayer network is the list of quintuples $(\alpha, i, \beta, j, w_{ij}^{[\alpha,\beta]})$ indicating that node (i, α) and node (j, β) are connected with weight $w_{ij}^{[\alpha,\beta]}$.

The multilayer edge list of an unweighted, directed multilayer network is a list of quadruples (α, i, β, j) indicating a link from node (i, α) to node (j, β).

The multilayer edge list of a weighted, directed multilayer network is a list of quintuples $(\alpha, i, \beta, j, w_{ij}^{[\alpha,\beta]})$ indicating a link from node (i, α) to node (j, β) with weight $w_{ij}^{[\alpha,\beta]}$.

5.6   Tensorial formalism for multilayer networks

In the previous sections we have seen that the information encoded in the interactions of a multilayer network goes beyond the information that it is possible to include in a single matrix. In fact, multilayer networks are described by a set of adjacency matrices indicating the interactions existing within and between the layers of the multilayer network. The supra-adjacency matrix apparently encodes all these interactions in a single matrix, but the supra-adjacency matrix cannot be fully interpreted if we do not have information about its block structure (i.e. how many nodes form each layer of the network). In order to fully address these issues, a tensorial formalism has been proposed in Refs [130, 195, 92]. This framework is able to describe rather efficiently the multilayer nature of multiplex networks, multi-slice networks and general multilayer networks in which there is a one-to-one mapping between the nodes.

Two main tensorial approaches have been considered. The first approach [130, 195] applies to multiplex networks and multi-slice networks and does not make use of interlinks. The second approach [92] applies to multilayer networks as long as there are replica nodes and allows for the explicit use of interlinks.

The idea common to both approaches is to proceed formally and introduce an N-dimensional vector space V associated with the nodes of the multiplex or multi-slice network and an M-dimensional vector space $\hat{V}$ associated with the layers of the multiplex or multi-slice network. The vector space V is spanned by the N-dimensional canonical covariant i-th vectors $\mathbf{e}_i$ of elements

$$e_i(j) = \delta(i,j)$$

with $i, j \in \{1, 2 \ldots, N\}$. The vector space $\hat{V}$ is spanned by the M-dimensional canonical covariant vectors $\hat{\mathbf{e}}_\alpha$ of elements

$$\hat{e}_\alpha(\beta) = \delta(\alpha, \beta)$$

with $\alpha, \beta \in \{1, 2, \ldots, M\}$. The dual spaces $V^\star$ and $\hat{V}^\star$ have canonical bases indicated by θ^i and $\hat{\theta}^\alpha$ respectively.

In the first approach [130, 195] the tensor $\mathcal{T}$ that characterizes formally the multiplex or the multi-slice network in the absence of multilinks is the 3-contravariant tensor belonging to the tensor product space

$$\mathcal{T} : V^\star \otimes V^\star \otimes \hat{V}^\star \tag{5.36}$$

with

$$T = \sum_{i,j=1,\ldots N} \sum_{\alpha=1,\ldots M} T_{ij\alpha} \theta^i \otimes \theta^j \otimes \hat{\theta}^\alpha. \tag{5.37}$$

This tensor identifies a multiplex network or a multi-slice network if we associate with the tensor elements $T_{ij\alpha}$ the same numerical values of the adjacency matrix elements $a_{ij}^{[\alpha]}$.

This tensorial formalism has been used both on temporal social networks [130] and on molecular networks [195] to extract valuable information from multiplex datasets. The algorithmic techniques used in these papers include non-negative tensor factorization analysis and combinatorial optimization problems.

In the second approach [92], aimed at representing a multiplex network with inter-links, the tensor $\mathcal{T}$ that characterizes formally the multilayer network with replica nodes is the 2-covariant, 2-contravariant tensor belonging to the tensor product space

$$\mathcal{T} : V \otimes \hat{V} \otimes V^\star \otimes \hat{V}^\star \tag{5.38}$$

with

$$T = \sum_{i,j=1,\ldots N} \sum_{\alpha,\beta=1,\ldots M} T^{j\beta}_{i\alpha} \mathbf{e}_j \otimes \hat{\mathbf{e}}_\beta \otimes \theta^i \otimes \hat{\theta}^\alpha. \tag{5.39}$$

This tensor can be interpreted as a map T

$$T : V \otimes \hat{V} \to V \otimes \hat{V} \tag{5.40}$$

such that

$$T \, \mathbf{e}_i \otimes \hat{\mathbf{e}}_\alpha = \sum_{j=1}^{N} \sum_{\beta=1}^{M} T^{j\beta}_{i\alpha} \mathbf{e}_j \otimes \hat{\mathbf{e}}_\beta. \tag{5.41}$$

The relation between this map and the multiplex (multi-slice) network is simply stated. To each canonical base $\mathbf{e}_i \otimes \hat{\mathbf{e}}_\alpha$ of the tensor product space $V \otimes \hat{V}$ we attribute a replica node (i,α). The map T associates with each replica node the set of the replica nodes connected to it.

The tensor elements $T^{j\beta}_{i\alpha}$ indicate the weight of the (eventually directed) interaction between the replica node (i,α) and the replica node (j,β). Therefore, the tensor elements $T^{j\beta}_{i\alpha}$ have the same numerical values of the elements of the supra-adjacency matrix $\mathcal{A}_{i\alpha j\beta}$. The relation between the tensor T and the supra-adjacency matrix $\mathcal{A}$ corresponds to the so-called *flattening* of the tensor.

Although this representation of multilayer networks in terms of tensors is formal, this formalism can provide advantages and can be a convenient way to express multilayer measures when the Einstein summation convention is adopted for the contraction of the indices. The convention is applied to repeated covariant and contravariant indices that are considered as dummy and summed indices. Therefore, we have for the generic tensors T, $\tilde{T}$ of the multiplex (multi-slice) tensor space

$$T^{i\beta}_{i\alpha} = \sum_{i=1}^{N} T^{i\beta}_{i\alpha},$$

$$T^{j\alpha}_{i\alpha} = \sum_{\alpha=1}^{M} T^{j\alpha}_{i\alpha},$$

$$T^{j\beta}_{i'\gamma} \, \tilde{T}^{i'\gamma}_{i\alpha} = \sum_{i'=1}^{N} \sum_{\gamma=1}^{M} T^{j\beta}_{i'\gamma} \, \tilde{T}^{i'\gamma}_{i\alpha}. \tag{5.42}$$

For example, consider an undirected multiplex network. The number of layers in which node i is connected to node j, i.e. the multiplicity of the overlap, is given by $T^{j\alpha}_{i\alpha}$. Similarly, the number of paths of length two connecting replica nodes (i,α) and (j,β) is given by $T^{j\beta}_{j'\gamma} \, T^{i'\gamma}_{i\alpha}$.

Although this formalism can provide a shortcut to characterize multilayer networks, in this book we have chosen to use the more intuitive matrix formalism where possible. The matrix formalism is more familiar to the interdisciplinary community of network scientists. Additionally, it is also more generic as it extends also to networks of networks and multilayer networks where the nodes in different layers are not mapped one-to-one.

6
Basic Structural Properties

6.1 The effect of multiplexity on network structure

Since the links of a multilayer network indicate different types of interactions, it is important not to make oversimplifications and to distinguish links of different nature. As a result, for multilayer networks the most fundamental structural properties of single networks must also be modified to take into account the multilayer nature of these structures. Here we give an account of the basic structural properties of multilayer networks including the degree, the clustering coefficient and the distance-dependent properties such as the average distance, the cross-betweenness and the interdependence.

6.2 Degree

When defining the degree of the nodes in multilayer networks it is important to distinguish the case of multiplex and multi-slice networks from the other types of multilayer network. Therefore, in the following we will define the degree of the nodes in these two different scenarios.

6.2.1 Multiplex degree

Let us first discuss the generalization of the degree of a node to undirected and unweighted multiplex and multi-slice networks with M layers. In these networks the *multiplex degree* $\mathbf{k}_i$ of node i is a vector and not a scalar. Specifically, the degree $\mathbf{k}_i$ of node i is an M-dimensional vector,

$$\mathbf{k}_i = \left(k_i^{[1]}, k_i^{[2]}, \ldots k_i^{[M]} \right) \tag{6.1}$$

with the element $k_i^{[\alpha]}$ indicating the degree of the node i within layer $\alpha = 1, 2, \ldots, M$, i.e.

$$k_i^{[\alpha]} = \sum_{j=1}^{N} a_{ij}^{[\alpha]}. \tag{6.2}$$

Multilayer Networks. Ginestra Bianconi, Oxford University Press (2018).
© Ginestra Bianconi. DOI: 10.1093/oso/9780198753919.001.0001

For directed and unweighted multiplex and multi-slice networks we distinguish between the *multiplex in-degree* $\mathbf{k}_{i,in}$ and the *multiplex out-degree* $\mathbf{k}_{i,out}$ of each node $i = 1, 2 \ldots, N$. In particular, we have

$$\mathbf{k}_{i,in} = \left(k_{i,in}^{[1]}, k_{i,in}^{[2]}, \ldots k_{i,in}^{[M]} \right),$$

$$\mathbf{k}_{i,out} = \left(k_{i,out}^{[1]}, k_{i,out}^{[2]}, \ldots k_{i,out}^{[M]} \right), \tag{6.3}$$

with the elements $k_{i,in}^{[\alpha]}, k_{i,out}^{[\alpha]}$ indicating the in/out-degree of the node i within layer α, i.e.

$$k_{i,in}^{[\alpha]} = \sum_{j=1}^{N} a_{ji}^{[\alpha]},$$

$$k_{i,out}^{[\alpha]} = \sum_{j=1}^{N} a_{ij}^{[\alpha]}. \tag{6.4}$$

The expression for the degree of the nodes in weighted multiplex or multi-slice networks can be obtained from the weighted adjacency matrices of elements $a_{ij}^{[\alpha]}$ by neglecting the weights of the links. Therefore, the multiplex degree $\mathbf{k}_i$ of a node i in a weighted multiplex or multi-slice network is given by Eq. (6.1) with the element $k_i^{[\alpha]}$ indicating the degree of the node i within layer α, i.e.

$$k_i^{[\alpha]} = \sum_{j=1}^{N} \theta(a_{ij}^{[\alpha]}), \tag{6.5}$$

where here $a_{ij}^{[\alpha]}$ can take real or integer-positive values and $\theta(x)$ is the Heaviside function. Similarly, the multiplex in-degree and the out-degree of a node i in a directed, weighted multiplex or multi-slice network are given by Eq. (6.3), with the elements $k_{i,in}^{[\alpha]}, k_{i,out}^{[\alpha]}$ indicating the in/out-degree of the node i within layer α, i.e.

$$k_{i,in}^{[\alpha]} = \sum_{j=1}^{N} \theta(a_{ji}^{[\alpha]}),$$

$$k_{i,out}^{[\alpha]} = \sum_{j=1}^{N} \theta(a_{ij}^{[\alpha]}). \tag{6.6}$$

6.2.2   Degree distributions of multiplex and multi-slice networks

In an undirected multiplex and multi-slice network we define the *multiplex degree distribution* $P(\mathbf{k})$ as the probability that a random node has multiplex degree $\mathbf{k}_i = \mathbf{k}$.

By indicating with $N(\mathbf{k})$ the number of nodes of multiplex degree $\mathbf{k}$ the multiplex degree distribution $P(\mathbf{k})$ is given by

$$P(\mathbf{k}) = \frac{N(\mathbf{k})}{N}.$$

(6.7)

The degree distributions in each layer α are instead indicated by $P^{[\alpha]}\left(k^{[\alpha]}\right)$.

For directed multiplex or multi-slice networks we distinguish between the multiplex in-degree distribution $P^{in}(\mathbf{k})$ and the out-degree distribution $P^{out}(\mathbf{k})$, indicating the probability that a random node has multiplex in-degree $\mathbf{k}_{i,in} = \mathbf{k}$ and the probability that it has multiplex out-degree $\mathbf{k}_{i,out} = \mathbf{k}$ respectively. The *multiplex in-degree distribution* $P^{in}(\mathbf{k})$ and the *multiplex out-degree distribution* $P^{out}(\mathbf{k})$ of a directed network are given by

$$P^{in}(\mathbf{k}) = \frac{N^{in}(\mathbf{k})}{N},$$
$$P^{out}(\mathbf{k}) = \frac{N^{out}(\mathbf{k})}{N},$$

(6.8)

where $N^{in}(\mathbf{k})$ and $N^{out}(\mathbf{k})$ indicate the total number of nodes of the network with multiplex in-degree $\mathbf{k}$ and multiplex out-degree $\mathbf{k}$ respectively.

6.2.3 Aggregated degree and aggregated strength

It is sometimes useful to consider the node properties in the aggregated network. The aggregated network of a multiplex network has been defined in Sec. 5.3.5 and is a single network between the nodes of the multiplex where all the interactions are considered on the same footing. The aggregated network has an adjacency matrix $\tilde{\mathbf{a}}$ of elements $\tilde{a}_{ij}$ given by Eq. (5.16) indicating whether the link (i,j) exists at least in one layer. It is also useful to associate with each link (i,j) in the projection network a weight called the *link multiplicity* v_{ij} indicating, in the case of an unweighted multiplex network, in how many layers the link (i,j) is present

$$v_{ij} = \sum_{\alpha=1}^{M} a_{ij}^{[\alpha]}.$$

(6.9)

In the case of an undirected multiplex or multi-slice network the *aggregated degree* K_i of node i is given by

$$K_i = \sum_{j=1}^{N} \tilde{a}_{ij}.$$

(6.10)

Therefore, the aggregated degree K_i of the node i is the degree of the node i in the network where we neglect the attributes of the links (the type of link for a multiplex network or the time at which the interaction occurs in the multi-slice networks). Similarly, in the undirected case the *aggregated strength* S_i (also called *overlapping degree*) [30] is the strength of node i in the projection network and is given by

$$S_i = \sum_{j=1}^{N} v_{ij}. \tag{6.11}$$

It can be shown that if the original multiplex or multi-slice network is unweighted, the aggregated strength is given by the sum of the degrees of node i across all the layers of the multiplex network,

$$S_i = \sum_{j=1}^{N} \sum_{\alpha=1}^{M} a_{ij}^{[\alpha]} = \sum_{\alpha=1}^{M} k_i^{[\alpha]}. \tag{6.12}$$

We note here that the sum of the degrees of a node across each layer is in general not equivalent to the aggregated degrees because some of the links in different layers might connect the same nodes. Instead, it is the aggregated strength that is equal to the sum of the degrees of nodes across different layers. The two quantities K_i and S_i are the same only if the links of each layer do not overlap with the links of any other layer, i.e. if every pair of nodes is at most linked in a single layer.

In the directed case, we distinguish between the *aggregated in-degree* and the *aggregated out-degree* of node i, given respectively by

$$K_{i,in} = \sum_{j=1}^{N} \tilde{a}_{ji},$$

$$K_{i,out} = \sum_{j=1}^{N} \tilde{a}_{ij}. \tag{6.13}$$

Similarly, it is possible to define the *aggregated in-strength* and the *aggregated out-strength* of node i, given respectively as

$$S_{i,in} = \sum_{j=1}^{N} v_{ji},$$

$$S_{i,out} = \sum_{j=1}^{N} v_{ij}. \tag{6.14}$$

6.2.4   Multilayer degree

In general multilayer networks each node i of layer α is assigned the *multilayer degree* $\mathbf{k}_{i\alpha}$ given by

$$\mathbf{k}_{i\alpha} = \left(k_i^{[\alpha,1]}, k_i^{[\alpha,2]}, \ldots, k_i^{[\alpha,M]} \right) \tag{6.15}$$

with $k_i^{[\alpha,\beta]}$ indicating the sum of links connecting node (i,α) to other nodes in layer β, i.e. for an unweighted multilayer network

$$k_i^{[\alpha,\beta]} = \sum_{j=1}^{N} a_{ij}^{[\alpha,\beta]}. \tag{6.16}$$

Given this definition, we note that $k_i^{[\alpha,\alpha]}$ plays a special role since it indicates the degree of node (i,α) within layer α.

In the case of a directed network of networks we distinguish between the *multilayer in-degree* $\mathbf{k}_{i\alpha,in}$ and the *multilayer out-degree* $\mathbf{k}_{i\alpha,out}$ of node i in layer α. These are given by

$$\mathbf{k}_{i\alpha,in} = \left(k_{i,in}^{[\alpha,1]}, k_{i,in}^{[\alpha,2]} \ldots, k_{i,in}^{[\alpha,M]} \right)$$
$$\mathbf{k}_{i\alpha,out} = \left(k_{i,out}^{[\alpha,1]}, k_{i,out}^{[\alpha,2]}, \ldots, k_{i,out}^{[\alpha,M]} \right), \tag{6.17}$$

with $k_{i,in/out}^{[\alpha,\beta]}$ indicating the sum of in-/outcoming links connecting node (i,α) to nodes in layer β, i.e. for an unweighted multilayer network

$$k_{i,in}^{[\alpha,\beta]} = \sum_{j=1}^{N} a_{ji}^{[\alpha,\beta]},$$
$$k_{i,out}^{[\alpha,\beta]} = \sum_{j=1}^{N} a_{ij}^{[\alpha,\beta]}. \tag{6.18}$$

The generalization of the definition of the multilayer degree for weighted multilayer networks is straightforward.

6.2.5   Multilayer degree distributions

For a general undirected multilayer network we define the *multilayer degree distribution* $P^{[\alpha]}(\mathbf{k})$ as the probability that a random node of layer α has multilayer degree $\mathbf{k}_{i\alpha} = \mathbf{k}$. By indicating with $N^{[\alpha]}(\mathbf{k})$ the number of nodes of layer α with multilayer degree $\mathbf{k}_{i\alpha} = \mathbf{k}$ the multilayer degree distribution $P^{[\alpha]}(\mathbf{k})$ is given by

$$P^{[\alpha]}(\mathbf{k}) = \frac{N^{[\alpha]}(\mathbf{k})}{N_\alpha}. \tag{6.19}$$

Similarly, for a directed multilayer network it is possible to define the *multilayer in-degree and out-degree distributions* $P^{in,[\alpha]}(\mathbf{k})$ and $P^{out,[\alpha]}(\mathbf{k})$ given by

$$P^{in,[\alpha]}(\mathbf{k}) = \frac{N^{in,[\alpha]}(\mathbf{k})}{N_\alpha},$$

$$P^{out,[\alpha]}(\mathbf{k}) = \frac{N^{out,[\alpha]}(\mathbf{k})}{N_\alpha}, \qquad (6.20)$$

where $N^{in,[\alpha]}(\mathbf{k})$, and $N^{out,[\alpha]}(\mathbf{k})$ indicate the number of nodes of layer α that have multilayer in-degree $\mathbf{k}_{i\alpha,in}$ or multilayer out-degree $\mathbf{k}_{i\alpha,out}$ given by $\mathbf{k}$ respectively.

6.3 Clustering coefficient

6.3.1 General remarks and definitions for multiplex networks

In a multiplex network the clustering coefficient can be generalized in several different ways. The major challenge in choosing an appropriate definition of the clustering coefficient in multiplex networks is that in these structures triangles (cycles of length 3) can include links belonging to different layers. In the majority of cases, this feature of the multiplex networks is exactly the property that we aim at capturing. For example, we might ask what is the probability that two of our colleagues do some sportive activity together, or we might ask what is the probability that our new acquaintance from the foreign language class knows one of our schoolmates. In one case we are asking if a triangle closes across two layers (work, sport), in the other case we are asking if a triangle closes across three layers (language course, common school, unknown situation). If we are interested in generalizing the clustering coefficient of a node i in an undirected multiplex network the first approach is to consider only triangles formed by three links, in the layers α, β, γ respectively. In this case, the definition of the local clustering coefficient $C_i^{[\alpha,\beta,\gamma]}$ is straightforward and given by

$$C_i^{[\alpha,\beta,\gamma]} = \frac{\sum_{r,s} a_{ir}^{[\alpha]} a_{rs}^{[\beta]} a_{si}^{[\gamma]}}{\mathcal{N}_{\alpha,\gamma}}, \qquad (6.21)$$

where $\mathcal{N}_{\alpha,\gamma}$ is the number of ordered pairs of neighbours of node i belonging respectively to layer α and layer γ, i.e.

$$\mathcal{N}_{\alpha,\gamma} = \begin{cases} k_i^{[\alpha]} k_i^{[\gamma]} & \text{for } \alpha \neq \gamma \\ k_i^{[\alpha]} (k_i^{[\alpha]} - 1) & \text{for } \alpha = \gamma. \end{cases} \qquad (6.22)$$

The above definition is valid as long as the normalization sum $\mathcal{N}_{\alpha,\gamma}$ is greater than zero. When $\mathcal{N}_{\alpha,\gamma}$ vanishes the local clustering coefficient $C_i^{[\alpha,\beta,\gamma]}$ is set to zero.

Although this definition is very immediate, it cannot account for situations in which one of the considered layers is unknown or undetermined. Moreover, this definition is restricted to given triplets of layers and as such it does not provide a unique estimation of the local organization of the triangles passing through node i across all the layers of the multiplex network.

In Ref. [30] the definitions of the local clustering coefficients $C_{i,1}$, $C_{i,2}$ of node i in an undirected multiplex network are given as

$$C_{i,1} = \frac{\sum_{\alpha=1}^{M} \sum_{\beta|\beta\neq\alpha} \sum_{r,s} a_{ir}^{[\alpha]} a_{rs}^{[\beta]} a_{si}^{[\alpha]}}{(M-1) \sum_{\alpha=1}^{M} k_i^{[\alpha]}(k_i^{[\alpha]}-1)},$$

$$C_{i,2} = \frac{\sum_{\alpha=1}^{M} \sum_{\gamma|\gamma\neq\alpha} \sum_{\beta|\beta\neq\alpha,\gamma} \sum_{r,s} a_{ir}^{[\alpha]} a_{rs}^{[\beta]} a_{si}^{[\gamma]}}{(M-2) \sum_{\alpha=1}^{M} \sum_{\gamma\neq\alpha} k_i^{[\alpha]} k_i^{[\gamma]}}, \qquad (6.23)$$

where $C_{i,1}(C_{i,2})$ evaluates the normalized number of triangles spanning two (three) layers and including node i. Note that also in this case, if the denominator of the above expressions vanishes the local clustering coefficient is assumed to be null. The local clustering coefficient can be averaged across the total number of nodes of the multiplex network, obtaining

$$C_1 = \frac{1}{N} \sum_{i=1}^{N} C_{i,1},$$

$$C_2 = \frac{1}{N} \sum_{i=1}^{N} C_{i,2}. \qquad (6.24)$$

Finally, the transitivity T_1 is defined as the ratio between the total number of 2-triangles and $M-1$ times the total number of 1-triads, and the transitivity T_2 is defined as the ratio between the total number of 3-triangles and $M-2$ times the total number of 2-triads in the multiplex network, i.e.

$$T_1 = \frac{\sum_{\alpha=1}^{M} \sum_{\beta|\beta\neq\alpha} \sum_{i,r,s} a_{ir}^{[\alpha]} a_{rs}^{[\beta]} a_{si}^{[\alpha]}}{(M-1) \sum_{\alpha=1}^{M} \sum_{i=1}^{N} k_i^{[\alpha]}(k_i^{[\alpha]}-1)},$$

$$T_2 = \frac{\sum_{\alpha=1}^{M} \sum_{\gamma|\gamma\neq\alpha} \sum_{\beta|\beta\neq\alpha,\gamma} \sum_{i,r,s} a_{ir}^{[\alpha]} a_{rs}^{[\beta]} a_{si}^{[\gamma]}}{(M-2) \sum_{\alpha=1}^{M} \sum_{\gamma\neq\alpha} \sum_{i=1}^{N} k_i^{[\alpha]} k_i^{[\gamma]}}. \qquad (6.25)$$

6.3.2 Parametrized clustering coefficient in multiplex networks

A series of alternative definitions of the clustering coefficient proposed in Ref. [83] uses a parametrization of the clustering coefficient in terms of two parameters describing the cost associated with a jump from one layer to the other (δ) and with a hop from a link to another one on the same layer (η) respectively. Let us define the weighted sum $t_{i\alpha}^{\star}$ and $q_{i\alpha}^{\star}$

respectively as the weighted sum of close triadic paths and the weighted sum of paths of length 2 passing through the replica node (i, α). The quantities $t_{i\alpha}^{\star}$ and $q_{i\alpha}^{\star}$ have a quite elegant expression in terms of $(N \cdot M) \times (N \cdot M)$ supra-matrices $\hat{A}$ and $\hat{C}$ defined as

$$
\hat{A} =
\begin{pmatrix}
\mathbf{a}^{[1]} & 0 & 0 & \cdots & 0 & 0 \\
0 & \mathbf{a}^{[2]} & 0 & \cdots & 0 & 0 \\
\vdots & \vdots & \vdots & \ddots & \vdots & \vdots \\
0 & 0 & 0 & \cdots & \mathbf{a}^{[M-1]} & 0 \\
0 & 0 & 0 & \cdots & 0 & \mathbf{a}^{[M]}
\end{pmatrix}
\tag{6.26}
$$

and

$$
\hat{C} = \eta \mathcal{I} + \delta \mathcal{C}
\tag{6.27}
$$

with

$$
\mathcal{I} =
\begin{pmatrix}
\mathbf{I} & 0 & 0 & \cdots & 0 & 0 \\
0 & \mathbf{I} & 0 & \cdots & 0 & 0 \\
\vdots & \vdots & \vdots & \ddots & \vdots & \vdots \\
0 & 0 & 0 & \cdots & \mathbf{I} & 0 \\
0 & 0 & 0 & \cdots & 0 & \mathbf{I}
\end{pmatrix}, \quad
\mathcal{C} =
\begin{pmatrix}
0 & \mathbf{I} & \mathbf{I} & \cdots & \mathbf{I} & \mathbf{I} \\
\mathbf{I} & 0 & \mathbf{I} & \cdots & \mathbf{I} & \mathbf{I} \\
\vdots & \vdots & \vdots & \ddots & \vdots & \vdots \\
\mathbf{I} & \mathbf{I} & \mathbf{I} & \cdots & 0 & \mathbf{I} \\
\mathbf{I} & \mathbf{I} & \mathbf{I} & \cdots & \mathbf{I} & 0
\end{pmatrix}.
\tag{6.28}
$$

The supra-matrix $\hat{A}$ fully characterizes the intralinks and the matrix $\mathcal{C}$ fully characterizes the interlinks of the multiplex network. Therefore, the weighted sum of closed triadic paths passing through replica node (i, α) is given by

$$
t_{i\alpha}^{\star} = 2 \left[\left(\hat{A}\hat{C} \right)^3 \right]_{i\alpha, i\alpha}.
\tag{6.29}
$$

This expression can be derived by observing that the number of closed triadic paths passing through a replica node (i, α) and belonging to a single layer can be expressed as

$$
\left[\hat{A}\hat{A}\hat{A} \right]_{i\alpha, i\alpha},
$$

the number of closed triadic paths passing through node (i, α) having two links in layer α and one link in another layer is given by

$$
\left[\hat{A}\hat{A}\mathcal{C}\hat{A}\mathcal{C} \right]_{i\alpha, i\alpha},
$$

and so on (see Fig. 6.1). By performing a weighted sum over all these types of triadic paths, when we associate a cost δ with every hop from one layer to another one and a

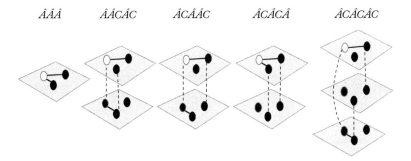

$$\hat{A}\hat{A}\hat{A} \qquad \hat{A}\hat{A}\hat{C}\hat{A}\hat{C} \qquad \hat{A}\hat{C}\hat{A}\hat{A}\hat{C} \qquad \hat{A}\hat{C}\hat{A}\hat{C}\hat{A} \qquad \hat{A}\hat{C}\hat{A}\hat{C}\hat{A}\hat{C}$$

Fig. 6.1 *Sketch of the closed triadic paths including different layers. The intralinks are the solid lines and the interlinks are the dotted lines.*

cost η with every pair of consecutive links belonging to the same layer, we obtain the expression in Eq. (6.29).

Similarly, by defining the supra-matrix $\mathcal{F} = \oplus_\alpha \mathbf{F}^{[\alpha]}$ where $\mathbf{F}^{[\alpha]}$ is the adjacency matrix of a complete graph on layer α, it is possible to express $q^\star_{i,\alpha}$ as

$$q^\star_{i\alpha} = 2\left(\hat{A}\hat{C}\mathcal{F}\hat{C}\hat{A}\hat{C}\right)_{i\alpha,i\alpha}. \tag{6.30}$$

We note that, thanks to their weights, $t^\star_{i\alpha}$ and $q^\star_{i\alpha}$ can be decomposed in contributions coming from closed triadic paths that belong to one, two and three layers respectively. In fact, by denoting by $t_{i\alpha,n}$ the total number of closed triadic paths belonging to n layers and by $q_{i\alpha,n}$ the total number of pairs of nodes belonging to n layers we have

$$\begin{aligned} t^\star_{i\alpha} &= \eta^3 t_{i\alpha,1} + \eta\delta^2 t_{i\alpha,2} + \delta^3 t_{i\alpha,3}, \\ q^\star_{i\alpha} &= \eta^3 q_{i\alpha,1} + \eta\delta^2 q_{i\alpha,2} + \delta^3 q_{i\alpha,3}. \end{aligned} \tag{6.31}$$

Using $t^\star_{i\alpha}, q^\star_{i\alpha}$ in Ref. [83] the following *local parametrized clustering coefficients* have been defined as

$$\begin{aligned} c^\star_{i\alpha} &= \frac{t^\star_{i\alpha}}{q^\star_{i\alpha}} \\ C^\star_i &= \frac{\sum_\alpha t^\star_{i\alpha}}{\sum_\alpha q^\star_{i\alpha}}, \end{aligned} \tag{6.32}$$

where it is assumed that $c^\star_{i\alpha} = 0$ if $q^\star_{i\alpha} = 0$ and $C^\star_i = 0$ if $\sum_\alpha q^\star_{i\alpha} = 0$. Note that while $c^\star_{i\alpha}$ characterizes the local loop structure associated with the replica node (i,α), $C^\star_i$ characterizes the local loop structure of node i across all layers. If we choose $\delta = 0$ the parametrized local clustering coefficients only depend on the triangles belonging to the same triangle. If we take $\eta = 0$ instead, only the triangles belonging to three different layers contribute. Finally, the *parametric transitivity* is given by

$$T^\star = \frac{\sum_i \sum_\alpha t_{i\alpha}^\star}{\sum_i \sum_\alpha q_{i\alpha}^\star}.$$ (6.33)

We observe that the definitions of the local parametrized clustering coefficients and the parametrized transitivity are invariant under the transformation $(\eta, \delta) \to (\eta', \delta')$ as long as $\eta'/\eta = \delta'/\delta$. Therefore, without loss of generality, as long as $\eta \neq 0$ we can always set $\eta' = 1$ and parametrize the clustering coefficient with the single parameter $\delta' = \delta/\eta$.

6.3.3   Cross-clustering coefficient in general multilayer networks

In a general multilayer network the clustering coefficient can also evaluate the density of triangles formed by two interlinks and one intralink. To this end, in Ref. [101] three structural measures have been defined: the local cross-clustering coefficient, the global cross-clustering coefficient and the cross-transitivity. The *local cross-clustering coefficient* $\tilde{C}_i^{[\alpha,\beta]}$ of a node (i,α) with respect to the layer β measures the density of links within any two neighbours of node (i,α) belonging to layer β. Therefore, it is a measure of the tendency of a node (i,α) to connect to nodes of layer β that belong to a close community. Mathematically, for any node (i,α) with $k_i^{[\alpha,\beta]} > 1$ the local cross-clustering coefficient is given by

$$\tilde{C}_i^{[\alpha,\beta]} = \frac{1}{k_i^{[\alpha,\beta]}(k_i^{[\alpha,\beta]} - 1)} \sum_{r,s,\alpha,\beta} a_{ir}^{[\alpha,\beta]} a_{rs}^{[\beta,\beta]} a_{si}^{[\beta,\alpha]},$$ (6.34)

where it is assumed that if $k_i^{[\alpha,\beta]} = 0$ or $k_i^{[\alpha,\beta]} = 1$ the local cross-clustering coefficient is zero, i.e. $\tilde{C}_i^{[\alpha,\beta]} = 0$. The *global cross-clustering coefficient* $\tilde{C}^{[\alpha,\beta]}$ estimates the probability for a node of layer α to have pairs of mutually connected neighbours in layer β and it is given by

$$\tilde{C}^{[\alpha,\beta]} = \frac{1}{N_\alpha} \sum_i \tilde{C}_i^{[\alpha,\beta]}.$$ (6.35)

Finally, the *cross-transitivity* $\tilde{T}^{[\alpha,\beta]}$ indicates the probability that two nodes in layer β are connected if they are both connected to a common neighbour in layer α,

$$\tilde{T}^{[\alpha,\beta]} = \frac{\sum_{i,r,s} \sum_{\alpha,\beta} a_{ir}^{[\alpha,\beta]} a_{rs}^{[\beta,\beta]} a_{si}^{[\beta,\alpha]}}{\sum_{i,r,s} \sum_{\alpha,\beta} a_{ir}^{[\alpha,\beta]} a_{si}^{[\beta,\alpha]}}.$$ (6.36)

6.4   Distance-dependent measures

6.4.1   Interdependence of a multiplex network

Multiplexity has a strong influence on the properties of shortest paths in a multiplex network. In fact, if we build a multiplex network starting from a single isolated layer, the inclusion of more layers can only reduce the shortest distance between any pair of two nodes. As a consequence, an important fraction of shortest paths between pairs of nodes includes links coming from different layers. In this context it is relevant to evaluate the added value that multiplexity has on the reachability of each node i of the multiplex network. In Ref. [217] the Authors introduced the *interdependence* which for every node is defined as $\lambda_i \in [0, 1]$ given by

$$\lambda_i = \sum_{i \neq j} \frac{\psi_{ij}}{\sigma_{ij}} \tag{6.37}$$

where ψ_{ij} is the number of shortest paths between node i and node j including links from more than a layer and σ_{ij} is the number of shortest paths between node i and node j including both paths travelling exclusively in a layer and paths including links from more than a layer. The average multiplex interdependence λ is computed as

$$\lambda = \frac{1}{N} \sum_{i=1}^{N} \lambda_i. \tag{6.38}$$

For $\lambda = 0$ all the shortest paths lie on just one layer, and for $\lambda = 1$ all the shortest paths include links of more than one layer.

6.4.2   Distance properties of general multilayer networks

In a general multilayer network it is important to evaluate the topological closeness of different layers. By indicating with $d_{i\alpha,j\beta}$ the shortest distance between node (i, α) and node (j, β), the *cross-average distance* $\tilde{\ell}^{[\alpha,\beta]}$ [101] between layer α and layer β is given by

$$\tilde{\ell}^{[\alpha,\beta]} = \left\langle d_{i\alpha,j\beta} \right\rangle_{i,j}, \tag{6.39}$$

where the average is performed over pairs of nodes (i, α) and (j, β), which are mutually reachable. The cross-average distance is large for two closely interwoven networks, indicating functional interdependence, while it can become low for two topologically distant layers that are likely to be functionally and dynamically independent.

Finally, the *cross-betweenness centrality* [101] of node (i, α) with respect to two layers β and γ indicates its role in mediating and controlling the communication between two layers β and γ. It is given by

$$B_{i,\alpha}^{[\beta,\gamma]} = \sum_{r \neq s} \frac{\sigma_{r\beta,s\gamma}(i,\alpha)}{\sigma_{r\beta,s\gamma}} \tag{6.40}$$

where $\sigma_{r\beta,s\gamma}$ indicates the total number of shortest paths from (r,β) to (s,γ) and $\sigma_{r\beta,s\gamma}(i,\alpha)$ counts the number of such paths that pass through node (i,α). The cross-betweenness $B_{i,\alpha}^{[\beta,\gamma]}$ evaluates the role of node (i,α) for the transmission of information between layer β and layer γ. A hub that has high cross-betweenness between two layers is more vulnerable with respect to damage of the interlinks between the two layers; a node of low cross-betweenness instead has high redundancy. From the cross-betweenness $B_{i,\alpha}^{[\beta,\gamma]}$ it is possible to obtain the betweenness centrality $B_{i,\alpha}$ simply by summing over all pairs of layers β, γ, i.e.

$$B_{i,\alpha} = \sum_{\beta,\gamma} B_{i,\alpha}^{[\beta,\gamma]}. \tag{6.41}$$

Although all these measures have been proposed for a general network of networks, they can also be applied to a multiplex network in which the interlinks are explicitly taken into account.

7
Structural Correlations of Multiplex Networks

7.1 Correlations in multiplex networks

In the vast majority of real cases multiplex networks have a highly correlated structure from which it is possible to extract more information than from the study of their layers taken in isolation. The correlations are typically induced by the identification (or the one-to-one mapping) among nodes in different layers, so it would be very misleading to consider a multiplex network as a single large network.

These correlations include:

- *Interlayer degree correlations*
  For every two layers, these correlations are able to indicate whether the hubs in one layer are also the hubs in the other layer, or whether, instead, they are the low-degree nodes.

- *Link overlap and multilinks*
  A multiplex network displays significant link overlap when a finite fraction of pairs of nodes are linked in multiple layers. For instance, typically we use different means of communication to interact with our social ties such as mobile phone, email and instant messaging, implying a significant overlap among the three layers. The different patterns of connections between two nodes, eventually including link overlap, are exhaustively characterized by the multilinks that extend the concept of links between two nodes in the multiplex network scenario.

- *Correlations in weighted multiplex networks*
  In weighted multiplex networks, weights might not be distributed homogeneously. For example, in a multiplex network formed by scientists that collaborate and cite each other, it is possible to observe that scientists tend to cite their collaborators in a significantly different way than other scientists with whom they have not collaborated. Weight–topology correlations existing in multiplex networks are revealed by the multistrengths and inverse multiparticipation ratio.

- *Node pairwise multiplexity*
  When the nodes are not all active (i.e. connected) in all layers, two nodes can have correlated activity patterns. For example, they can be active on the same or on different layers. These correlations are captured by the *node pairwise multiplexity*.

Multilayer Networks. Ginestra Bianconi, Oxford University Press (2018).
© Ginestra Bianconi. DOI: 10.1093/oso/9780198753919.001.0001

- *Layer pairwise multiplexity*

  When the nodes are not all active (i.e. connected) in all layers, two layers can have correlated activity patterns. For example, they can contain the same active nodes or different active nodes. These correlations are captured by the *layer pairwise multiplexity*.

7.2   Interlayer degree correlations in multiplex networks

The degrees of the same node in different layers can be correlated. Given any two layers, it is possible to distinguish between positive degree correlations, indicating that a node tends to have either high degree or low degree in both layers, and negative degree correlations, describing the tendency of high-degree nodes of one layer to have low degree in the other. For instance, the citation/collaboration multiplex network formed by nodes representing scientists that cite each other and collaborate with each other displays positive degree correlations, indicating that typically a hub of the citation network is also a scientist with many collaborations. On the contrary, the multiplex networks between airports connected by low-cost airline companies display negative degree correlations because, due to their competition, low-cost airlines tend to have different hub airports.

From a theoretical perspective, valid, for instance, when discussing network models, multiplex networks do not display degree correlations if the multiplex degree distribution $P(\mathbf{k})$ can be expressed in terms of the degree distributions $P^{[\alpha]}\left(k^{[\alpha]}\right)$ of each single layer α as

$$P(\mathbf{k}) = \prod_{\alpha=1}^{M} P^{[\alpha]}\left(k^{[\alpha]}\right). \tag{7.1}$$

On the contrary, when this relation does not hold, the multiplex network displays degree correlations.

When we need to characterize real multiplex network data different measures can be useful for evaluating the degree correlations between a layer α and a layer β. These measures describe the correlations by performing a different level of coarse graining. Here we describe exhaustively these measures, highlighting the pros and cons of each of them.

- *Full characterization of the degree correlations across the two layers*

  Given a multiplex network, the probability that a node has degree $k^{[\alpha]}$ in layer α and degree $k^{[\beta]}$ in layer β is given by

$$P\left(k^{[\alpha]}, k^{[\beta]}\right) = \frac{N\left(k^{[\alpha]}, k^{[\beta]}\right)}{N}, \tag{7.2}$$

where $N\left(k^{[\alpha]}, k^{[\beta]}\right)$ is the number of nodes that have degree $k^{[\alpha]}$ in layer α and degree $k^{[\beta]}$ in layer β. From this matrix, the full pattern of interlayer degree correlations between layer α and layer β can be determined. Specifically, by plotting this two-dimensional function it is possible to investigate whether there are assortative or disassortative correlations between the two layers, mainly if the degrees of the same node in the two layers are positively or negatively correlated. Moreover, it is possible to calculate the Mutual Information $I_{[\alpha,\beta]}$ given by

$$I_{[\alpha,\beta]} = \sum_{k^{[\alpha]}, k^{[\beta]}} P(k^{[\alpha]}, k^{[\beta]}) \ln \frac{P(k^{[\alpha]}, k^{[\beta]})}{P^{[\alpha]}(k^{[\alpha]}) P^{[\beta]}(k^{[\beta]})}. \tag{7.3}$$

The Mutual Information is zero if there are no degree correlations, and typically higher values of the mutual information indicate more correlated degree sequences in the two layers. This technique aims at fully characterizing the correlations between the two layers, however the network might be in some cases too small to provide a solid statistic for all the matrix entries $P(k^{[\alpha]}, k^{[\beta]})$. In these cases the values of the degrees $k^{[\alpha]}$ and $k^{[\beta]}$ can be binned, or else a more coarse-grained measure of correlations can be used.

- *Average degree in layer α conditioned on the degree of the node in layer β*
  A more coarse-grained measure of correlation is $\langle k^{[\alpha]} | k^{[\beta]} \rangle$, i.e. the average degree of a node in layer α conditioned on the degree of the same node in layer β:

$$\left\langle k^{[\alpha]} | k^{[\beta]} \right\rangle = \sum_{k^{[\alpha]}} k^{[\alpha]} P(k^{[\alpha]} | k^{[\beta]}) = \frac{\sum_{k^{[\alpha]}} k^{[\alpha]} P(k^{[\alpha]}, k^{[\beta]})}{\sum_{k^{[\alpha]}} P(k^{[\alpha]}, k^{[\beta]})}. \tag{7.4}$$

If this function does not depend on $k^{[\beta]}$, the degrees in the two layers are uncorrelated. If this function is increasing (decreasing) in $k^{[\beta]}$, the degrees of the nodes in the two layers are positively (negatively) correlated. When this function is not monotonically increasing or monotonically decreasing, the interpretation is less straightforward. This method of calculating the degree correlations is particularly useful for scale-free networks where the possible values of the degree of each node can span over a wide range, while it is less efficient if the degree distribution is more homogeneous.

- *Pearson, Spearman and Kendall correlation coefficients*
  Even more coarse-grained correlation measures are the Pearson, the Spearman and the Kendall correlation coefficients. These measures provide a single scalar number to evaluate the interlayer degree correlations between two layers globally across the entire set of nodes of the multiplex network.
  The Pearson correlation coefficient $r_{\alpha\beta}$ is given by

$$r_{\alpha\beta} = \frac{\left\langle k^{[\alpha]} k^{[\beta]} \right\rangle - \left\langle k^{[\alpha]} \right\rangle \left\langle k^{[\beta]} \right\rangle}{\sigma_\alpha \sigma_\beta}, \tag{7.5}$$

where

$$\left\langle k^{[\alpha]} \right\rangle = \frac{1}{N} \sum_{i=1}^{N} k_i^{[\alpha]},$$

$$\left\langle k^{[\alpha]} k^{[\beta]} \right\rangle = \frac{1}{N} \sum_{i=1}^{N} k_i^{[\alpha]} k_i^{[\beta]},$$

$$\sigma_\alpha = \frac{1}{N} \sum_{i=1}^{N} \left[\left(k_i^{[\alpha]} - \left\langle k^{[\alpha]} \right\rangle \right)^2 \right].$$

The Pearson correlation coefficient can be dominated by the correlations of the high-degree nodes if the degree distribution of the network is broad.

The Spearman correlation coefficient $\rho_{\alpha\beta}$ between the degree sequences $\{k_i^{[\alpha]}\}$ and $\{k_i^{[\beta]}\}$ in the two layers α and β is given by

$$\rho_{\alpha\beta} = \frac{\left\langle x^{[\alpha]} x^{[\beta]} \right\rangle - \left\langle x^{[\alpha]} \right\rangle \left\langle x^{[\beta]} \right\rangle}{\hat{\sigma}_\alpha \hat{\sigma}_\beta}, \tag{7.6}$$

where $x_i^{[\alpha]}$ is the rank of the degree $k_i^{[\alpha]}$ in the sequence $\{k_i^{[\alpha]}\}$, $x_i^{[\beta]}$ is the rank of the degree $k_i^{[\beta]}$ in the sequence $\{k_i^{[\beta]}\}$ and $\hat{\sigma}_\alpha$ is given by

$$\left\langle x^{[\alpha]} \right\rangle = \frac{1}{N} \sum_{i=1}^{N} x_i^{[\alpha]},$$

$$\left\langle x^{[\alpha]} x^{[\beta]} \right\rangle = \frac{1}{N} \sum_{i=1}^{N} x_i^{[\alpha]} x_i^{[\beta]},$$

$$\hat{\sigma}_\alpha = \frac{1}{N} \sum_{i=1}^{N} \left[\left(x_i^{[\alpha]} - \left\langle x^{[\alpha]} \right\rangle \right)^2 \right]. \tag{7.7}$$

The Spearman coefficient has the problem that the ranks of the nodes according to their degree in a given layer are not uniquely defined because degree sequences typically include some degeneracy (not all the nodes have different degree). Therefore, the Spearman correlation coefficient is not a uniquely defined number.

The Kendall τ correlation coefficient between the degree sequences $\{k_i^{[\alpha]}\}$ and $\{k_i^{[\beta]}\}$ is a measure that takes into account possible degeneracy of the ranks. A pair of nodes i and j is concordant if their degrees have the same order in the two sequences, i.e. $(k_i^{[\alpha]} - k_j^{[\alpha]})(k_i^{[\beta]} - k_j^{[\beta]}) > 0$, and discordant if $(k_i^{[\alpha]} - k_j^{[\alpha]})(k_i^{[\beta]} - k_j^{[\beta]}) < 0$. The Kendall's τ is defined in terms of the number n_c of concordant pairs and the number n_d of discordant pairs and is given by

$$\tau = \frac{n_c - n_d}{\sqrt{(n_0 - n_1)(n_0 - n_2)}}. \tag{7.8}$$

In the equation above, $n_0 = N(N-1)/2$ and the terms n_1 and n_2 account for the degeneracy of the ranks and are given by

$$n_1 = \frac{1}{2} \sum_{k^{[1]}} P^{[1]}\left(k^{[1]}\right)\left[P^{[1]}\left(k^{[1]}\right) - 1\right],$$

$$n_2 = \frac{1}{2} \sum_{k^{[2]}} P^{[2]}\left(k^{[2]}\right)\left[P^{[2]}\left(k^{[2]}\right) - 1\right]. \tag{7.9}$$

The degree correlations between two layers of a multiplex network can be tuned by modifying the one-to-one mapping between their nodes. This can be achieved by reshuffling the interlinks and changing the identity of the connected replica nodes among the different layers as suggested in Ref. [213]. This tuning of the degree correlations can be used to generate either Maximally Positively correlated (MP) multiplex networks or Maximally Negatively correlated (MN) multiplex networks. Given a duplex network having two layers with degree sequence given by $\left\{k_i^{[1]}\right\}_{i=1,2,...,N}$ and $\left\{k_i^{[2]}\right\}_{i=1,2,...,N}$ respectively, let us rank the degrees of each layer in descending order. The Maximally Positively correlated (MP) multiplex network has the same two layers of the original duplex network, but the pairs of replica nodes are different from the ones of the original network. In fact, replica nodes are nodes having the same rank in the two degree

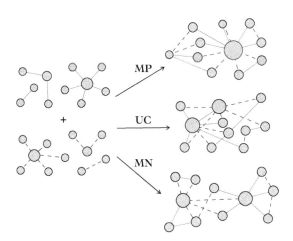

Fig. 7.1 *Schematic illustration of three kinds of correlated multiplex network, Maximally Positive (MP), Uncorrelated (UC) and Maximally Negative (MN). Each layer of the networks has different types of links, indicated by solid and dashed links, respectively. Reprinted figure with permission from [213] ©2014 by the American Physical Society.*

sequences (see Fig. 7.1). In this way, the Maximally Positively correlated (MP) multiplex network acquires the maximal positive degree correlation among all the multiplex networks having the same layers.

Conversely, if the first-degree sequence is ordered in ascending order and the second-degree sequence is sorted in descending order, by forming pairs of replica nodes associating nodes of the same rank in the two sequences we generate a Maximally Negatively correlated (MN) multiplex network (see Fig. 7.1). In fact, the Maximally Negatively correlated (MN) multiplex network is the one with the largest negative degree correlations among all the multiplex networks with the same layers.

Uncorrelated (UC) multiplex networks are typically generated by randomly associating pairs of nodes of the two layers for forming replica nodes (see Fig. 7.1).

7.2.1   Other measures to evaluate the heterogeneity of the degrees of the nodes

A different set of measures has been introduced to evaluate the correlations existing between the degree of the same node across more than two layers [30]. These quantities are intended to measure the heterogeneity between the degrees $k_i^{[\alpha]}$ of the same node i across the different layers α. The first one is the entropy H_i associated with the distribution of the degree of each node across different layers defined as

$$H_i = -\sum_\alpha \frac{k_i^{[\alpha]}}{S_i} \ln\left(\frac{k_i^{[\alpha]}}{S_i}\right), \tag{7.10}$$

where $S_i = \sum_\alpha k_i^{[\alpha]}$ is the aggregated strength of node i. Therefore, H_i evaluates the heterogeneity of the degrees of node i across the different layers α. It is minimal ($H_i = 0$) if node i is connected only in one layer (maximum heterogeneity) and is maximal ($H_i = \ln M$) if node i has the same degree in all the layers (minimum heterogeneity). The second measure proposed to evaluate the heterogeneity between the degrees of the same node across all the layers of the multiplex network is the participation coefficient P_i given by

$$P_i = \frac{M}{M-1}\left[1 - \sum_{\alpha=1}^{M}\left(\frac{k_i^{[\alpha]}}{S_i}\right)^2\right]. \tag{7.11}$$

The participation coefficient takes values in the interval $[0, 1]$. If all the links of node i are distributed uniformly across the layers we have $P_i = 1$. If the links of node i belong to a single layer we have $P_i = 0$.

The third measure that can be considered, the *inverse participation ratio* y_i of the degrees of a node i, is a traditional measure used in statistical mechanics. The participation y_i is defined as

$$y_i = \left[\sum_{\alpha=1}^{M} \left(\frac{k_i^{[\alpha]}}{S_i} \right)^2 \right]^{-1} \tag{7.12}$$

and indicates the effective number of layers where node i is connected. The quantity y_i spans the interval $[1, M]$. If node i is connected in one layer only $y_i = 1$, whereas if i has the same degree in all M layers $y_i = M$.

7.3   Overlap, multilinks and multidegrees

A relevant feature of many multiplex networks is their link overlap. This means that a significant number of pairs of nodes are connected in multiple layers. Multiplex networks with significant link overlap are ubiquitous and include multiplex airport networks, on-line social games, collaboration and citation networks [290, 71, 209]. Take, for example, the airport network where the different layers are the flight connections of different airline companies [71]. As most of the airline companies aim at providing the most popular flight connections, the multiplex network displays a significant link overlap (see Fig. 4.5). Another example of a multiplex network with significant overlap is the duplex network between scientists connected in one layer by a citation network (who cites whom) and in the other layer by the collaboration network (who collaborates with whom) [209]. The two layers in this dataset display significant overlap because two coauthors are also usually citing each other in their papers. Finally also in the *in silico* social network constituted by the *Pardus* online game [290], where avatars have different types of interactions, a significant overlap is observed between different layers (see Fig. 7.2).

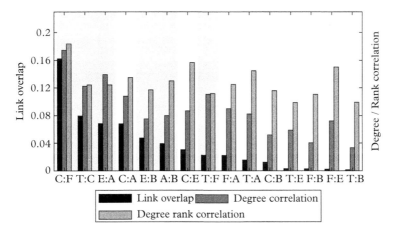

Fig. 7.2 *In the* Pardus *online game the layers indicate friendship (F), communication (C), trade (T), enmity (E), attack (A) and bounty (B). Here the correlations encoded in the multiplex network structure are shown: the layers display a significant overlap of the links, correlations between the degrees of the nodes and correlation between their ranks. Reprinted figure from Ref. [290].*

7.3.1   Overlap

Two measures [41, 290] of overlap have been recently proposed to characterize the link overlap across two layers: the *total overlap* and the *local overlap*. The *total overlap* $O^{[\alpha,\beta]}$ between two layers α and β [41, 290] is defined as the total number of links that are in common between layer α and layer β, i.e. for an undirected network

$$O^{[\alpha,\beta]} = \sum_{i<j} a_{ij}^{[\alpha]} a_{ij}^{[\beta]}, \tag{7.13}$$

where $\alpha \neq \beta$.

The *local overlap* $o_i^{[\alpha,\beta]}$ between two layers α and β [41] is defined as the total number of neighbours of node i that are neighbours in both layer α and layer β:

$$o_i^{[\alpha,\beta]} = \sum_{j=1}^{N} a_{ij}^{[\alpha]} a_{ij}^{[\beta]}. \tag{7.14}$$

These quantities are sometimes also normalized in order to obtain a measure of the significance of the overlap in the multiplex network. The normalized total overlap $\hat{O}^{[\alpha,\beta]}$ and the normalized local overlap $\hat{o}_i^{[\alpha,\beta]}$ are the Jaccard indices given by

$$\hat{O}^{[\alpha,\beta]} = \frac{\sum_{i<j} a_{ij}^{[\alpha]} a_{ij}^{[\beta]}}{\sum_{i<j} \left(a_{ij}^{[\alpha]} + a_{ij}^{[\beta]} - a_{ij}^{[\alpha]} a_{ij}^{[\beta]} \right)},$$

$$\hat{o}_i^{[\alpha,\beta]} = \frac{\sum_{j=1}^{N} a_{ij}^{[\alpha]} a_{ij}^{[\beta]}}{\sum_{j=1}^{N} \left(a_{ij}^{[\alpha]} + a_{ij}^{[\beta]} - a_{ij}^{[\alpha]} a_{ij}^{[\beta]} \right)}. \tag{7.15}$$

7.3.2   Multilinks and multidegrees

In a multiplex network with M layers a significant link overlap can be achieved in multiple ways, since any two nodes can be linked across the different layers of the multiplex network in multiple ways.

Multilinks [41, 209] are a comprehensive way to characterize the pattern of connections between any two nodes. A multilink between any given two nodes i and j indicates in which layers of the multiplex network nodes i and j are connected. Consider, for example, the multiplex with $M = 2$ layers, i.e. the network shown in Fig. 7.3. Nodes 1 and 2 are connected by one link in the first layer and one link in the second. Thus, we say that the nodes are connected by a multilink $(1, 1)$. Similarly, nodes 2 and 3 are connected by one link in the first layer and no link in layer 2. Therefore, they are connected by a multilink $(1, 0)$.

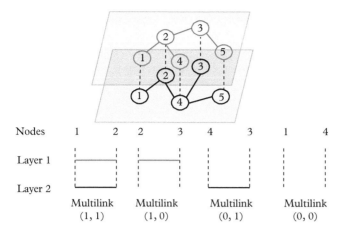

Fig. 7.3 *Example of all possible multilinks in a multiplex network with* M = *2 layers and* N = *5 nodes. Nodes* i *and* j *are linked by one multilink* $\vec{m}$ = (m$^{[1]}$, m$^{[2]}$). *Reprinted figure from Ref. [209] ©2014 Menichetti et al.*

In general, for a multiplex network formed by M layers we can define *multilinks* in the following way. Let us consider the vector

$$\vec{m} = \left(m^{[1]}, m^{[2]}, \ldots, m^{[\alpha]}, \ldots m^{[M]} \right) \tag{7.16}$$

of elements $m^{[\alpha]} \in \{0, 1\}$. Any pairs of nodes i and j are connected by a *multilink* $\vec{m}$ if and only if they are connected in every layer α in which $m^{[\alpha]} = 1$ and they are disconnected in every layer α in which $m^{[\alpha]} = 0$.

It follows that for an unweighted multiplex network every two nodes i and j are connected by a multilink $\vec{m}_{ij}$ given by

$$\vec{m}_{ij} = \left(a_{ij}^{[1]}, a_{ij}^{[2]}, \ldots, a_{ij}^{[M]} \right), \tag{7.17}$$

while for a weighted multiplex network

$$\vec{m}_{ij} = \left(\theta \left(a_{ij}^{[1]} \right), \theta \left(a_{ij}^{[2]} \right), \ldots, \theta \left(a_{ij}^{[M]} \right) \right). \tag{7.18}$$

Therefore, every pair of nodes is connected by a single multilink. Note that the trivial multilink $\vec{m} = \vec{0}$ indicates that there is no layer connecting the two considered nodes. All the other multilinks with $\vec{m} \neq \vec{0}$ indicate the different patterns of connections that can exist between any two nodes of a multiplex network and are therefore called the non-trivial multilinks.

In a single network any two nodes can be either connected or non-connected. In a general multiplex network of M layers, there are many more possibilities accounted for by the multilinks. Since in each layer the two nodes are either connected or non-connected, the overall number of possible multilinks is 2^M.

Let us now introduce the *multi-adjacency matrices* $A^{\vec{m}}$ with elements $A_{ij}^{\vec{m}}$ equal to 1 if there is a multilink $\vec{m}$ between node i and node j, and 0 otherwise:

$$A_{ij}^{\vec{m}} = \begin{cases} 1 & \text{if } \vec{m}_{ij} = \vec{m} \\ 0 & \text{otherwise} \end{cases}$$

The multi-adjacency matrices can be expressed in terms of the adjacency matrices of the unweighted multiplex networks as

$$A_{ij}^{\vec{m}} = \prod_{\alpha=1}^{M} \left[a_{ij}^{[\alpha]} m^{[\alpha]} + \left(1 - a_{ij}^{[\alpha]}\right)\left(1 - m^{[\alpha]}\right) \right]. \tag{7.19}$$

For weighted multiplex networks we have instead

$$A_{ij}^{\vec{m}} = \prod_{\alpha=1}^{M} \left[\theta\left(a_{ij}^{[\alpha]}\right) m^{[\alpha]} + \left(1 - \theta\left(a_{ij}^{[\alpha]}\right)\right)\left(1 - m^{[\alpha]}\right) \right]. \tag{7.20}$$

Thus, multi-adjacency matrices satisfy the condition

$$\sum_{\vec{m}} A_{ij}^{\vec{m}} = 1 \tag{7.21}$$

for every fixed pair of nodes (i,j). Multi-adjacency matrices are a way to describe the same information encoded in the M adjacency matrices of each layer, and while they do not encode more information, they can be useful auxiliary tools to derive the general expression of the multidegrees as a function of the adjacency matrices $\mathbf{a}^{[\alpha]}$. The *multidegree* $k_i^{\vec{m}}$ of a node i is the total number of multilinks $\vec{m}$ incident to node i,

$$k_i^{\vec{m}} = \sum_{j|j \neq i} \delta(\vec{m}, \vec{m}_{ij}), \tag{7.22}$$

where $\delta(\vec{m}, \vec{m}_{ij}) = 1$ for $\vec{m} = \vec{m}_{ij}$ and $\delta(\vec{m}, \vec{m}_{ij}) = 0$ otherwise. Considering the fact that every pair of nodes is connected by a given multilink $\vec{m}$, we have that the multidegrees $\vec{m}$ of each node satisfy

$$\sum_{\vec{m}} k_i^{\vec{m}} = N - 1 \tag{7.23}$$

where this expression is derived in the absence of tadpoles, i.e. in each layer there are no links connecting a node with itself.

In terms of the multi-adjacency matrices we can express the multidegree $k_i^{\vec{m}}$ as

$$k_i^{\vec{m}} = \sum_{j=1}^{N} A_{ij}^{\vec{m}} \tag{7.24}$$

and consequently using Eq. (7.19) we obtain for the unweighted multiplex network

$$k_i^{\vec{m}} = \sum_{j=1}^{N} \prod_{\alpha=1}^{M} \left[a_{ij}^{[\alpha]} m^{[\alpha]} + \left(1 - a_{ij}^{[\alpha]}\right)\left(1 - m^{[\alpha]}\right) \right], \tag{7.25}$$

while for weighted multiplex networks we have

$$k_i^{\vec{m}} = \sum_{j=1}^{N} \prod_{\alpha=1}^{M} \left[\theta\left(a_{ij}^{[\alpha]}\right) m^{[\alpha]} + \left(1 - \theta\left(a_{ij}^{[\alpha]}\right)\right)\left(1 - m^{[\alpha]}\right) \right]. \tag{7.26}$$

As an example, here we list the non-trivial multidegrees $\vec{m}$, with $\vec{m} \neq \vec{0}$, for a given duplex network in terms of the adjacency matrix $\mathbf{a}^{[1]}$ of layer 1 and $\mathbf{a}^{[2]}$ of layer 2 given by

$$k_i^{(1,0)} = \sum_{j=1}^{N} a_{ij}^{[1]}\left(1 - a_{ij}^{[2]}\right),$$

$$k_i^{(0,1)} = \sum_{j=1}^{N} \left(1 - a_{ij}^{[1]}\right) a_{ij}^{[2]},$$

$$k_i^{(1,1)} = \sum_{j=1}^{N} a_{ij}^{[1]} a_{ij}^{[2]}. \tag{7.27}$$

The trivial multidegree $k_i^{(0,0)}$ can be found by using Eq. (7.23). Considering the example of a social duplex, formed by the same set of agents interacting by mobile phone in layer 1 and by email in layer 2, $k_i^{(1,0)}$ indicates the number of acquaintances of node i that only communicate with it via mobile phone, $k_i^{(0,1)}$ indicates the number of acquaintances of node i that only communicate with it via email and $k_i^{(1,1)}$ indicates the number of acquaintances of node i that communicate with it via both mobile phone and email.

Sometimes it might be convenient to represent the type of multilink $\vec{m}$ with an integer number $\bar{q} \in \{0, 1, 2, \ldots, 2^M - 1\}$ whose binary representation is $\vec{m}$, i.e.

$$\bar{q} = \sum_{\alpha=1}^{M} 2^{\alpha-1} m^{[\alpha]}. \tag{7.28}$$

Using this notation we can construct a *weighted aggregated multi-adjacency matrix* $\bar{\mathbf{a}}$ of elements

$$\bar{a}_{ij} = \bar{q}_{ij} = \sum_{\alpha=1}^{M} 2^{\alpha-1} m_{ij}^{[\alpha]}. \tag{7.29}$$

The weighted aggregated multi-adjacency matrix is encoding all the information of the multi-adjacency matrices as we have

$$A_{ij}^{\vec{m}} = \delta \left(\bar{a}_{ij}, \sum_{\alpha=1}^{M} 2^{\alpha-1} m^{[\alpha]} \right) \tag{7.30}$$

where $\delta(x,y)$ is the Kronecker delta. Finally, the multidegree $k_i^{\vec{m}}$ can be also expressed in terms of the weighted aggregated multi-adjacency matrix $\bar{\mathbf{a}}$ as

$$k_i^{\vec{m}} = \sum_{j=1}^{N} \delta \left(\bar{a}_{ij}, \sum_{\alpha=1}^{M} 2^{\alpha-1} m^{[\alpha]} \right). \tag{7.31}$$

When the number of layers M is very large, the classification into multilinks can become computationally demanding. In this case, a partial information about the overlap of the links across different layers is given by the *link multiplicity* v_{ij} defined in Sec. 6.2.3. This quantity can be related to the multilinks as

$$v_{ij} = \sum_{\alpha=1}^{M} m_{ij}^{[\alpha]}, \tag{7.32}$$

where the nodes i and j are linked by a multilink $\vec{m}_{ij}$.

In the context of financial networks it has been shown that the multidegree can be an extremely powerful tool. Specifically, in a recent study [218] a multiplex network between assets has been constructed by considering four layers, each one formed by the network of significant correlations between the assets as indicated respectively by the Pearson, Kendall, Tail and Partial dependencies. While the Pearson dependence evaluates the linear correlations, the other measures go beyond the linear correlations. By analysing the financial market during the 2007–8 crisis, the Authors show evidence that the multidegree corresponding to links existing simultaneously in the Kendall, Tail and Partial layers but not existing in the Pearson layers reveals the roles of different industrial sectors in financial crises.

Recently, multilinks have also been used as motifs in order to characterize the connectome of the nematode *C. elegans*. In Ref. [37] the identity of the over- and under-represented multilinks are shown in the multiplex network with $M = 3$ layers formed by the monoamide extra-synaptic layer (directed), electric gap junction connections (bidirectional) and synaptic connections (directed). The analysis reveals the location of the network where aminergic signalling modulates synaptic activity.

7.3.3   Multidegree distribution

The multidegree distribution $P(\{k^{\vec{m}}\})$ indicates the probability that a random node of a multiplex network has multidegrees $k_i^{\vec{m}} = k^{\vec{m}}$ for every non-trivial multidegree $\vec{m} \neq \vec{0}$ and is given by

$$P(\{k^{\vec{m}}\}) = \frac{N(\{k^{\vec{m}}\})}{N}, \tag{7.33}$$

where $N(\{k^{\vec{m}}\})$ indicates the number of nodes with non-trivial multidegrees $\{k^{\vec{m}}\}$. The multidegrees can be either uncorrelated or correlated. For uncorrelated multidegree distributions we have

$$P(\{k^{\vec{m}}\}) = \prod_{\vec{m} \neq \vec{0}} P^{\vec{m}}(k^{\vec{m}}) \tag{7.34}$$

where $P^{\vec{m}}(k^{\vec{m}})$ is the distribution of the multidegrees $k^{\vec{m}}$, i.e.

$$P^{\vec{m}}(k^{\vec{m}}) = \frac{N^{\vec{m}}(k^{\vec{m}})}{N} \tag{7.35}$$

where $N^{\vec{m}}(k^{\vec{m}})$ is the number of nodes with $k_i^{\vec{m}} = k^{\vec{m}}$. If the relation defined in Eq. (7.34) does not hold, the multidegrees are correlated. For a duplex network, having correlated multidegree can indicate that nodes having many multilinks $(1, 1)$ characterizing interactions occuring on both layers might also have many multilinks of the type $(1, 0)$ or $(0, 1)$ characterizing single-layer interactions. In this case, the nodes with high multidegree $k^{(1,1)}$ will also tend to have high multidegree $k^{(1,0)}$ and $k^{(0,1)}$.

7.4   Correlations in weighted multiplex networks

In single networks the interplay between weights and topology is revealed by studying the strength and the inverse participation ratio as a function of the node degrees (see Sec. 2.5.3). In weighted multiplex networks the statistical properties of the weights can be strongly affected by link overlap. In order to capture this phenomenon the *multistrengths* and the *inverse multiparticipation ratio* have been introduced in Ref. [209]. For every layer α the multistrength $s_i^{\vec{m},[\alpha]}$ and the inverse multiparticipation ratio $Y_i^{\vec{m},[\alpha]}$ of node i are defined as

$$s_i^{\vec{m},[\alpha]} = \sum_{j=1}^{N} a_{ij}^{[\alpha]} A_{ij}^{\vec{m}},$$

$$Y_i^{\vec{m},[\alpha]} = \sum_{j=1}^{N} \left(\frac{a_{ij}^{[\alpha]} A_{ij}^{\vec{m}}}{\sum_r a_{ir}^{[\alpha]} A_{ir}^{\vec{m}}} \right)^2. \tag{7.36}$$

The multistrength $s_i^{\vec{m},[\alpha]}$ indicates the total weight of the links incident to node i in layer α that participate in a multilink of type $\vec{m}$. The inverse multiparticipation ratio $Y_i^{\vec{m},[\alpha]}$ is a measure of the inhomogeneity of the weights of the nodes that are incident to node i in layer α and at the same time are part of a multilink $\vec{m}$.

The average multistrength of nodes with a given multidegree, i.e. $s^{\vec{m},[\alpha]}(k^{\vec{m}}) = \left\langle s_i^{\vec{m},[\alpha]} \delta(k_i^{\vec{m}}, k^{\vec{m}}) \right\rangle$, and the average inverse multiparticipation ratio of nodes with a given multidegree, $Y^{\vec{m},[\alpha]}(k^{\vec{m}}) = \left\langle Y_i^{\vec{m},[\alpha]} \delta(k_i^{\vec{m}}, k^{\vec{m}}) \right\rangle$, are typically scaling as

$$s^{\vec{m},[\alpha]}(k^{\vec{m}}) \propto (k^{\vec{m}})^{\hat{\beta}_{\vec{m},\alpha}}$$

$$Y^{\vec{m},[\alpha]}(k^{\vec{m}}) \propto (k^{\vec{m}})^{-\hat{\lambda}_{\vec{m},\alpha}}, \qquad (7.37)$$

with exponents $\hat{\beta}_{\vec{m},\alpha} \geq 1$ and $\hat{\lambda}_{\vec{m},\alpha} \leq 1$. In Ref. [209] the weighted multilink properties of multiplex networks between scientists collaborating (in layer 1) and citing each other (in layer 2) have been analysed starting from the APS data set. In Fig. 7.4, the multistrength

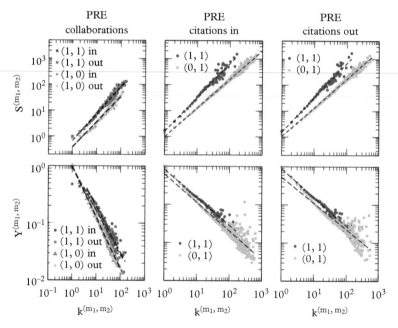

Fig. 7.4 *Properties of multilinks in the weighted collaboration (layer 1)/citation (layer 2) multiplex network based on the PRE publications analysed in Ref. [209]. In the case of the collaboration network, the distributions of multistrengths versus multidegrees always have the same exponent, but the average weight of multilinks (1,1) is larger than the average weight of multilinks (1,0). Moreover, the exponents $\lambda_{(1,0),col,in}$, $\lambda_{(1,0),col,out}$ are larger than exponents $\lambda_{(1,1),col,in}$, $\lambda_{(1,1),col,out}$. In the case of the citation layer, both the incoming multistrengths and the outgoing multistrengths have a functional behaviour that varies depending on the type of multilink. Conversely, the average inverse multiparticipation ratio in the citation layer does not show any significant change of behaviour when compared across different multilinks. Reprinted figure from Ref. [209] ©Menichetti et al.*

and inverse multiparticipation ratio for the collaboration and citation networks of PRE authors are shown.

In this case the citation network is also directed, and the multistrength $s_{i,1}^{(1,0),[2],in}$ indicates the number of times Author i is cited by scientists that are not his coauthors, while $s_{i,1}^{(1,1),[2],in}$ indicates the number of times he is cited by his coauthors. Similarly, $s_{i,1}^{(1,0),[2],out}$ indicates the number of times Author i cites scientists that are not his coauthors while $s_{i,1}^{(1,1),[2],in}$ indicates the number of times he cites his coauthors. These multistrengths display a non-linear scaling with $\hat{\beta}_{\vec{m},\alpha,in} > 1$ and $\hat{\beta}_{\vec{m},\alpha,in} > 1$. Interestingly, this analysis shows that the statistical properties of the weights belonging to multilinks $\vec{m} = (0,1)$ and $\vec{m} = (1,1)$ are distinct (the exponent $\hat{\beta}_{(1,0),\alpha,in/out}$ is significantly smaller than the exponent $\hat{\beta}_{(1,1),\alpha,in/out}$). In this way it is possible to reveal a pattern that cannot be inferred by looking at the single layer taken in isolation, quantifying the trend according to which *authors tend to cite significantly more the hub authors with whom they have collaborated than the authors with whom they have not collaborated.*

7.5 The activities of the nodes and pairwise multiplexity

Any multiplex network of N nodes and M layers has an underlying bipartite network describing which nodes are active (connected) in which layer (see Fig. 7.5). Let us indicate by $\mathbf{b}$ the incidence matrices of this undirected bipartite network. For an undirected multiplex network, the incidence matrix $\mathbf{b}$ of the bipartite network has the generic element $b_{i\alpha}$ indicating whether node i is connected in layer α, i.e.

$$b_{i\alpha} = 1 - \delta\left(0, k_i^{[\alpha]}\right). \tag{7.38}$$

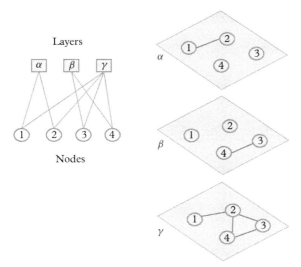

Fig. 7.5 *Schematic representation of the bipartite network between nodes and layers characterizing the activity of the nodes in a multiplex network.*

For a directed multiplex network, node i is inactive in layer α if both its in-degree and its out-degree in layer α are zero. Therefore, we can define the matrix elements $b_{i\alpha}$ as

$$b_{i\alpha} = 1 - \delta\left(0, k_{i,in}^{[\alpha]}\right)\delta\left(0, k_{i,out}^{[\alpha]}\right). \tag{7.39}$$

Note that for directed multiplex networks it is also possible to extract a directed bipartite network indicating whether a given node has positive in-degree and/or positive out-degree in any given layer (see its use in the context of centrality measure discussed in Sec. 9.5). The *activity* B_i of a node i has been defined in Ref. [230] and is given by the number of layers in which node i is active:

$$B_i = \sum_{\alpha=1}^{M} b_{i\alpha}. \tag{7.40}$$

In Ref. [230] several real multiplex networks were analysed, showing that the activity of the nodes has a broad distribution with unbounded fluctuations in most cases.

The layer activity $N^{[\alpha]}$ has been defined in Ref. [230] and is given by the number of nodes active in layer α:

$$N^{[\alpha]} = \sum_{i=1}^{N} b_{i\alpha}. \tag{7.41}$$

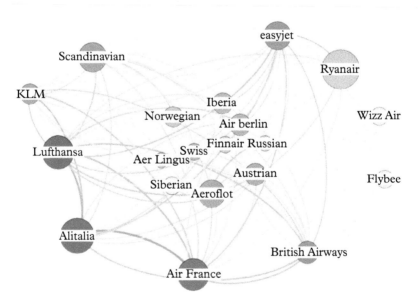

Fig. 7.6 *The network between 20 European airline companies in which each link is weighted with its layer pairwise multiplexity. This method can reveal non-trivial correlations between the activities of the nodes in the different layers. Reprinted figure with permission from [230] ©(2015) by the American Physical Society.*

In Ref. [230] the Authors studied the *layer pairwise multiplexity* $Q_{\alpha\beta}$, measuring the correlation between the layers. The layer pairwise multiplexity is defined as

$$Q_{\alpha\beta} = \frac{1}{N} \sum_{i=1}^{N} b_{i\alpha} b_{i\beta}, \qquad (7.42)$$

and quantifies the fraction of nodes that are active in layer α as well as in layer β. Similarly, one can use the *node pairwise multiplexity* Q_{ij} measuring the correlation of activities between two nodes. The *node pairwise multiplexity*, introduced in Ref. [84], is defined as

$$Q_{ij} = \frac{1}{M} \sum_{\alpha=1}^{M} b_{i\alpha} b_{j\beta} \qquad (7.43)$$

and quantifies the fraction of layers in which both node i and node j are active.

In Fig. 7.6 the network of European airline companies is plotted, where the links between two different companies are weighted with the layer pairwise multiplexity. This method can reveal non-trivial correlations between the activities of the nodes in different layers.

8
Communities

8.1 The relevance of communities in multilayer networks

Communities of multilayer networks might span across different layers. For example, in scientific collaboration networks communities might form around similar scientific topics. Therefore, the communities detected in the different layers of the scientific collaboration networks can be correlated. Similarly, in multi-slice networks, political coalitions identified by politicians having similar voting behaviours in the US Congress might have a continuity over time that spans several legislatures. Interestingly, communities in multi-slice networks might also be recurrent. For example, if we consider the communities formed by contact networks of children in primary school it can be observed that the communities depend on the time of the day, and that the communities forming during class activities can form in the morning, dissolve during lunch and playground activities and form again in the afternoon. In this scenario we will indicate the communities spanning different layers (which might have a temporal connotation or not) as multilayer communities. In order to evaluate the multilayer nature of communities, different methods have been explored. A wide class of algorithms deals directly with the problem of multilayer community detection. Another series of studies instead evaluates the similarity between the mesoscale structure of different layers. Finally, some works are starting to address the related question of whether or not the information of a multilayer network can be compressed by aggregating different layers.

8.2 Multilayer community detection

In this section we will describe several approaches proposed for community detection in multilayer networks. These approaches can be classified in five major classes: maximization of multilayer modularity; consensus clustering; clustering based on the properties of multilayer random walks; multilink community detection; and approaches based on tensor decomposition and tensor computation.

Multilayer Networks. Ginestra Bianconi, Oxford University Press (2018).
© Ginestra Bianconi. DOI: 10.1093/oso/9780198753919.001.0001

8.2.1 Community detection algorithms using a generalized modularity measure

The first paper that has proposed a multilayer community detection algorithm is Ref. [219]. The proposed algorithm applies to multilayer networks in which there is a one-to-one map between the nodes in each layer and interlinks are placed between corresponding nodes. Therefore, this algorithm applies to multiplex networks and to multi-slice temporal networks (see Fig. 4.1 panels (a) and (b)). A central element of this algorithm is the explicit treatment of the interlinks among replica nodes that are used to favour the persistence of the communities across different layers.

Specifically, in Ref. [219] the multilayer communities are detected by a greedy algorithm based on the optimization of the generalized modularity $\mathcal{Q}^M$. The generalized modularity $\mathcal{Q}^M$ aims at quantifying how significant the multilayer communities are with respect to a random hypothesis. A suitable random hypothesis for the multilayer network structure is that the generic layer α is an uncorrelated random network. In this case, the probability $p_{ij}^{[\alpha]}$ that node i and node j are connected by a link in layer α is given by

$$p_{ij}^{[\alpha]} = \frac{k_i^{[\alpha]} k_j^{[\alpha]}}{\langle k^{[\alpha]} \rangle N}, \tag{8.1}$$

where $k_i^{[\alpha]}$ indicates the degree of node i in layer α. The multilayer community assignment is determined when each replica node (i, α) is associated with the community $g_i^{[\alpha]} \in \{1, 2, \ldots, P\}$. The generalized modularity quantifies the quality of this multilayer community assignment by measuring how tightly connected the multilayer communities are with respect to the random hypothesis given by Eq. (8.1). Specifically, the generalized modularity $\mathcal{Q}^M$ is taken to be

$$\mathcal{Q}^M = \frac{1}{\mu} \sum_{i,j,\alpha,\beta} \left\{ \left(A_{i\alpha,j\beta} - \gamma^{[\alpha]} \frac{k_i^{[\alpha]} k_j^{[\alpha]}}{\langle k^{[\alpha]} \rangle N} \right) \delta_{\alpha,\beta} + \omega A_{i\alpha,j\beta} \delta_{ij} \right\} \delta \left(g_i^{[\alpha]}, g_j^{[\beta]} \right), \tag{8.2}$$

where $A_{i\alpha,j\beta}$ is the supra-adjacency matrix, $\mu = \sum_{i,j,\alpha} A_{i\alpha,j\alpha} + \omega \sum_{i,\alpha,\beta} A_{i\alpha,i\beta}$ and $\delta(x, y)$ indicates the Kronecker delta. The multilayer modularity can be optimized for finding the optimal community assignment of the multilayer network as a function of the parameters ω and $\gamma^{[\alpha]}$, also called the resolution parameters. Specifically, the generalized modularity $\mathcal{Q}^M$ is maximized according to a greedy optimization which follows similar steps to those used by the Louvain algorithm [53].

For $\omega = 0$ and $\gamma^{[\alpha]} = 1$, the generalized modularity $\mathcal{Q}^M$ is proportional to the average of the modularities $\mathcal{Q}^{[\alpha]}$ of any given layer α, i.e.

$$\mathcal{Q}^M = \frac{1}{\mu} \sum_{\alpha=1}^{M} \mathcal{Q}^{[\alpha]} \tag{8.3}$$

where

$$Q^{[\alpha]} = \frac{1}{\langle k^{[\alpha]} \rangle N} \sum_{ij} \left(A_{i\alpha,j\alpha} - \frac{k_i^{[\alpha]} k_j^{[\alpha]}}{\langle k^{[\alpha]} \rangle N} \right) \delta \left(g_i^{[\alpha]}, g_j^{[\alpha]} \right). \tag{8.4}$$

Interestingly, in biological applications when considering a multiplex network formed by the protein–protein interaction network and the regulatory network one might just use this measure to find the multilayer communities [36].

By setting the parameter ω to a non-zero value $\omega \neq 0$, in the definition of Q^M (Eq. (8.2)) it is possible to favour the persistence of the communities involving the same node in different layers. Finally, the parameter $\gamma^{[\alpha]}$ can also be tuned to bias the algorithm toward bigger or smaller communities. In fact, by taking $\gamma^{[\alpha]} = 1$, $\forall \alpha \in \{1, 2 \ldots, M\}$, the resulting partition may not reflect the true community structure of the network. Consequently, the method might favour the detection of smaller communities. By setting $\gamma^{[\alpha]} < 1$ larger communities are resolved, whereas for $\gamma^{[\alpha]} > 1$ more communities containing fewer nodes are detected. Therefore, exploring the communities detected by the algorithm for different values of the resolution parameters might provide important additional insights into the mesoscale structure of the multilayer dataset under consideration.

This community detection algorithm has been successfully used [219] for characterizing a variety of datasets including the temporal network between US Senators with similar voting profiles from the first to the 110th US Congresses, covering the years 1789–2008.

Interesting results are obtained in Ref. [28] where this algorithm is applied to multi-slice brain functional networks. In particular, the Authors show that the *flexibility* of the brain functional network correlates to the learning activity. The flexibility f_i of a node i evaluates the fraction of times that a node changes multilayer community assignment in a given experimental session of length M. The flexibility F is the average of the flexibility of the nodes i, i.e.

$$F = \frac{1}{N} \sum_{i=1}^{N} f_i. \tag{8.5}$$

In Ref. [28] brain functional networks of individuals performing simple motor tasks are considered. It is found that the flexibility measured at a given session of the experiment is correlated with the amount of learning in the subsequent session (see Fig. 8.1).

8.2.2 Consensus clustering

A large number of community detection algorithms used for single networks provides results that are not deterministic and that depend on random seeds and other contingencies. In Ref. [182] consensus clustering has been proposed to generate stable, accurate results from the stochastic partitions provided by this class of community detection

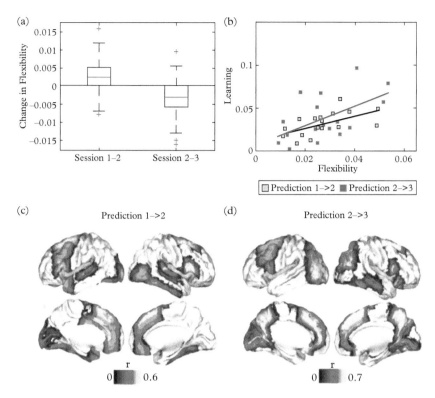

Fig. 8.1 *The flexibility of brain functional networks measured during three experimental sessions in which a given pool of individuals are asked to learn a motor task. The increase in flexibility from experimental session 1 to session 2 and the magnitude of the decrease in flexibility from session 2 to session 3 is significantly greater than zero as shown in panel (a). Panel (b) demonstrates the significant predictive correlations between flexibility in session 1 and learning in session 2 and between flexibility in session 2 and learning in session 3. In panel (c) the brain regions whose flexibility in session 1 predicted learning in session 2 are shown. In panel (d) the brain regions whose flexibility in session 2 predicted learning in session 3 are shown. Reprinted from Ref. [28].*

algorithms. The method applies to single layers and multilayer networks as well, including multiplex and temporal networks [70].

Let us first briefly discuss this algorithm on single layers and then its application to temporal and multiplex networks. The algorithmic steps required for finding the consensus clustering on single networks are described in the following [182]:

(i) Make a choice of a community detection algorithm that can cluster weighted networks and that generates at each run a stochastic partition such as the Louvain or the Infomap algorithm.

(ii) Run the algorithm n_P times, obtaining n_P stochastic partitions.

(iii) Construct the weighted consensus graph with weighted adjacency matrix **D** with elements

$$D_{ij} = \frac{n_{ij}}{n_P} \tag{8.6}$$

where n_{ij} is the number of partitions in which node i and node j are assigned to the same cluster (see Fig. 8.2).

(iv) Set to zero all the entries below a certain threshold τ, as long as the network remains connected.

(v) Apply the chosen community detection algorithm to the weighted consensus graph n_P times to yield n_P partitions.

(vi) If the partitions are all equal, stop, otherwise go back to point (iii).

In Ref. [182] the Authors have shown that the consensus clustering improves both the stability and the accuracy of the detected partitions.

This method can be applied both to temporal networks and to multiplex networks.

When considering a temporal network formed by M slices, it is possible to apply the consensus clustering to the layers belonging to a sliding window of r temporal slices such that the first window includes the slices for 1 to r, the second window includes the slices from 2 to $r + 1$ and so on. There are two sources of fluctuations: the fluctuations generated by the stochastic community detection algorithm associated with each single layer and the fluctuations due to the intrinsic dynamics of the temporal network. Once the consensus clustering is determined for each different sliding window, the clusters detected at different timeframes are identified using the Jaccard index given by the number of replica nodes in both clusters divided by the number of replica nodes that are in either one or the other cluster. In Ref. [182] this method has been applied to the APS citation network studied as a function of time as a multi-slice network. This technique

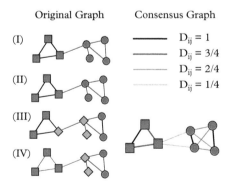

Fig. 8.2 *Schematic description of the construction of the consensus graph with weighted adjacency matrix* **D** *out of four stochastic partitions of a single network. Reprinted by permission from Macmillan Publishers Ltd: Scientific Reports [182]. ©2012.*

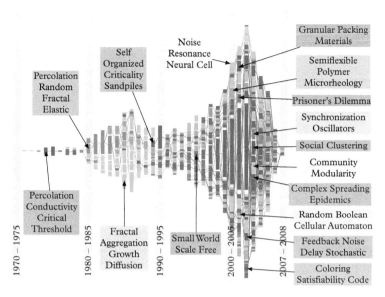

Fig. 8.3 *The clusters detected by consensus clustering in the APS citation network that have the keyword 'Network(s)' among the top 15 most frequent words appearing in the title of the papers, are displayed. A colour uniquely identifies a module, while the width of the links between clusters is proportional to the number of papers they have in common. A rapid growth of the field 'Complex Networks' is observed which eventually splits in to a number of smaller subtopics like Community Structure, Epidemic Spreading, Robustness, etc. Adapted by permission from Macmillan Publishers Ltd: Scientific Reports [182]. ©2012. Courtesy of the Authors.*

has been shown to be able to characterize the temporal evolution and the emergence of scientific topics over time. As an example of the results obtained Fig. 8.3 shows the multilayer clusters found where only papers including the keyword 'Network' in the title are considered, revealing the temporal evolution of this scientific field.

In Ref. [70], a consensus algorithm has been applied to multiplex biological networks of $M = 4$ layers to detect cancer driver genes. In this case, first the consensus algorithm was found for each single layer. Subsequently, the consensus clustering was applied to the full multiplex network for merging the information coming from the partitions found on the different layers.

8.2.3   Detecting mesoscale structure using random walks

Random walks on single networks are known to reveal their mesoscale structure. In fact, the transient to the steady state is characterized by a dynamical flow that remains localized on network clusters, revealing mesoscale communities and functional modules. One of the most successful community detection algorithms on single networks, the Infomap, exploits the properties of the random walk and, using information theory techniques, determines the modular structures that capture the constraints on the random walk flow

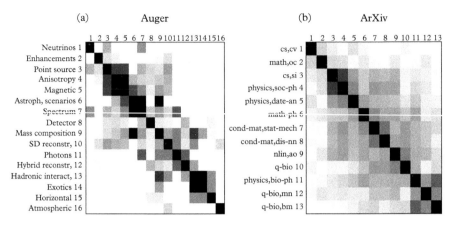

Fig. 8.4 *Multilayer communities detected in Ref. [89] are found to span across different network layers. The heat maps show the similarities between network layers, measured as the fraction of replica nodes in different network layers that are assigned to the same communities. Reprinted figure from Ref. [89].*

dynamics (see Sec. 2.6.4). In particular, by minimizing the length of the string encoding the dynamics of the random walk on a network with a modular partition, the method finds the dynamical regularities that identify the functional modules of the multiplex network and identifies the best community structure for describing the random walk dynamics. Recently this technique has been extended [89] to multiplex networks by allowing the random walk to jump from one layer to another along their interlinks with a tunable diffusion parameter.

Interestingly, this method can capture overlapping communities, i.e. the situation in which single nodes participate in multiple communities at the same time. In fact, by clustering replica nodes it is possible that each single node has different replica nodes belonging to different clusters. This is a simple mechanism that can explain the rather common appearance of community overlap as a clear manifestation of the relevance of the multilayer nature of complex networks.

The Pierre Auger Collaboration network, including collaboration in the observatory of ultra-high-energy cosmic rays, and the arXiv collaboration network of scientists working on networks have been analysed in Ref. [89], finding highly overlapping modules. The multilayer community structure can be exploited to construct a similarity between the different layers of the two datasets. This similarity is equal to the number of scientists whose replica nodes are assigned to the same community (see Fig. 8.4).

8.2.4   Multilink community detection

In multiplex networks it might be misleading to associate a single community to a replica node. Consider, for instance, a multiplex network formed by several online social network platforms, say Twitter and Facebook. It might be unrealistic to assume that an individual or even a single account belongs just to a single community. In fact, influential Twitter or Facebook individuals/accounts tend to reach more than one community in both social network platforms. It turns out that an individual can be a broker of information in a community defined only on Twitter or in a community that extends to both social network platforms.

The multilink community detection algorithm [215] addresses this concern and uncovers the rich community structure of multiplex networks by clustering the multilinks instead of nodes (or replica nodes), in this way providing a multiplex network equivalent to the link communities of single layers (see Sec. 2.6.5). The algorithm is a hierarchical clustering of the multilinks that operates starting from a similarity matrix between multilinks. The similarity between two incident multilinks evaluates how clustered the multiplex network is locally around the two multilinks, including additionally a cost for local paths of lengths three and four that pass through multiple layers.

This community detection method reveals the rich interplay between the mesoscale structure of the multiplex networks and their multiplexity. For instance, some nodes can belong to many layers and few communities, while others can belong to few layers but many communities. Moreover, the multilink communities can be formed by a different number of relevant layers.

Let us outline the main properties of the similarity matrix between multilinks. Every pair of multilinks connecting nodes i and r and nodes j and s is associated with the similarity $S_{ir,js}$. The similarity $S_{ir,js}$ is non-zero only between incident multilinks (i.e. for $s = r$) and is a function of two parameters: ϵ and z. The parameter $\epsilon \in (0, 1)$ can be tuned depending on the role assigned to the similarity among the two incident multilinks (in terms of the presence of interactions in the same layers) with respect to the similarity and the clustering of their local neighbourhood. The parameter $z \in (0, 1)$ evaluates the role of multiplexity and represents the cost attributed to incident multilinks of different composition.

Starting from the similarity matrix between the multilinks it is possible to perform a hierarchical clustering of clusters formed by multilinks. The resulting dendrogram can be cut according to the selected score function. In Ref. [215] multilink community detection has been applied to the Florentine Families Multiplex Network [236] formed by $M = 2$ layers, one layer describing the business dealings between $N = 16$ Florentine families in the fifteenth century, the other layer their alliances due to marriages (see Fig. 8.5).

For each family the *layer activity* indicating in how many layers it is active is compared with its *community activity* indicating in how many multilink communities it participates (Fig. 8.5(d)). The families with high community activity emerge as the powerful brokers between different communities, and we note that the Medici, belonging to three multilink communities, play a pivotal role in the network.

The multilink community detection algorithm has also been applied to the European Multiplex Air Transportation Network. The algorithm reveals that the main multilink communities of this multiplex network have very different composition in terms of single layers. In fact, the largest community is formed by connections existing in many layers, while the second-largest community has a very different structure, with only a few airlines contributing to this community. In order to capture this difference for

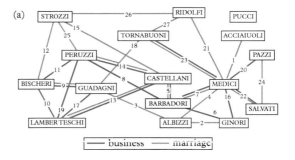

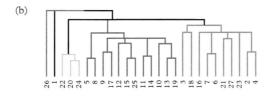

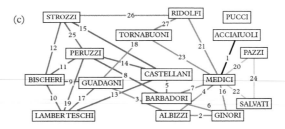

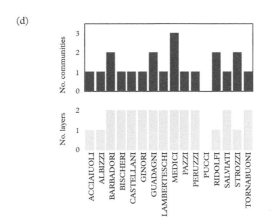

Fig. 8.5 *The Florentine Families Multiplex Network describing the business and marriage alliances of the fifteenth-century Florentine families (panel (a)). A heat map displaying the multilink similarity matrix and its relative dendrogram (panel (b)). Partition of the Florentine Families Multiplex Network into five multilink communities (panel (c)). Layer and community activity of the different families (panel (d)). The Medici family is characterized by achieving the maximum community activity. The multilink communities are detected using $\epsilon = 0.4, z = 0.6$. Reprinted figure from Ref. [215].*

each community it is possible to define the layer specificity given by the fraction of multilinks of the community having a link in the chosen layer. Fig. 8.6(a) shows the very different composition of the two largest communities as indicated by the layer specificities. Additionally, when comparing the airports and their community activity, we observe (see Fig. 8.6(b)) that a large layer activity seems to be correlated with high community activity, i.e. airports serving multiple airline companies typically belong to many multilink communities. However, we note that there is significant variability in the community activity of airports that are active in many layers. For example, Vienna (VIE) and Amsterdam (AMS) have a comparable layer activity but very different community activity. Similarly, there are airports with small layer activity but significant community activity, for example Luton (LTN) and Bergamo (BGY) airports.

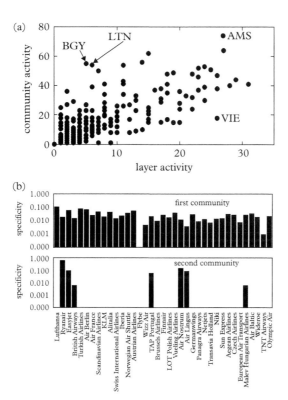

Fig. 8.6 *Community vs layer activity of the European airports (panel (a)). While the layer activity appears to have a positive correlation with the layer activity, large difference in community activity can be observed between airports with large layer activity (compare, for instance, Amsterdam (AMS) and Vienna (VIE)). Specificity for the first- and second-largest communities for the European Multiplex Air Transportation Network (panel (b)). Both panels indicate a large variability in the layer composition of different communities. The multilink communities are detected using $\epsilon = 0.4, z = 0.6$. Reprinted figure from Ref. [215].*

Multilink communities reveal that the mesoscale structure of a multiplex network can be organized via communities containing links in many different layers and, at the same time, communities having one predominant layer. This suggests that the mesoscale organization of multiplex networks has a rich structure that is not captured by methods that aim at compressing the information on a few single layers.

8.2.5   Detecting mesoscale organization using tensor computation

The tensorial description of multiplex and multi-slice networks has allowed network scientists to extend methods coming from tensor computation to analyse the mesoscale structure of these networks.

In Ref. [130] a non-negative tensor decomposition is used to detect the community structure in multi-slice (multiplex) networks. Although the method proposed by the Authors is very general, in Ref. [130] it is applied specifically to the study of temporal multi-slice networks. The temporal multi-slice network is specified by a three-dimensional tensor $\mathcal{T} \in \mathbb{R}^{N \times N \times M}$ of elements $\mathcal{T}_{ij\alpha}$, indicating the interaction between node i and node j in layer α (temporal slice) with $i = 1, 2, \ldots, N$ and $\alpha = 1, 2, \ldots, M$. Any such tensor can be factorized in the product of three rank-1 tensors according to the decomposition

$$\mathcal{T} = \sum_{r=1}^{\mathcal{R}_T} \mathbf{a}_r \otimes \mathbf{b}_r \otimes \mathbf{c}_r, \tag{8.7}$$

where $\mathbf{a}_r, \mathbf{b}_r$ and $\mathbf{c}_r$ are three vectors in $\in \mathbb{R}^N$. The smaller value of $\mathcal{R}_T$ for which the decomposition in Eq. (8.7) holds is the rank of the tensor. A visual representation of this decomposition, also known as Kruskal decomposition, is shown in Fig. 8.7. The rank-1 tensors $\mathbf{a}_r, \mathbf{b}_r$ and $\mathbf{c}_r$ are called the *components* of the temporal network and for each value of r they indicate respectively two sets of interacting nodes $(\mathbf{a}_r, \mathbf{b}_r)$ and a set of typical times $(\mathbf{c}_r)$ at which these interactions occur.

Although the 3-rank tensor can always be decomposed exactly according to Eq. (8.7), the exact decomposition might represent too fine a decomposition of communities. In order to decompose the tensor into fewer and more relevant components, the so-called

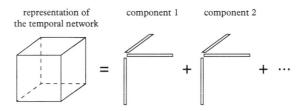

representation of   component 1   component 2
the temporal network

Fig. 8.7 *Pictorial representation of the Kruskal decomposition. The original three-way tensor represented in the left as a cube is decomposed into the sum of several components (rank-1 tensors) each one generated by the outer product of three one-dimensional vectors. Reprinted figure from Ref. [130].*

PARAFAC decomposition can be used. This method minimizes the difference between the tensor $\mathcal{T}$ and the sum of R outer products of three vectors

$$\min_{\{\mathbf{a}_r, \mathbf{b}_r, \mathbf{c}_r\}} \left\| \mathcal{T} - \sum_{r=1}^{R} \mathbf{a}_r \otimes \mathbf{b}_r \otimes \mathbf{c}_r \right\|_F^2, \tag{8.8}$$

where $||\tilde{\mathcal{T}}||_F$ indicates the Frobenius norm $||\tilde{\mathcal{T}}||_F^2 = \sqrt{\sum_{ij,\alpha} |\tilde{T}_{ij\alpha}|^2}$. In Ref. [130] the Authors focus on non-negative PARAFAC decomposition, imposing that the rank-1 tensors have non-negative elements. This is a customary condition ensuring a straight-forward interpretation of the resulting decomposition as a purely additive representation of the tensor. The resulting factorization of the tensor depends on the number of components R. In order to find the best factorization, in Ref. [130] the Authors propose a method of assessing the quality of the factorization and achieving very robust results of this tensor-based community detection algorithm.

This procedure has been applied to temporal network data describing face-to-face interactions between children in a primary school collected by the Sociopatterns project [289]. The detected communities simultaneously cluster students and interaction times. Most of the detected clusters correlate with the division of students in different classes; nevertheless, this technique is also able to detect components corresponding to mixed class activities such as lunch and playground activities (see Fig. 8.8). Interestingly, it has been found that some communities such as class activities have a non-trivial temporal pattern and during a typical day can form, dissolve and reform again.

A distinct study [195] uses the tensor representation of a multiplex network together with a combinatorial optimization algorithm to detect the so-called *recurrent heavy sub-graphs* (RHS). The recurrent heavy subgraphs represent meaningful biological modules from a large dataset of gene co-expression networks. The experimental microarray data about gene expression in different biological tissues is represented as a collection of co-expression networks forming a multiplex network. The multiplex network is described by a tensor $T_{ij\alpha}$ indicating the weights of the links between gene i and gene j in tissue α. This representation of the dataset allows the Authors of the paper to use continuous optimization methods to analyse the entire dataset. In particular, the recurrent heavy subgraphs (RHV) are identified by two membership vectors:

a) the gene membership vector $\mathbf{x} = (x_1, x_2, \ldots x_i, \ldots, x_N)$ indicating whether node i belongs to RHS ($x_i = 1$) or not ($x_i = 0$)

b) the network membership vector $\mathbf{y} = (y_1, y_2, \ldots y_\alpha, \ldots, y_M)$ indicating whether the RHS appears in layer α ($y_\alpha = 1$) or not ($y_\alpha = 0$).

The detection of the RHS is turned into a combinatorial optimization problem: the optimization problem identifies the heaviest among all RHS. The weight of an RHS is determined by the function $H(\mathbf{x}, \mathbf{y})$ given by the sum of the weights of the link in the RHS, i.e.

$$H(\mathbf{x}, \mathbf{y}) = \frac{1}{2} \sum_{i=1}^{N} \sum_{j=1}^{N} \sum_{\alpha=1}^{M} T_{ij\alpha} x_i x_j y_\alpha. \tag{8.9}$$

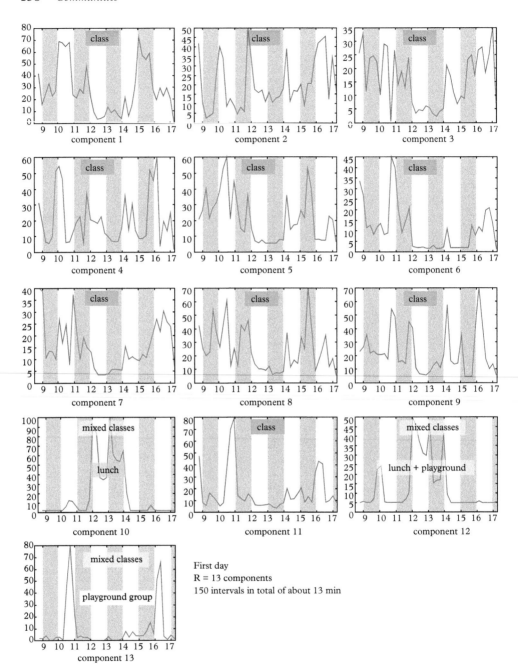

Fig. 8.8 *The non-negative tensor decomposition algorithm proposed in Ref. [130] is applied to the Sociopattern dataset of school children, detecting 13 temporal communities. Most of them correspond to class-specific activities and correlate with the division of students into classes. Three of them instead correspond to mixed class activities such as lunch and playground activities. Reprinted figure from Ref. [130].*

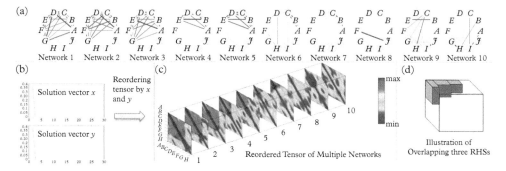

Fig. 8.9 *Illustration of the recurrent heavy subgraphs (RHS) (panel (a)) and its characterization in terms of the gene membership vector **x** and the tissue (network) membership vector **y**, indicating the relevance of the RHV respectively for the genes and the tissues (networks)(panel (b)). The reordering of the tensor according to the descending rank of the membership vectors reveals a compact cluster of gene and tissue (networks) involved in the RHS (panels (c), (d)). Reprinted from Ref. [195]*

Nevertheless, this problem has high computational complexity as long as the vectors $\mathbf{x}, \mathbf{y}$ have binary elements taking exclusively the values $0, 1$ (the problem has been proved to be NP-hard). Therefore, in Ref. [195] the Authors have turned the problem into a continuous optimization problem by allowing the membership vectors to take real values. Specifically, they consider the following optimization problem to detect the RHS:

$$\max_{\mathbf{x}\in\mathbb{R}_+^N, \mathbf{y}\in\mathbb{R}_+^M} H(\mathbf{x}, \mathbf{y})$$
$$\text{subject to} \begin{cases} f(\mathbf{x}) = 1 \\ g(\mathbf{y}) = 1 \end{cases} \tag{8.10}$$

where $\mathbb{R}_+$ is the non-negative real space and $f(\mathbf{x}), g(\mathbf{y})$ are vector norms. By tailoring the choice of the vector norms to their problem and taking a combination of vector p-norms, the Authors are able to detect the relevant RHS for the large database of gene expressions and to relate these subgraphs to meaningful biological modules (see Fig. 8.9).

8.3 Correlations in the community structure of multiplex networks

The communities in different layers of multiplex networks often present significant similarities. This phenomenon has been observed in a variety of multilayer network datasets including scientific collaboration networks and actor collaboration networks. Importantly, the mesoscale similarity between the layers of a multiplex network can be used to extract the network of effective interactions between layers.

Already in the context of single layers, several measures have been proposed to characterize the similarity between two community assignments. These include the Normalized Mutual Information [85] and the information theory indicator function Θ, [50] based on the entropy of network ensembles. It is therefore natural to use extensions of these measures to characterize the similarity of the communities found in different layers of a multiplex network.

8.3.1   Normalized Mutual Information

The measure of Normalized Mutual Information (NMI) [85] between two layers α and β quantifies to which extent groups of nodes belong to the same community in both levels of the multiplex. Therefore, NMI is a measure of the similarity of the community structure between two layers. A high value of NMI indicates that the community structure of the two layers are quite similar. Conversely, a low value is attributed to dissimilar structures. Let us consider two community assignments, $\{g_i^{[\alpha]}\}$ and $\{\tilde{g}_i^{[\beta]}\}$, indicating respectively the clusters $\sigma = g_i^{[\alpha]}$ to which the replica node (i, α) belongs, with $\sigma = 1, 2 \ldots P^{[\alpha]}$, and the clusters $\sigma' = \tilde{g}_i^{[\beta]}$ to which the replica node (i, β) belongs, with $\sigma' = 1, 2 \ldots P^{[\beta]}$. The similarity in the community structure can be captured by the NMI, which is given by

$$
NMI(\{g_i^{[\alpha]}\}, \{\tilde{g}_i^{[\beta]}\}) = \frac{-2 \sum_{\sigma=1}^{P^{[\alpha]}} \sum_{\sigma'=1}^{P^{[\beta]}} N_{\sigma,\sigma'}^{[\alpha,\beta]} \log \left(\frac{N_{\sigma,\sigma'}^{[\alpha,\beta]} N}{N_{\sigma}^{[\alpha]} N_{\sigma'}^{[\beta]}} \right)}{\sum_{\sigma=1}^{P^{[\alpha]}} N_{\sigma}^{[\alpha]} \log \left(\frac{N_{\sigma}^{[\alpha]}}{N} \right) + \sum_{\sigma'=1}^{P^{[\beta]}} N_{\sigma'}^{[\beta]} \log \left(\frac{N_{\sigma'}^{[\beta]}}{N} \right)}
\tag{8.11}
$$

where $N_{\sigma}^{[\alpha]}, N_{\sigma'}^{[\beta]}$ are the number of nodes in the community σ in layer α and in the community σ' in layer β respectively, whereas $N_{\sigma,\sigma'}^{[\alpha,\beta]}$ is the number of nodes that have their replica node in layer α in the community σ and their replica node in layer β in the community σ'. In mathematical terms

$$
N_{\sigma}^{[\alpha]} = \sum_{i=1}^{N} \delta \left(g_i^{[\alpha]}, \sigma \right),
$$

$$
N_{\sigma'}^{[\beta]} = \sum_{i=1}^{N} \delta \left(\tilde{g}_i^{[\beta]}, \sigma' \right),
$$

$$
N_{\sigma,\sigma'}^{[\alpha,\beta]} = \sum_{i=1}^{N} \delta \left(g_i^{[\alpha]}, \sigma \right) \delta \left(\tilde{g}_i^{[\beta]}, \sigma' \right).
\tag{8.12}
$$

The multilayer nature of the communities can be established by measuring the similarities of the communities between any two layers of a multiplex network dataset.

For a large number of multilayer network datasets it is found that communities of different layers display significant variation of the NMI between their layers despite having rather homogeneous structural properties, sometimes achieving significant similarity. An example of this phenomenon is reported in Ref. [29] in the context of multiplex scientific collaboration networks (APS datasets where scientists collaborate on different topics) and actor collaboration networks (the IMDb dataset where actors play in movies of different genres) (see Fig. 8.10 and Table 8.1).

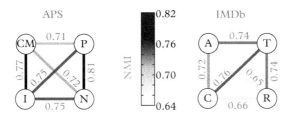

Fig. 8.10 *Illustration of the similarity between four layers of the APS (left) and the IMDb (right) multiplex collaboration networks. These include the Condensed Matter (CM), the Particles (P), the Nuclear (N) and the Interdisciplinary (I) layers of the APS multiplex collaboration network and the Action (A), Thriller (T), Romance (R) and Crime (C) layers of the IMDb multiplex collaboration network. The links between the layers indicate the similarity of the community structures (from dark gray—most similar to light gray—most dissimilar). Reprinted from Ref. [29].*

Table 8.1 *Structural properties of collaboration layers in the APS and IMDb datasets. The number of nodes N, the average degree ⟨k⟩ and the clustering coefficient C are shown for each layer in a subset of the APS and IMDb datasets. The layers are the same as in Fig. 8.10. Data from Ref. [29].*

APS	Code	N	$\langle k \rangle$	C
Nuclear	N	1238	4.75	0.27
Particles	P	1238	4.66	0.30
Cond. Matt. I	CM	1238	10.29	0.24
Interdisciplinary	I	1238	7.37	0.26
IMDb	**Code**	N	$\langle k \rangle$	C
Action	A	55797	83.56	0.61
Crime	C	55797	82.30	0.58
Romance	R	55797	86.00	0.59
Thriller	T	55797	77.75	0.56

8.3.2   Similarity index based on information theory

An alternative measure to evaluate the similarity between the mesoscale structure of different layers of a multiplex network is the $\tilde{\Theta}^S$ indicator. This indicator is based on information theory tools and specifically on the network entropy. In comparison to the NMI this method has the advantage that it applies also to multiplex networks where nodes have different activity, as it estimates the similarity between the layers also accounting for the fact that eventually some nodes are present only in a few layers.

The indicator $\tilde{\Theta}^S$ is a variation of the Θ indicator proposed in Ref. [50] to quantify the significance of a partition induced by a given feature of the nodes $\{q_i\}$ (such as a node property or the node community assignment) with respect to the network structures in single networks. The information theory measure Θ makes explicit use of the entropy of network ensembles (see Sec. 2.8.5). Firstly, the block model resulting from the division of the nodes into blocks formed by nodes of the same degree and the same community is considered. The block model is defined by the number of links within each block and between every pair of different blocks. The entropy of network ensembles $\Sigma_{k,q}$ evaluates how many networks can be constructed with the given block structure. The smaller the entropy of the network ensemble, the better the community structure reproduces the partition of the network into blocks. To evaluate the significance of the partition with respect to the network structure, in Ref. [50] it has been proposed to compare the obtained value $\Sigma_{k,q}$ to the typical value $\Sigma_{k,\pi(q)}$ that one obtains for a random permutation $\pi(q)$ of the features of the nodes. To this end, the significance of a community with respect to the network structure is given by the indicator function Θ which is the zeta-score of the entropy of network ensemble with respect to the randomization of the features of the nodes given by

$$\Theta = \frac{|\Sigma_{k,q} - \langle \Sigma_{k,\pi(q)} \rangle |}{\sigma\left(\Sigma_{k,\pi(q)}\right)} \tag{8.13}$$

with

$$\sigma\left(\Sigma_{k,\pi(q)}\right) = \left\langle \Sigma_{k,\pi(q)}^2 \right\rangle - \left\langle \Sigma_{k,\pi(q)} \right\rangle^2$$

and $\langle \ldots \rangle$ indicating the average of the permutations $\pi(q)$.

Given a multiplex network, this method can be adapted to evaluate how similar the community structures of two layers are. The technique is summarized in Fig. 8.11. Let q_i^α and q_i^β indicate the community assignment of node i in layer α and in layer β respectively. These communities can be obtained by running a single-layer community detection algorithm as the Louvain or the Infomap algorithm on each layer separately.

The indicator function $\Theta_{k^\alpha, q^\beta}$ measures the specificity of the layer α with respect to the particular community assignment q_i^β derived from the mesoscale structure of layer β.

This quantity can be normalized to obtain an unbiased (non-symmetric) indicator $\tilde{\Theta}_{\alpha\beta}$ which quantifies how much the community assignment of layer β is reflected in the

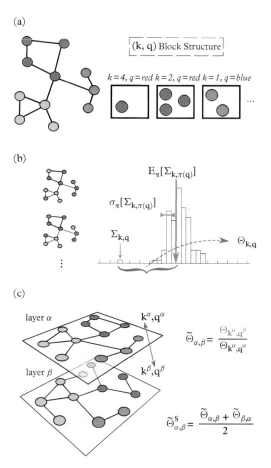

Fig. 8.11 *Diagram showing the method for the extraction of the similarity matrix $\Theta^S_{\alpha\beta}$ between the mesoscale structure of the layers of a multiplex network. As a first step (panel (a)) the nodes are divided into classes (k, q) of nodes, where k indicates their degree and q their characteristic. These classes induce a block structure of the network indicating the number of links between the nodes of each class and the links connecting the nodes in different classes. Subsequently (panel (b)), the entropy $\Sigma_{k,q}$ given by Eq. (10.15) is calculated and compared with the entropy distribution obtained in a random hypothesis. Indeed, the entropies $\Sigma_{k,\pi(q)}$ calculated after a uniform random permutation $\pi(q)$ of the characteristics of the nodes are calculated. The mean $E(\Sigma_{k,\pi(q)})$ and standard deviation $\sigma(\Sigma(k, \pi(q)))$ of the entropy distribution $P(\Sigma_{k,\pi(q)})$ is calculated. The indicator function Θ measures the difference between $\Sigma_{k,q}$ and $E(\Sigma_{k,\pi(q)})$ in units of $\sigma(\Sigma(k, \pi(q)))$. Finally (panel (c)), in a multiplex network $\tilde{\Theta}_{\alpha,\beta}$ characterizes the similarities between layer α and layer β. This quantity is found by normalizing with respect to a null hypothesis and symmetrizing the indicator function $\Theta_{k^\alpha,q^\beta}$ that quantifies the information about structure in layer α, carried by the community structure in layer β. Reprinted figure from Ref. [165].*

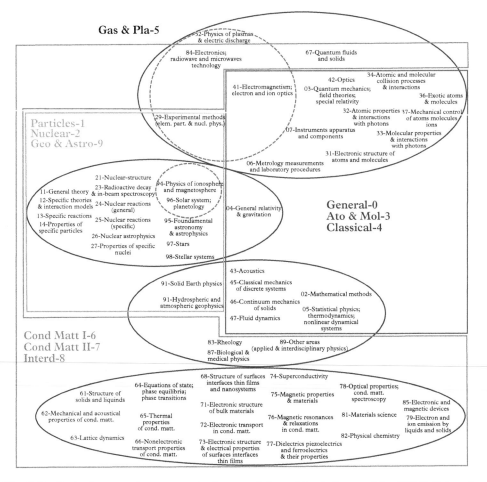

Fig. 8.12 *Optimal community structure in the network formed by the layers of the APS Collaboration Network in which each layer represents a collaboration network in a specific area of physics, as described by the second hierarchical level of the PACS code. The four communities using the $\widetilde{\Theta}^S_{\alpha,\beta}$ matrix contructed starting from the community assignment provided by Infomap are represented by solid line ovals. The partition obtained from the $\widetilde{\Theta}^S_{\alpha,\beta}$ matrix constructed starting from the Louvain communities includes a finer division into two sub-communities indicated by dashed-line ovals. These communities form the coarse-grained partition of the three blocks found at the first hierarchical level of the PACS code (rectangles). Reprinted from Ref. [165].*

network structure of layer α. The indicator $\widetilde{\Theta}^S_{\alpha,\beta}$ symmetrizes $\widetilde{\Theta}_{\alpha\beta}$ and quantifies how similar layer α and layer β are with respect to their community structures. The complete set of similarities $\widetilde{\Theta}^S_{\alpha,\beta}$ forms an $M \times M$ matrix $\widetilde{\Theta}^S$ between the layers of the multiplex network. From this matrix it is possible to extract the network between the layers of the multiplex structure.

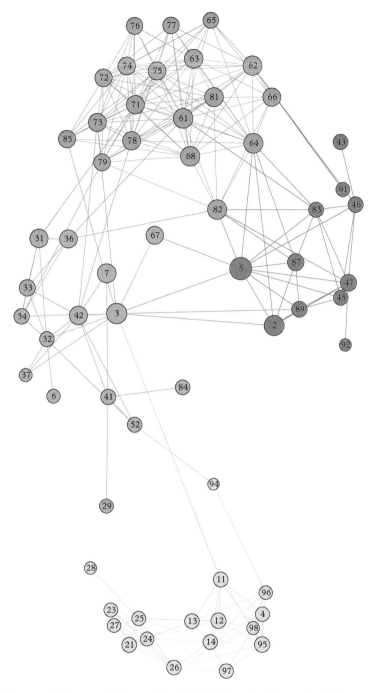

Fig. 8.13 *The network between the layers of the APS Collaboration Network with layers corresponding to the PACS code at the second level of the PACS hierarchy is shown. The network is extracted from the* $\widetilde{\Theta}^S$ *similarity matrix constructed starting from the Infomap partitions of each separate layer. The colour code indicates the four different detected communities of the network among the layers of the multiplex collaboration network. Reprinted from Ref. [165].*

When comparing different layers of a multiplex network, the nodes that are active in one layer might not be active in another layer. Nevertheless, the information carried by the activity of the node might be significant. For example, if two layers have very different activity patterns it might occur that the nodes that are inactive in one layer form a well-defined cluster in the other layer, resulting in a very significant structural property of the multiplex network that it is important to capture. Therefore, to distinguish between active and inactive nodes in a layer it is very convenient to classify all the 'inactive' nodes in one layer with a given common characteristic q.

The similarity matrix $\widetilde{\Theta}^S$ has been used to analyse the American Physical Society (APS) Collaboration Multiplex Network, in which physicists collaborate on different scientific subjects determined by the subject classification known as PACS (Physics and Astronomy Classification Scheme). The APS Collaboration Multiplex Network considered in Ref. [165] is formed by $M_2 = 66$ layers, each one describing the collaboration network in a specific field of physics (second level of the PACS code hierarchy).

From the similarity matrix $\widetilde{\Theta}^S$ it is possible to propose an alternative hierarchy between the PACS numbers. This method allows to characterize with a bottom-up approach how the organization of knowledge in physics is effectively perceived by scientists while shaping their collaboration network (see Fig. 8.12). We observe that while the PACS hierarchy clearly captures main features of the collaboration network, the analysis of Collaboration Multiplex Networks at the second level of the PACS hierarchy clearly suggests a hierarchical organization of these PACS numbers that is not equivalent to the first level of the PACS hierarchy. Finally, the information gained by this analysis has been used to construct the network of networks between the layers of the Collaboration Multiplex Networks. To this end, the weighted network determined by imposing an opportune threshold on the similarity matrix $\widetilde{\Theta}^S$ has been constructed (see Fig. 8.13). The threshold is given by the minimum value of the similarity matrix $\widetilde{\Theta}^S$ that ensures that each layer is connected to at least one other layer of its own cluster. From these networks, it is possible to appreciate that, although the network between the layer of the Collaboration Multiplex Networks is highly interconnected, the clusters found correspond to layers much more similar to each other than to the layers of the other clusters.

8.4   To aggregate or to disaggregate?

8.4.1   Reducibility of multilayer networks

When links have different connotations and indicate interactions of different types, there are in general multiple ways to represent the raw data as a multilayer network. In fact, different multilayer networks can be obtained by aggregating interactions of a similar nature into single layers. Since the analysis of a multilayer network usually includes this preprocessing step, one crucial problem is to identify for any given dataset which is the ideal number of layers M that allows us to find a tradeoff between keeping all the relevant multilayer information and having an efficient representation of the dataset including the smallest number of relevant layers. While this question has been addressed first by studying local properties and centrality measures of multilayer networks [71], recently this question has been tackled by looking at the mesoscale structure of networks. The method proposed in Ref. [90] starts with a multilayer structure including a large number of layers. At each step the most similar layers are aggregated, forming a dendrogram describing the subsequent aggregation of layers. Finally, the best aggregation procedure is found by cutting the dendrogram at a level corresponding to the best score function, characterizing the best aggregation (see Fig. 8.14).

This procedure is very general, and one can imagine using several measures of similarities between the layers to construct the dendrogram describing the aggregation

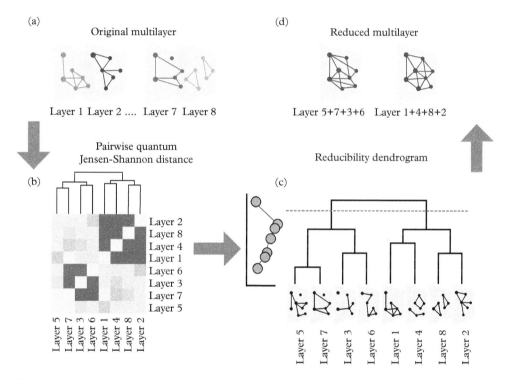

Fig. 8.14 *The figure describes the method used in Ref. [90] for determining the multiplex network reducibility. Panel (a) displays the original multilayer (multiplex) network. Panel (b) describes the aggregation procedure based on the similarity distance between layers and the construction of the hierarchical dendrogram. Panel (c) describes the choice of the best aggregation according to a global score function that is used to cut the dendrogram at a given level of the hierarchy. Panel (d) shows the reduced multilayer (multiplex) network. Reprinted by permission from Macmillan Publishers Ltd: Nature Communication [90] ©2015.*

procedure. In Ref. [90] the Authors consider a specific choice for these functions and use network measures inspired by quantum information.

This method is applied to a large variety of datasets in biological, transportation and social networks. These results suggest that multiplex networks might exhibit high redundancy that allows the merging of up to 75% of layers with some notable exceptions such as non-redundant efficient transportation networks.

8.4.2   Revealing the hidden multilayer structure of single networks

Interestingly, not only has the problem of finding the best aggregation procedure attracted the scientific interest of network scientists, but also the opposite theoretical problem has been explored in the literature. In this case, the goal is to uncover the hidden multilayer structure of an aggregated network that does not distinguish between links of different types. In Ref. [298] an approximated method for detecting the best partition of an aggregated network into two different layers is proposed using stochastic block models. It is shown that by assuming that a given single network is the result of the aggregation of different layers of a hidden multilayer structure, it is possible to make more reliable predictions of missing links and more accurate predictions of spurious interactions existing in real-world networks. The method has been applied to a large variety of datasets including molecular, neuronal, social and transportation networks.

8.4.3   Assessing the multiplex nature of communities

In the precedent paragraphs we have discussed works that aim i) at finding the best aggregation of multiplex networks ii) at disaggregating in an optimal way a single network. A recent work addresses both questions using inference methods and stochastic block models. In particular, in Ref. [247] the Author has proposed two generative stochastic block models for layered networks and for each considered dataset has inferred the parameters of these models using maximum likelihood. The first model considers the layers as edge covariates. In this case, the aggregated network is generated first and subsequently the layer membership of each edge is drawn randomly. In the second model, each layer is generated independently of the others. These two generative models are called *edge covariates* and *independent layers* (see Fig. 8.15). These two models represent two opposite assumptions on the nature of communities of the multilayer network. In the first case, the communities are essentially communities of the aggregated network; in the second case their nature is intrinsically multilayer, as every layer has its own independent community structure.

Performing an inference study on a real dataset allows the investigation of the intrinsic character of its community structure. If the first model is the one that best fits the data, the communities of the multilayer networks are the communities of the aggregated network. On the contrary, if the second model best fits the data the communities of the dataset have a multilayer nature.

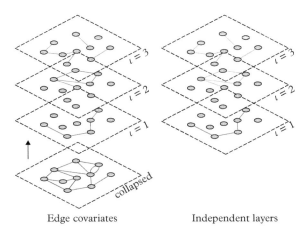

Fig. 8.15 *Schematic representation of two processes that can be used to generate multilayer networks. In the* edge covariates *model the aggregated network is generated first, and conditioned on it, the edges are distributed among the layers. In the* independent layers *model the layers are formed independently of each other. Reprinted figure with permission from Ref. [247] ©2015 by the American Physical Society.*

Interestingly, this study provides two clear examples in which the model that best fits the data is different. The first example is a social network of $N = 241$ physicians that were asked to identify their social connection with respect to the following three questions:

(1) *When you need information or advice about questions of therapy where do you usually turn?*

(2) *And who are the three or four physicians with whom you most often find yourself discussing cases or therapy in the course of an ordinary week? Last week for instance?*

(3) *Would you tell me the first names of your three friends whom you see most often socially?*

This multilayer network of physicians is best fitted by the edge covariates model and the independent layers model is rejected with the posterior odds ratio $\Lambda \sim 10^{-51}$. Therefore, in this case the communities do not have an intrinsic multilayer nature, revealing that the questions asked do not allow us to distinguish clearly between the different layers.

The second example is the network of vote correlations among federal deputies of the Brazilian national congress between the years 1999 and 2006 including the two consecutive terms (1999–2002 and 2003–6) forming the two layers of the network. In this case the independent layer model is the one that best fits the data and the edge covariates model can be rejected, with $\Lambda \sim 10^{-111}$ indicating that the multilayer network is very informative of the network structure.

In conclusion, this work shows that there is no general answer to the question: *to aggregate or to disaggregate?* and the suitable answer depends on the dataset.

9

Centrality Measures

9.1 Centrality measures and multiplexity

In multilayer networks as well as in single layer networks it is often the case that a ranking of the nodes is necessary for summarizing the information encoded in the network structure. Centrality measures originally defined on single networks have been used extensively in social sciences and technological and biological networks. In multilayer networks these measures can be extended in different ways. A set of centrality measures is especially tailored to multiplex networks without interlinks. In these settings it is possible to either evaluate the effect that the centrality of a node in one layer has on its centrality on the other layer (Multiplex PageRank) or to associate a different *influence* with the links of different layers that weights their contribution to the centrality of the nodes (Multiplex Eigenvector Centralities, Functional Multiplex PageRank and MultiRank). Other algorithms allow us to tackle the centrality of nodes on general multilayer networks directly as they include an explicit treatment of interlinks (Versatility, Communicability). Finally, suitable centrality measures for multi-slice networks have also been explored that take into account the temporal sequence of the layers.

9.2 Multiplex PageRank

The Multiplex PageRank [147] evaluates the centrality of the nodes of multiplex networks. The main effect that the Multiplex PageRank aims at capturing is the influence of the centrality of a node in one layer on its centrality in another layer. Consider for example a very central actor in the movie collaboration network. If the famous actor takes part in social causes his/her centrality in the actor–movie collaboration network might influence his/her centrality in socio-political causes. This is the case, for example, with famous actors such as Angelina Jolie, who is also a UN Goodwill ambassador. Therefore, in the Multiplex PageRank the centrality of a node in one layer might affect the centrality of the same node in other layers.

In order to capture this phenomenon a master layer α with adjacency matrix $a_{ij}^{[\alpha]}$ can be chosen and the corresponding PageRank centrality can be calculated. The PageRank x_i of a node i in a network is given by

Multilayer Networks. Ginestra Bianconi, Oxford University Press (2018).
© Ginestra Bianconi. DOI: 10.1093/oso/9780198753919.001.0001

$$x_i = \mu \sum_{j=1}^{N} a_{ji}^{[\alpha]} \frac{x_j}{\hat{\kappa}_j} + \omega, \tag{9.1}$$

where $\hat{\kappa}_j = \max(\tilde{\kappa}_j, 1)$, with $\tilde{\kappa}_j = \sum_{r=1}^{N} a_{jr}^{[\alpha]}$. Here $\delta(x, y)$ indicates the Kronecker delta and ω, the teleportation parameter, enforces the normalization of the PageRank centralities.

Given the rank x_i of node i in the layer α, the Multiplex PageRank algorithm measures the centrality of this node in any other layer β with adjacency matrix $a_{ij}^{[\beta]}$ using a random walker, as in the usual PageRank. At each time step the random walker placed on a given node of the network either hops to a neighbouring node or jumps to a random node of the network (teleportation jump). Moreover, the random walker of the Multiplex PageRank has a probability of visiting any of the nodes that is affected by their rank x_i in layer α. In other words, the Multilayer PageRank evaluates the rank of the nodes in layer β by determining the stationary probability of a random walk in layer β, biased by the PageRank of the nodes in layer α. The Multiplex PageRank $X_i(q, n)$ of node i depends on two parameters $q, n \in \{0, 1\}$ and satisfies

$$X_i(q, n) = \mu \sum_{j=1}^{N} (x_i)^q a_{ji}^{[\beta]} \frac{X_j(q, n)}{\kappa_j} + \omega \left(\frac{x_i}{\langle x \rangle} \right)^n, \tag{9.2}$$

where κ_j and ω are given by

$$\kappa_j = \max \left(\sum_{r=1}^{N} a_{jr}^{[\beta]} (x_r)^q, 1 \right)$$

$$\omega = \frac{1}{N} \sum_{j=1}^{N} \left[1 - \mu + \mu \delta \left(\sum_{r=1}^{N} a_{jr}^{[\beta]} (x_r)^q, 0 \right) \right] X_j(q, n). \tag{9.3}$$

For $(q, n) = (0, 0)$ the Multiplex PageRank $X_i(q, n)$ reduces to the PageRank of layer β which is independent of the ranking on the layer α. However, for $(q, n) \neq (0, 0)$ we distinguish the following non-trivial cases of Multiplex PageRank:

(1) *Additive Multiplex PageRank:*
    The Additive Multiplex PageRank is obtained for $(q, n) = (0, 1)$. In this case, when performing the teleportation jump the random walker chooses his destination node i according to his centrality x_i in layer α.
    In the Additive Multiplex PageRank each node in layer β derives an added benefit by being central in network α, regardless of the relevance of the nodes that point to it in layer β.

(2) *Multiplicative Multiplex PageRank:*
    The Multiplicative Multiplex PageRank is obtained for the parameter values $(q, n) = (1, 0)$. In this case, the random walker hopping from node j to a

neighbouring node i chooses the node i from the neighbours of node j with a probability proportional to its PageRank centrality x_i in layer α.

In the Multiplicative Multiplex PageRank each node in layer β derives an added benefit by being central in network α, but this benefit is contingent upon the connections that a node receives from central nodes in network β.

(3) *Combined Multiplex PageRank*:

The Combined Multiplex PageRank is obtained for the parameter values $(q, n) = (1, 1)$. In this case, the random walker hopping from node j to a neighbour node i chooses a node i from the neighbours of node j with a probability proportional to its PageRank centrality x_i in layer α. Additionally, when the random walker jumps to a random node it chooses the random node with a probability proportional to x_i.

In the Combined Multiplex PageRank the effect of network α on network β is a combination of the effects of an additive and of a Multiplicative PageRank.

In multiplex networks with more than two layers the Multiplex PageRank can also be applied repeatedly to different layers of a given multiplex network once the most opportune order between the layers has been chosen [163].

9.3 Multiplex Eigenvector Centralities

In multiplex networks links in different layers can contribute differently to the centrality of the nodes. In order to capture this phenomenon in Ref. [278] the Authors have introduced an $M \times M$ *influence matrix* **w** of non-negative elements $w_{\alpha\beta}$ describing the influence of the generic layer β on layer α. The Multiplex Eigenvector Centralities proposed are called *local and global heterogeneous eigenvector-like centralities* and rank replica nodes in the multiplex network neglecting the effect of interlinks. Ranking replica nodes responds to the need to tailor the ranking to a given aspect captured by a layer of the multiplex network. For example, the centrality of an actor in the actor collaboration network might be different from his centrality in social media. The local and the global heterogeneous eigenvector-like centrality take into account all the layers of the multiplex network, weighting their contribution as predetermined by a given influence matrix.

The *global heterogeneous eigenvector centrality* $x_{i\alpha}^{\star}$ of a replica node (i, α) is given by the eigenvector corresponding to the maximum eigenvalue of the supra-weighted matrix $\mathcal{W}$ of dimension $(N \cdot M) \times (N \cdot M)$ given by

$$\mathcal{W}_{i\alpha,j\beta} = \sum_{\beta} w_{\alpha\beta} a_{ij}^{[\beta]}. \qquad (9.4)$$

The *local heterogeneous eigenvector centrality* instead is the eigenvector of the block diagonal supra-weighted matrix $\hat{\mathcal{W}}$ of dimension $(N \cdot M) \times (N \cdot M)$ given by

$$\hat{W}_{i\alpha,j\beta} = \delta(\alpha,\beta) \sum_{\gamma} w_{\alpha\gamma} a_{ij}^{[\gamma]}. \tag{9.5}$$

The idea of considering the influence matrix is brilliant and meaningful for a number of real-world multiplex networks. However, the shortcoming of these definitions of centrality is that they require the knowledge of the influences $w_{\alpha\beta}$ which are M^2 parameters typically not accessible to the network scientist. We will see how the concept of influences of the layers has subsequently been developed in the Functional Multiplex PageRank and in the MultiRank algorithms.

9.4 Functional Multiplex PageRank

9.4.1 A function as centrality measure

Characterizing the centrality of nodes in multiplex networks is a challenging task because links of different types can contribute differently to the centrality of the nodes. Take, for example, the multiplex airport network. A hub airport like Heathrow might gain important centrality thanks to British Airways connections, while a hub airport like Frankfurt might gain a relevant role thanks to its important function in the Lufthansa network.

As different patterns of connections contribute differently to the centrality of a node, in Ref. [164] an *influence* $z^{\vec{m}}$ is associated with each multilink $\vec{m}$. The Functional Multiplex PageRank centrality determines the centrality of the node as a function of the influences $\mathbf{z} = \{z^{\vec{m}} | \vec{m} \neq \vec{0}\}$ associated with its non-trivial multilinks in this way, describing exhaustively the contribution to the centrality of the node coming from every possible type of connection between two nodes of a multiplex network.

The Functional Multiplex PageRank captures the fact that different multilinks contribute differently to the centrality of each node and associates with each node i an entire function $X_i(\mathbf{z})$. For each node i the function $X_i(\mathbf{z})$ is also called the *pattern to success* because from this function it is possible to extract the information about which types of connections contribute the most to the centrality of the node.

The Functional Multiplex PageRank describes the steady state of a random walker that hops from a node j to a neighbour node i with probability μ if this is possible, and otherwise it jumps to a random node that is not isolated. When the random walker hops to a random neighbour it follows each multilink $\vec{m} \neq \vec{0}$ with a probability proportional to $z^{\vec{m}}$ (see Fig. 9.1). As a function of the parameters $\mathbf{z}$ the Functional Multiplex PageRank $X_i(\mathbf{z})$ of node i can be reduced to:

(a) PageRank on each separate layer;

(b) PageRank on the aggregated network;

(c) PageRank on the network formed by the links (i,j) present at the same time in every layer $\alpha = 1, 2, \ldots M$.

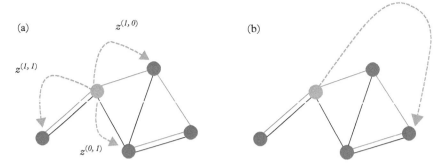

Fig. 9.1 *The Functional Multiplex PageRank defines the stationary state of a random walker that either hops to a neighbour node connected by a non-trivial multilink $\vec{m}$ with a probability proportional to the influence $z^{\vec{m}}$, or jumps to a random node of the multiplex network.*

9.4.2   Definition

The Functional Multiplex PageRank $X_i(\mathbf{z})$ of node i is defined as the steady state of the random walker described in the previous section. Mathematically, $X_i(\mathbf{z})$ is given by the solution of the following recursive equation,

$$X_i(\mathbf{z}) = \mu \sum_{j=1}^{N} A_{ji}^{\vec{m}^{ji}} z^{\vec{m}^{ij}} \frac{1}{\kappa_j} X_j + \omega v_i, \tag{9.6}$$

where $z^{\vec{0}} = 0$ and where κ_j and ω are normalization factors and the vector $\mathbf{v}$ determines which nodes are not isolated. Specifically, by defining

$$\tilde{\kappa}_j = \sum_{i=1}^{N} A_{ji}^{\vec{m}^{ji}} z^{\vec{m}^{ji}}$$

we have that κ_j, ω and v_i are given by

$$\kappa_j = \tilde{\kappa}_j + \delta(\tilde{\kappa}_j, 0),$$

$$\omega = \sum_{i=1}^{N} \left[(1 - \mu) + \delta(0, \tilde{\kappa}_i) \right] X_i,$$

$$v_i = \frac{1}{N} \theta \left(\sum_{j=1}^{N} A_{ij}^{\vec{m}^{ij}} z^{\vec{m}^{ij}} + \tilde{\kappa}_j \right), \tag{9.7}$$

where $\delta(x, y)$ indicates the Kronecker delta and $\theta(x)$ indicates the Heaviside step function. Using the definition of the Functional Multiplex PageRank given by Eq. (9.6), by making an opportune choice of the influences **z** it is possible to recover the desired limiting cases.

(a) The Functional Multiplex PageRank reduces to the PageRank in the layer α for

$$z^{\vec{m}} = \begin{cases} z^\star > 0 & \text{if } m^{[\alpha]} = 1, \\ 0 & \text{if } m^{[\alpha]} = 0. \end{cases} \tag{9.8}$$

(b) The Functional Multiplex PageRank reduces to the PageRank in the aggregated network obtained for

$$z^{\vec{m}} = z^\star > 0 \tag{9.9}$$

as long as $\vec{m} \neq \vec{0}$.

(c) The Functional Multiplex PageRank reduces to the PageRank fully overlapping network for

$$z^{\vec{m}} = \begin{cases} z^\star > 0 & \text{if } \vec{m} = \vec{1}, \\ 0 & \text{if } \vec{m} \neq \vec{1}. \end{cases} \tag{9.10}$$

The Functional Multiplex PageRank can describe non-linear effects due to the overlap between the links. For instance, in a duplex network we can have

$$z^{(1,1)} > z^{(1,0)} + z^{(0,1)}, \tag{9.11}$$

indicating that the multilinks with link overlap have more influence than the sum of influences attributed to multilinks having connections exclusively in the first and second layers. Unlike when

$$z^{(1,1)} < z^{(1,0)} + z^{(0,1)}, \tag{9.12}$$

we are attributing less influence to multilinks having link overlap than the sum of the influences attributed to multilinks having connections exclusively in the first and second layers. From the definition of the Functional Multiplex PageRank, one observes that the ranking $\mathbf{X}(\mathbf{z})$ is invariant under the transformation

$$\mathbf{z} = \gamma \mathbf{z} \tag{9.13}$$

for $\gamma > 0$. Therefore, by considering **z** as a vector in a $(2^M - 1)$-dimensional space, with elements $z^{\vec{m}}$ for every $\vec{m} \neq \vec{0}$, the Functional Multiplex PageRank only depends on the direction of this vector and not on its normalization. Therefore, the general definition of the Functional Multiplex PageRank depends on $(2^M - 2)$-independent parameters.

9.4.3   Correlations between the Functional Multiplex PageRank of different nodes

By evaluating the Functional Multiplex PageRank on a given $(2^M - 2)$-dimensional grid with N_g points, we can calculate the Pearson correlation ρ between the Functional Multiplex PageRank of the generic nodes i and j as

$$\rho = \frac{\overline{X_i X_j} - \overline{X_i}\,\overline{X_j}}{\sigma(X_i)\sigma(X_j)},\tag{9.14}$$

where $\overline{Y(\mathbf{z})}$ is given by

$$\overline{Y(\mathbf{z})} = \frac{1}{N_g}\sum_{n=1}^{N_g} Y(\mathbf{z}_n)\tag{9.15}$$

and $\sigma(Y) = \sqrt{\overline{Y^2} - \overline{Y}^2}$. Note that $\rho \in [-1,1]$ where negative values $\rho < 0$ indicate anticorrelations, while $\rho > 0$ indicates positive correlations.

9.4.4   Absolute Functional Multiplex PageRank

From the Functional Multiplex PageRank it is also possible to define an absolute ranking of nodes by assigning to each node i the maximum of the Functional Multiplex PageRank over all the space in which the vector $\mathbf{z}$ varies. To this end, the Absolute Multiplex PageRank $X_i^\star$ of node i can be defined as

$$X_i^\star = \max_{\mathbf{z}} X_i(\mathbf{z}).\tag{9.16}$$

Another interesting possibility for determining an absolute ranking from the Functional Multiplex PageRank is to take the absolute ranking induced by the average of the Functional Multiplex PageRank, i.e.

$$\hat{X}_i = \langle X_i(\mathbf{z})\rangle_{\mathbf{z}}.\tag{9.17}$$

9.4.5   Application to duplex networks

The Functional Multiplex PageRank of a duplex network depends on the values of the influences $\mathbf{z} = (z^{(1,0)}, z^{(0,1)}, z^{(1,1)})$. As has been discussed in the previous paragraphs, the Functional Multiplex PageRank only depends on the direction of $\mathbf{z}$ interpreted as a three-dimensional vector in $\mathbb{R}^3$. In a duplex network, changing only the direction of the vector $\mathbf{z}$ within the three-dimensional region where all the components of $\mathbf{z}$ are either positive or null is sufficient to span all cases. Accordingly, the different directions

Table 9.1 *Top-ranked airports according to the Abso-*
lute Multiplex PageRank in the duplex airport network
formed by Lufthansa and British Airways flight connec-
tions. Here, in order to find the absolute Functional Mul-
tiplex PageRank we evaluated the Functional Multiplex
PageRank for angles (θ, ϕ) *chosen on a grid with spacing*
$\delta\theta = \delta\phi = \pi/80.$ *Data from Ref. [164].*

Rank	Airport
1	Heathrow Airport (LHR)
2	Munich Airport (MUC)
3	Frankfurt Airport (FRA)
4	Gatwick Airport (LGW)

of **z** can be parametrized by using just two parameters. Therefore, the influences $\mathbf{z} = (z^{(1,0)}, z^{(0,1)}, z^{(1,1)})$ can be expressed in spherical coordinates as

$$z^{(1,0)} = \sin\theta\cos\phi,$$
$$z^{(0,1)} = \sin\theta\sin\phi,$$
$$z^{(1,1)} = \cos\theta, \tag{9.18}$$

with $\theta, \phi \in [0, \pi/2]$.

An example of a duplex network on which the Functional Multiplex PageRank has been applied is the airport duplex network formed by the European flight connections of Lufthansa (layer 1) and British Airways (layer 2).

The angle ϕ modulates the influence of the multilinks $(1,0)$ (exclusively Lufthansa flight connections) with respect to multilinks $(0,1)$ (exclusively British Airways flight connections). For $\phi = 0, \theta = \pi/2$ the influence of exclusively Lufthansa connections is maximized, for $\phi = \pi/2, \theta = \pi/2$ the influence of exclusively British Airways is maximized. The angle θ measures the influence of multilinks $(1,1)$ corresponding to flight connections existing in both airline companies with respect to the other two types of multilinks corresponding to flight connections existing in a single airline company. For $\theta = 0$ the influence of multilinks $(1,1)$ is maximized, while for $\theta = \pi/2$ it is minimized.

The Absolute Multiplex PageRank of this duplex network ranks its four top central airports according to the rank shown in Table 9.1. The major airports display a very different Functional Multiplex PageRank revealed by their distinct *pattern to success*. In Fig. 9.2 the pattern of success of four exemplary hub airports are shown. Frankfurt Airport (FRA) shows a pattern of success that establishes the airport as a central hub for Lufthansa. In fact, its Functional Multiplex PageRank displays a maximum for smaller values of ϕ and decreases as θ decreases toward zero, showing that Frankfurt Airport takes most of its centrality from flight connections operated exclusively by Lufthansa. On the contrary, Düsseldorf Airport (DUS) acquires significant centrality also by including connections existing in both layers, although it constitutes an important Lufthansa hub. Therefore, it has a Functional Multiplex PageRank that is as ϕ decreases, and also as θ approaches zero. By calculating the Functional Multiplex PageRank of Heathrow Airport (LHR) and Gatwick Airport (LGW) one can see that they are both British Airways hub airports but Heathrow acquires important centrality also by including connections existing in both layers.

Therefore, this result shows that the links that determine the centrality of the nodes can be of different types for two different nodes of the multiplex network. It is therefore interesting to measure the correlation between the Functional Multiplex PageRanks (pattern to success) of two different nodes. In Table 9.2 we report the correlations ρ existing between the Functional Multiplex PageRanks shown in Fig. 9.2, showing both positive (Heathrow/Gatwick, Frankfurt/Düsseldorf but also Heathrow/Düsseldorf) and negative values (Heathrow/Frankfurt, Gatwick/Frankfurt, Gatwick/Düsseldorf).

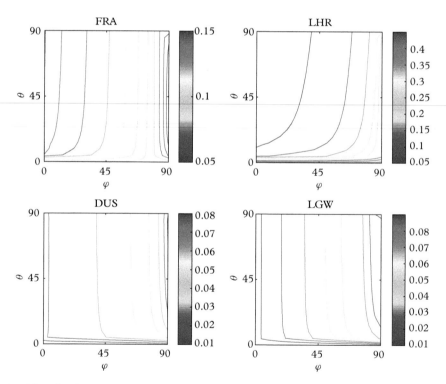

Fig. 9.2 *Functional Multiplex PageRanks for important airports such as Frankfurt Airport (FRA), Heathrow Airport (LHR), Düsseldorf Airport (DUS) and Gatwick Airport (LGW) in the duplex network formed by Lufthansa and British Airways flights. Reprinted from Ref. [164].Copyright ©EPLA, 2016.*

Table 9.2 *Correlation ρ between the Functional Multiplex PageRank of the airports Heathrow (LHR), Frankfurt (FRA), Gatwick (LGW) and Düsseldorf (DUS). Data from Ref. [164].*

ρ	LHR	FRA	LGW	DUS
LHR	1	-0.797	0.484	0.351
FRA	-0.797	1	-0.983	0.275
LGW	0.484	-0.983	1	-0.729
DUS	0.351	0.2758	-0.729	1

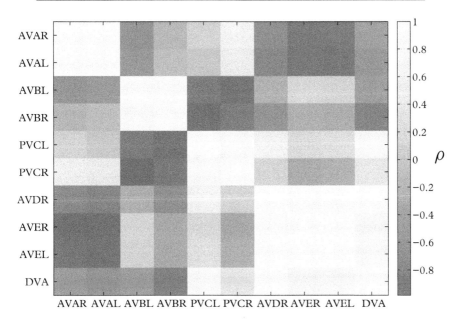

Fig. 9.3 *Correlation ρ between the Functional Multiplex PageRank of the top-ranked neurons in the duplex brain network of the nematode* C. elegans. *Reprinted from Ref. [164]. Copyright ©EPLA, 2016.*

As a second example, the multiplex connectome (brain network) of the nematode *C. elegans* has been analysed. This dataset includes all the connections existing between the 302 neurons of the animal. These connections can be chemical (synaptic connections forming layer 1) or electrical (gap junctions forming layer 2).

The Pearson correlation coefficient analysis performed on the top ten ranked nodes (see Fig. 9.3) shows that neurons of the same type have highly correlated Functional Multiplex PageRank.

9.5 MultiRank

9.5.1 Ranking nodes and layers

The Functional Multiplex PageRank, without assigning any predefined influence to the multilinks, has great flexibility and has been proved to be very useful for understanding duplex networks. However, its applicability to multiplex networks is practically limited to multiplex networks with few layers, as it requires the analysis of a function, the Functional Multiplex PageRank, with a number of variables that increases fast with the number of layers.

In order to address this limitation of the Functional Multiplex PageRank, a ranking algorithm, MultiRank, has been proposed in Ref. [258]. MultiRank works efficiently on multiplex networks with many layers and simultaneously ranks the nodes and the layers of the multiplex networks.

According to this algorithm, each layer α has a centrality $z^{[\alpha]}$ called the *influence* of the layer and each node i has a centrality X_i. The centrality of each node strongly depends on which nodes are pointing to it and in which layer, with links reaching the node from influent layers contributing more to its centrality. On their turn, layers including central nodes acquire a greater influence than layers including less central nodes.

Therefore, the algorithm consists of a set of coupled equations for the centrality of the nodes and the influence of the layers.

The MultiRank algorithm interprets the multiplex network as a combination between a coloured network and a bipartite network. The coloured network is the weighted aggregated network where different types of links are assigned a different weight (influence). On the other side, the algorithm also exploits the properties of the weighted, directed bipartite network formed by nodes and layers that can be constructed from the multiplex network. This bipartite network provides information about the activity of the nodes in each layer, indicating whether they are present (and therefore connected) in the layer. Additionally, the weighted and directed links of this bipartite network indicate the in-strength and out-strength of each node in any given layer. In Fig. 9.4 the bipartite network and the coloured networks extracted from any given multiplex network are represented.

Specifically, MultiRank uses the following two sets of matrices extracted from the multiplex network. The first matrix is the $(N \times N)$-weighted matrix $\mathbf{C}$ of the coloured network, where the links of each layer $\alpha = 1, 2, \ldots, M$ are weighted with the influences $z^{[\alpha]}$ associated with it. Therefore, the elements C_{ij} of the matrix $\mathbf{C}$ are given by

$$C_{ij} = \sum_{\alpha=1}^{M} a_{ij}^{[\alpha]} z^{[\alpha]}. \tag{9.19}$$

The second set of matrices are the incidence matrices of the bipartite network constructed from the multiplex network by considering the connectivity of each node i in layer α. For directed multiplex networks we distinguish two $M \times N$ incidence matrices $\mathbf{B}^{in}$ and $\mathbf{B}^{out}$ of elements

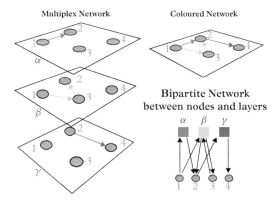

Multiplex Network

Coloured Network

Bipartite Network between nodes and layers

Fig. 9.4 *The schematic representation of a multiplex network formed by three different layers is shown together with the construction of the directed bipartite network formed by nodes and layers and the multiplex representation as a coloured network. The directed bipartite network indicates for each node in which layers the node is connected. Additionally, as is explained in the text, it gives direct information about in-strength and out-strength of each node in any given layer. Reprinted from Ref. [258].*

$$B_{\alpha i}^{in} = \frac{\sum_{j=1}^{N} a_{ji}^{[\alpha]}}{W^{[\alpha]}},$$

$$B_{\alpha i}^{out} = \frac{\sum_{j=1}^{N} a_{ij}^{[\alpha]}}{W^{[\alpha]}}, \tag{9.20}$$

where

$$W^{[\alpha]} = \sum_{i=1}^{N} \sum_{j=1}^{N} a_{ij}^{[\alpha]} \tag{9.21}$$

indicates the total weight of the links in layer α. Therefore, $B_{\alpha i}^{in}$ indicates the normalized in-strength of node i in layer α and $B_{\alpha i}^{out}$ indicates the normalized out-strength of node i in layer α. For undirected multiplex networks the matrices $\mathbf{B}^{in}$ and $\mathbf{B}^{out}$ are identical. Note that if node i is not connected in layer α we have $B_{\alpha i}^{in} = B_{\alpha i}^{out} = 0$ and that the node is *inactive* in layer α.

9.5.2 The definition

The MultiRank algorithm depends on three parameters: $\gamma > 0$, s taking values $s \in \{1, -1\}$ and a taking values $a \in \{1, 0\}$.

The MultiRank algorithm assigns a centrality X_i to each node i and an influence $z^{[\alpha]}$ to each layer α. The centrality X_i of node i is higher if central nodes in influent layers point to node i. Therefore, given the influences $z^{[\alpha]}$ the centrality X_i is determined by the generalized PageRank algorithm

$$X_i = \mu \sum_{j=1}^{N} \frac{C_{ji}}{\kappa_j} X_j + \omega v_i \qquad (9.22)$$

where μ is taken to be $\mu = 0.85$, and κ_j, v_i and ω are given by

$$\kappa_j = \max\left(1, \sum_{i=1}^{N} C_{ji}\right),$$

$$v_i = \theta\left(\sum_{j=1}^{N}[C_{ij} + C_{ji}]\right),$$

$$\omega = \frac{1}{\sum_{i=1}^{N} v_i} \sum_{j=1}^{N}\left[1 - \mu\,\theta\left(\sum_{i=1}^{N} C_{ji}\right)\right] X_j. \qquad (9.23)$$

These equations for X_i are then coupled with another set of equations specifying the values of the influences $z^{[\alpha]}$ given the centralities of the nodes. These equations depend on the three parameters a, s and γ and read

$$z^{[\alpha]} = \frac{1}{\mathcal{N}}\left[W^{[\alpha]}\right]^a \left[\sum_{i=1}^{N} B_{\alpha i}^{in}\,(X_i)^{s\gamma}\right]^s \qquad (9.24)$$

where $\mathcal{N}$ is a normalization constant.

- For $a = 1$ the influence of a layer is larger the larger the total weight $W^{[\alpha]}$ of the links in layer α.
- For $a = 0$, the influence of a layer is normalized with respect to $W^{[\alpha]}$.
- For $s = 1$, layers have larger influence if they include more central nodes. In this case the parameter γ can be tuned to either suppress ($\gamma > 1$) or enhance ($\gamma > 1$) the contribution of low-centrality nodes in determining the centrality of the layers in which they are active.
- For $s = -1$, however, layers have larger influence if they include *fewer* highly influential nodes. In other words, this algorithm awards *elite layers*. In this case, the parameter γ can be tuned to either enhance ($\gamma > 1$) or suppress ($\gamma > 1$) the contribution of low-centrality nodes in determining the centrality of the layers in which they are active.
- For $a = 0$ and $\gamma = 0$, all the layers have the same influence and the MultiRank reduces to the PageRank of the aggregated network.

9.5.3 Possible variations of the MultiRank algorithm

The MultiRank algorithm is fully defined by Eqs (9.22) and (9.24). However, this algorithm can be modified by changing the first equation (i.e. Eq. (9.22)) determining the centrality of the nodes. In fact, instead of adopting a PageRank algorithm for the centrality of the nodes, it is possible to consider either the eigenvector centrality or the Katz centrality. In the first case Eq. (9.22) could be substituted by the normalized eigenvector centrality, satisfying the system of equations

$$\lambda_1 X_i = \sum_{j=1}^{N} C_{ji} X_j,$$

$$\sum_{i=1}^{N} X_i = 1, \tag{9.25}$$

where λ_1 is the maximum eigenvalue of the matrix $\mathbf{C}$. In the second case Eq. (9.22) could be substituted by

$$X_i = \mu \sum_{j=1}^{N} C_{ji} X_j + \omega, \tag{9.26}$$

where ω ensures the normalization condition and $\mu > 0$ is suitably chosen to ensure the convergence of the algorithm.

9.5.4 Applications to real datasets

In Ref. [258] the MultiRank algorithm has been applied to a large variety of multiplex network datasets including the European Multiplex Air Transportation Network with $M = 37$ layers, the Food and Agriculture Organization Multiplex World Trade Network with $M = 364$ layers and the Pierre Auger Multiplex Collaboration Network with $M = 16$ layers finding very informative results. In Fig. 9.5 we visualize the results obtained by running the MultiRank on the European Multiplex Air Transportation Network for different values of the parameters (s, a) and γ. From these maps the very relevant but different roles of airports such as Stansted and Frankfurt are apparent as they are the top-ranked airports for different parameter values.

9.6 Versatility

The *versatility* of the nodes [94] of a multilayer network is a centrality measure that identifies the nodes that play the most central role in the cohesion of the multilayer structure. The most versatile nodes are in fact the nodes that keep the multilayer network together, connecting and bridging between the interaction existing in different layers.

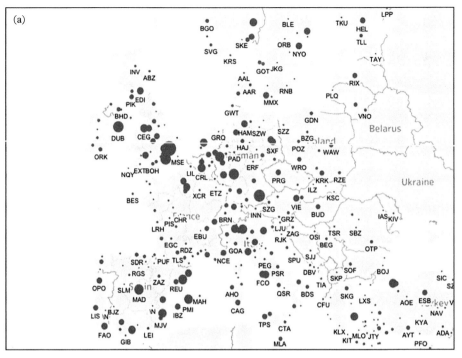

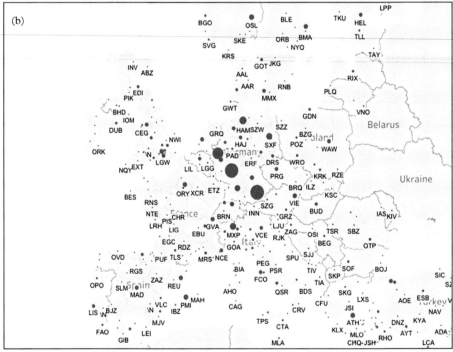

Fig. 9.5 *The maps representing the centrality* X_i *of European airports in the European Air Transportation Multiplex Network according to the MultiRank algorithm are shown for different values of the parameters* $s = 1, a = 1, \gamma = 1$ *(panel (a)) and* $s = 1, a = 1, \gamma = 3$ *(panel (b)). By comparing the results for* $\gamma = 1$ *and* $\gamma = 3$ *it is possible to observe that for* $\gamma = 3$ *few airports acquire a centrality much higher than the others. Reprinted from Ref. [258].*

This centrality algorithm provides a global ranking of every node belonging to any layer of the multilayer structure by using random walks that diffuse on intralinks and interlinks as well. This centrality measure can be applied to general multilayer networks but also to multiplex networks. In multiplex networks it is found that the versatility is strongly dependent on the activity of the node. In Ref. [94] the Authors distinguish between the eigenvector versatility and the PageRank versatility. The eigenvector versatility $\Theta_{i\alpha}$ of the node (i, α) is given by the solution to the eigenvector problem

$$\sum_{j=1}^{N}\sum_{\beta=1}^{M} \left(\mathcal{A}^T\right)_{i\alpha,j\beta} \Theta_{j\beta} = \lambda_1 \Theta_{i\alpha} \tag{9.27}$$

where $\mathcal{A}^T$ is the transpose of the supra-adjacency matrix of the multiplex network and λ_1 is the maximum eigenvalue of the supra-adjacency matrix $\mathcal{A}$. The PageRank versatility $\Theta_{i\alpha}$ of the node (i, α) is given by the stationary state of a random walk that hops between neighbour nodes and connected layers but that performs also additional *teleportation* jumps with probability ω. Therefore, $\Theta_{i\alpha}$ is given by the solution of the equation

$$\Theta_{i\alpha} = \mu \sum_{i=1}^{N}\sum_{\alpha=1}^{M} \frac{\left(\mathcal{A}^T\right)_{i\alpha,j\beta}}{\kappa_{j\beta}} \Theta_{j\beta} + \omega, \tag{9.28}$$

where $\kappa_{j\beta} = \max\left(1, \sum_i \sum_\alpha A_{i\alpha,j\beta}^T\right)$ and where the constant ω is fixed by the normalization condition

$$\sum_{i=1}^{N}\sum_{\alpha=1}^{M} \Theta_{i\alpha} = 1. \tag{9.29}$$

The PageRank versatility can be obtained by solving Eq. (9.28) by iteration, or, for relatively small matrices, it can be obtained by matrix inversion using the solution

$$\Theta_{i\alpha} = \omega \sum_{j=1}^{N}\sum_{\beta=1}^{M} \left[1 - \mu \mathcal{A}^T \mathcal{D}^{-1}\right]_{i\alpha,j\beta}^{-1}, \tag{9.30}$$

where $\mathcal{D}$ is the supra-matrix of elements $\mathcal{D}_{i\alpha,j\beta} = \delta(i,j)\delta(\alpha,\beta)\kappa_{i,\alpha}$.

Using a similar approach, it is also possible to define the Katz versatility $\Theta_{i\alpha}$ of the replica nodes given by the solution to the equation

$$\Theta_{i\alpha} = \mu \sum_{i=1}^{N}\sum_{\alpha=1}^{M} A_{i\alpha,j\beta}^T \Theta_{j\beta} + \omega \tag{9.31}$$

where the constant ω is fixed by the normalization condition, Eq. (9.29).

Table 9.3 *Comparison between the PageRank versatility and the PageRank of the aggregated Wikipedia multiplex network. Data from Ref. [94].*

Name	PageRank Versatility	PageRank Aggregated network
Milton Friedman	1	15 (+14)
Hilary Putnam	2	33 (+31)
Edward Osborne Wilson	3	331 (+328)
Harlond Clayton Urey	4	536 (+532)
Kurt Gödel	5	42 (+37)
Charles Stark Draper	9	1195 (+1186)
Aristotele	11	2 (-9)
Immanuel Kant	13	1 (-12)
Albert Einstein	23	9 (-14)
Plato	24	3 (-21)

On multilayer networks the nodes (i, α) belonging to different layers α are replica nodes. In this case, the node versatility X_i of the node i can be found from the set of replica node versatilities $\Theta_{i\alpha}$, as

$$X_i = \sum_{\alpha=1}^{M} \Theta_{i\alpha}. \tag{9.32}$$

This choice corresponds to a maximum entropy assumption that each replica node (i, α) of node i provides a contribution to the centrality of node i of equal weight.

In Ref. [94] the node versatility of philosophers, chemists and physicists in Wikipedia has been compared with their centrality in the aggregated multiplex network. Interestingly, significant differences have been found, indicating that the versatility of a node favours individuals with an impact on multiple disciplines. For example, the versatility ranking of Einstein and Kant is smaller than the versatility of Gödel whose research on logic has impact in different fields ranging from pure mathematics to physics. Maybe more surprisingly, the versatility attributes to Einstein and Kant a lower rank than the one of Edward Osborne Wilson, the founding father of sociobiology, and Harold Clayton Urey, the Nobel Prize-winner in Chemistry known for the theory of development of organic life from non-living matter (see Table 9.3).

9.7 Multilayer Communicability

The *multilayer communicability* [115] is a centrality measure which quantifies the number of paths taking both intralinks and interlinks that join a given node of a given layer to the other nodes of the multilayer structure. It is therefore a centrality measure that naturally extends the definition of communicability proposed in Refs [116, 117] for single layers (see Sec. 2.7.7). Let us consider the supra-matrix $\tilde{A}$ constructed from the supra-adjacency matrix by assigning a weight q to the interlinks, i.e.

$$
\tilde{A} =
\begin{pmatrix}
\begin{array}{c|c|c|c}
\mathbf{a}^{[1,1]} & 0 & \cdots & 0 \\
\hline
0 & \mathbf{a}^{[2,2]} & \cdots & 0 \\
\hline
\vdots & \vdots & \ddots & \vdots \\
\hline
0 & 0 & \cdots & \mathbf{a}^{[M,M]}
\end{array}
\end{pmatrix}
+ q
\begin{pmatrix}
\begin{array}{c|c|c|c}
0 & \mathbf{a}^{[1,2]} & \cdots & \mathbf{a}^{[1,M]} \\
\hline
\mathbf{a}^{[2,1]} & 0 & \cdots & \mathbf{a}^{[2,M]} \\
\hline
\vdots & \vdots & \ddots & \vdots \\
\hline
\mathbf{a}^{[M,1]} & \mathbf{a}^{[M,2]} & \cdots & 0
\end{array}
\end{pmatrix}. \tag{9.33}
$$

The weighted multiplicity of walks of length L between the representation of a node (i, α) and node (j, β) is given by different entries of the Lth power of the supra-adjacency matrix, $\tilde{A}^L$. In many instances, we are interested in assigning more importance to shorter walks than to longer ones. In this way, one defines the communicability matrix $\mathbf{G}$ as

$$
\mathcal{C} = \mathbf{I} + \tilde{A} + \frac{\tilde{A}^2}{2!} + \cdots = \sum_{L=0}^{\infty} \frac{\tilde{A}^L}{L!} = e^{\tilde{A}}. \tag{9.34}
$$

In a multilayer network in which each layer has the same number of nodes, i.e. $N_\alpha = N$ for every $\alpha = 1, 2, \ldots, M$, we can further express the communicability matrix as

$$
\mathcal{C} = e^{\tilde{A}} =
\begin{pmatrix}
\mathbf{c}^{[1,1]} & \mathbf{c}^{[1,2]} & \cdots & \mathbf{c}^{[1,M]} \\
\mathbf{c}^{[2,1]} & \mathbf{c}^{[2,2]} & \cdots & \mathbf{c}^{[2,M]} \\
\vdots & & \ddots & \vdots \\
\mathbf{c}^{[M,1]} & \mathbf{c}^{[M,2]} & \cdots & \mathbf{c}^{[M,M]}
\end{pmatrix}, \tag{9.35}
$$

where $\mathbf{c}^{[\alpha,\alpha]}$ is an $N \times N$ matrix containing the communicability between the representations of every pair of nodes within layer α of the multiplex and $\mathbf{c}^{[\alpha,\beta]}$ is the $N \times N$ matrix containing the communicability between pairs of nodes belonging to layer α and layer β respectively. It is important to note that $\mathbf{c}^{[\alpha,\alpha]} \neq \exp(\mathbf{a}^{[\alpha,\alpha]})$, where $\mathbf{a}^{[\alpha,\alpha]}$ is the (eventually weighted) adjacency matrix of each layer. Every node (i, α) of the network can therefore be ranked according to its total communicability with the other nodes of the same layer. Specifically, in directed networks we might be interested in ranking according to $x_{i\alpha}^{receive}$ which depends on the incoming paths or according to $x_{i\alpha}^{broadcast}$ which depends on the outgoing paths, i.e.

$$
x_{i\alpha}^{receive} = \sum_{j=1}^{N} \mathcal{C}_{j\alpha,i\alpha},
$$

$$
x_{i\alpha}^{broadcast} = \sum_{j=1}^{N} \mathcal{C}_{i\alpha,j\alpha}. \tag{9.36}
$$

Table 9.4 *Comparison between the communicability of the multiplex network formed by the six European airline companies (British Airways, Lufthansa, AirFrance, Ryanair, Easyjet, AirBerlin) and communicability for the aggregated network. Data from Ref. [115].*

Rank	$q = 0.0$	$q = 0.1$	$q = 1.0$	Aggregated
1	Paris CdG	London Stansted	London Stansted	Frankfurt
2	Barcelona	Madrid	Dublin	Munich
3	Venice	Barcelona	Madrid	London Stansted
4	Amsterdam	ParisCdG	Palma de Mallorca	London Gatwick

Finally, if the multilayer network is a multiplex, it is possible to attribute to each node i a centrality given by the inverse of the harmonic mean of the centralities of its replica nodes, i.e.

$$X_i^{type} = \left[\frac{1}{M} \sum_{\alpha=1}^{M} \frac{1}{x_{i\alpha}^{type}} \right]^{-1},$$ (9.37)

where 'type' indicates either the receive or the broadcast centrality measure. In Ref. [115] this centrality measure has been applied to a social multiplex network within a company and to a subset of the undirected European Air Transportation Multiplex Network [71]. The top five ranked airports of the multiplex network are indicated in Table 9.4 for different values of q and compared to the communicability of the aggregated network. Note that since this multiplex network is undirected there is no distinction between the received and the broadcast centrality.

9.8   Centrality of multi-slice networks

For multi-slice networks a new ranking algorithm has been proposed in Ref. [141] based on a modification of the Katz centrality of single networks.

The Katz centrality of a single network attributes to each node a centrality that is equal to the number of paths that reach the node starting from every node of the network, when the contribution of longer paths is discounted by a factor μ^n, where n is the length of the path and $\mu < 1$. Specifically, the Katz centrality of a node i in a layer with adjacency matrix $\mathbf{a}$ is given by

$$x_i = \sum_{j=1}^{N} (\mathbf{I} - \mu \mathbf{a})_{ji}^{-1} = \sum_{j=1}^{N} \sum_{n=0}^{\infty} \mu^n \mathbf{a}_{ji}^n.$$ (9.38)

In Ref. [141] it has been proposed to extend this definition to multi-slice temporal networks as in the following. Instead of considering the paths which can traverse a single layer, the Authors of Ref. [141] consider paths that can travel in multiple layers of the multi-slice networks. These paths can perform several steps in the same layer and can move from one layer to any subsequent layer, but they can never go back to an older layer. To this end, a series of $N \times N$ matrices $\hat{\mathcal{K}}^{[\ell]}$ with $\ell \in \{1, 2, \ldots, M\}$ is defined

$$\hat{\mathcal{K}}^{[\ell]} = \frac{\hat{\mathcal{K}}^{[\ell-1]}(\mathbf{I} - \mu \mathbf{a}^{[\ell]})^{-1}}{||\hat{\mathcal{K}}^{[\ell-1]}(\mathbf{I} - \mu \mathbf{a}^{[\ell]})^{-1}||} \tag{9.39}$$

where $\hat{\mathcal{K}}^{[0]} = \mathbf{I}$ and $|| \ldots ||$ denotes the Euclidean norm. Therefore, we have that $\hat{\mathcal{K}}^{[M]}$ is, up to a rescaling constant equal to the weighted sum of all the directed paths described above,

$$\hat{\mathcal{K}}^{[M]} \propto (\mathbf{I} - \mu \mathbf{a}^{[1]})^{-1} (\mathbf{I} - \mu \mathbf{a}^{[2]})^{-1} \ldots (\mathbf{I} - \mu \mathbf{a}^{[M]})^{-1}$$

$$= \sum_{n_1=0}^{\infty} \sum_{n_2=0}^{\infty} \ldots \sum_{n_M=0}^{\infty} \mu^{n_1+n_2+\ldots+n_M} (\mathbf{a}^{[1]})^{n_1} (\mathbf{a}^{[2]})^{n_2} \ldots (\mathbf{a}^{[M]})^{n_M}. \tag{9.40}$$

The broadcast centrality characterizes how well a node can broadcast a message, the receive centrality characterizes how central a node is with respect to receiving information. These centralities are given by

$$X_i^{receive} = \sum_{j=1}^{N} \hat{\mathcal{K}}_{ji}^{[M]},$$

$$X_i^{broadcast} = \sum_{j=1}^{N} \hat{\mathcal{K}}_{ij}^{[M]}. \tag{9.41}$$

In Ref. [141] these centrality measures have been applied to the Enron email datasets and the MIT telecommunication dataset showing that the multi-slice network centrality provides relevant information that cannot be extracted by considering the layer in isolation or by simply evaluating the degrees of the nodes.

10

Multilayer Network Models

10.1 Different approaches to multilayer network modelling

Multilayer network modelling allows us to artificially construct multilayer networks with given structural properties or obeying simple dynamical rules. The ability to construct multilayer networks from scratch or to randomize a given multilayer network is central in network theory. On the one side, it is important to test which dynamical rules are responsible for given multilayer network structures, on the other, comparing a given real multilayer network structure with null models of multilayer networks is essential to investigate which structural aspects are over- (or under-) represented in real datasets.

As in single networks, for multilayer networks there are two main classes of models. The non-equilibrium models usually imply network growth and include simple dynamical rules that are responsible for emergent structural properties. The equilibrium models instead are the least biased models, satisfying a set of constraints, and constitute the topologies of reference to which real multilayer networks are usually compared. Multi-slice networks are modelled using different approaches and can be classified depending on whether the models do or do not reproduce the burstiness of real datasets.

10.2 Growing multiplex network models

10.2.1 Growing multiplex networks

Growing multiplex network models are inspired by the widely used growing network models of single networks which most notably include the Barabási–Albert (BA) model [15]. The non-equilibrium framework of growing network models allows us to explore the role of simple dynamical rules in promoting the emergence of complexity in the large network limit. For instance, the BA model shows that preferential attachment plays a major role in the generation of scale-free networks. Similarly, here growing multiplex network models will determine some basic rules responsible for the emergence of positive and negative degree correlations, community structures and degree distributions in multiplex networks.

Multilayer Networks. Ginestra Bianconi, Oxford University Press (2018).
© Ginestra Bianconi. DOI: 10.1093/oso/9780198753919.001.0001

10.2.2   Generalized linear preferential attachment model

As in single networks, in multiplex networks scale-free layers are also ubiquitous. For instance multiplex networks with scale-free layers include, among others, collaboration multiplex networks between scientists or between movie actors, social multiplex networks formed by several online platforms and multiplex molecular networks.

It is therefore natural to discuss growing multiplex network models with generalized preferential attachment in which new links are attached preferentially to nodes that have high degrees. Here we will see that this mechanism not only produces scale-free layers but also generates positive degree correlations among the layers of the multiplex. For simplicity, we will consider the growth of a duplex network (multiplex network with $M = 2$ layers). We will assume that when new links are attached to a given layer, the target nodes are chosen with a probability that is larger, the larger the linear combination of their degrees in layers 1 and 2.

The model is simply defined. Starting at time $t = 1$ from a duplex network with $n_0 \geq m$ nodes (with a replica node in each of the two layers) connected by m_0 links in each layer, the model proceeds as follows:

- *Growth:* At each time $t > 1$ a node represented by a a replica node in each of the two layers is added to the multiplex. Each newly added replica node is connected to the other nodes of the same layer by m links.

- *Generalized preferential attachment:* Each new link in layers $\alpha \in \{1, 2\}$ is attached to node i with probability $\Pi_i^{[\alpha]}$ proportional to a linear combination of the degree $k_i^{[1]}$ of node i in layer 1 and the degree $k_i^{[2]}$ of node i in layer 2, i.e.,

$$\Pi_i^{[1]} \propto c_{1,1} k_i^{[1]} + c_{1,2} k_i^{[2]},$$
$$\Pi_i^{[2]} \propto c_{2,1} k_i^{[1]} + c_{2,2} k_i^{[2]}, \tag{10.1}$$

where $c_{\alpha,\beta} \in [0, 1]$ with $c_{1,1} + c_{1,2} = c_{2,1} + c_{2,2} = 1$.

In the case in which $c_{1,1} = c_{2,2} = 1$, the model reduces to two apparently decoupled BA models. Nevertheless, in the multiplex network two replica nodes have the same age, therefore this characteristic of the model is responsible for generating degree–degree correlations in the layers of the duplex. In fact, it is well known that in the BA model the most connected nodes are also the first nodes arrived in the network. Therefore, in the multiplex network model with $c_{1,1} = c_{2,2} = 1$, in which the two layers evolve as single BA networks, an old node that arrived early in the multiplex network will acquire large degree in both networks, whereas a young node will have small degree in both networks yielding a multiplex network with positive degree correlations. These results can easily be obtained in the mean-field approximation and extend also to the case $(c_{1,1}, c_{2,2}) \neq (1, 1)$.

Let us derive the degree correlations in the framework of the mean-field approximation. The degrees $\mathbf{k}_i = \left(k_i^{[1]}, k_i^{[2]} \right)^T$ of each node i evolve according to the mean-field equation

$$\frac{d\mathbf{k}_i}{dt} = m\mathbf{\Pi}_i \tag{10.2}$$

where $\mathbf{\Pi}_i = \left(\Pi_i^{[1]}, \Pi_i^{[2]}\right)^T$ with $\Pi_i^{[\alpha]}$ given by Eq. (10.1). In the large multiplex network limit, since $\sum_i k_i^{[\alpha]} \simeq 2mt$ for both $\alpha = 1, 2$, we have

$$\frac{d\mathbf{k}_i}{dt} = \frac{1}{2t}\mathbf{C}\mathbf{k}_i, \tag{10.3}$$

with the matrix $\mathbf{C}$ given by

$$\mathbf{C} = \begin{pmatrix} c_{1,1} & 1 - c_{1,1} \\ c_{2,1} & 1 - c_{2,1} \end{pmatrix}. \tag{10.4}$$

If $(c_{1,1}, c_{2,2}) \neq (1, 1)$, the matrix $\mathbf{C}$ has eigenvalues $\lambda_1 = 1$ and $\lambda_2 = (c_{1,1} - c_{2,1})$ with corresponding eigenvectors $\mathbf{u}_1 = (1, 1)$ and $\mathbf{u}_2 = (1 - c_{1,1}, -c_{2,1})$. Therefore, we can decompose the vector of degrees $\mathbf{k}_i(t) = D_1(t)\mathbf{u}_1 + D_2(t)\mathbf{u}_2$ and we can solve Eqs (10.3) obtaining

$$\mathbf{k}_i(t) = D_1(t_i)\left(\frac{t}{t_i}\right)^{\lambda_1/2}\mathbf{u}_1 + D_2(t_i)\left(\frac{t}{t_i}\right)^{\lambda_2/2}\mathbf{u}_2, \tag{10.5}$$

where t_i indicates the time when node i is arrived in the network. By imposing the initial condition $\mathbf{k}_i(t_i) = m\mathbf{u}_1$, implying $D_1(t_i) = m$ and $D_2(t_i) = m$, we obtain

$$\mathbf{k}_i(t) = m\left(\frac{t}{t_i}\right)^{1/2}\mathbf{u}_1 \tag{10.6}$$

and therefore in the mean-field approximation we have

$$k_i^{[1]} = k_i^{[2]} = m\left(\frac{t}{t_i}\right)^{1/2}. \tag{10.7}$$

In the case where $(c_{1,1}, c_{2,2}) = (1, 1)$ the system of equations decouples and it is easily derived that the solution does not change and Eq. (10.7) is recovered again. Therefore, we find that in the framework of the mean-field approximation the multiplex network displays positive degree correlations (see Fig. 10.1) as the conditioned average degree $\langle k^{[1]}|k^{[2]}\rangle$ defined in Sec. 7.2 is given by

$$\left\langle k^{[1]}|k^{[2]}\right\rangle = k_i^{[2]}. \tag{10.8}$$

Moreover, given this mean-field solution it is possible to show that both layers have scale-free degree distribution with exponent $\gamma = 3$ (see Fig. 10.1).

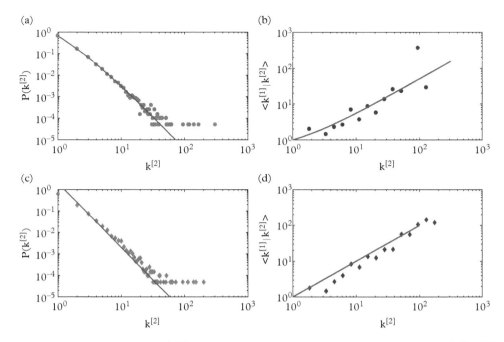

Fig. 10.1 *Degree distribution* $P\left(k^{[2]}\right)$ *in layer 2, and the interlayer degree–degree correlations* $\left(k^{[1]}|k^{[2]}\right)$ *for the growing multiplex network model of Ref. [228] in the cases* $c_{1,1} = c_{2,2} = 1$ *(panels (a) and (b)) and* $c_{1,1} = c_{2,2} = 0.5$ *(panels (c) and (d)). The solid lines indicate the theoretical results obtained with the master-equation approach (panels (a) and (b)) and with the mean-field approach (panels (c) and (d)).*

Interestingly, this model admits an explicit exact analytical solution for $c_{1,1} = c_{2,2} = 1$ obtained using the master equation approach [228]. The joined degree distribution $P(k,q)$ of having a node with degree $k^{[1]} = k$ in layer 1 and degree $k^{[2]} = q$ in layer 2, is given by (see Appendix C for the derivation)

$$P(k,q) = \frac{2\Gamma(2+2m)\Gamma(k)\Gamma(q)\Gamma(k+q-2m+1)}{\Gamma(m)\Gamma(m)\Gamma(k+q+3)\Gamma(k-m+1)\Gamma(q-m+1)}. \qquad (10.9)$$

The conditioned average degree $\left(k^{[1]}|k^{[2]}\right)$ can be calculated exactly from the joint degree distribution given by Eq. (10.9) as

$$\left(k^{[1]}|k^{[2]}\right) = \sum_{k^{[1]}} k^{[1]} P(k^{[1]}|k^{[2]} = q) = \frac{m(k^{[2]} + 2)}{1 + m}. \qquad (10.10)$$

This result agrees with simulation results (see Fig. 10.1) and is consistent with the result obtained in the mean-field approximation (Eq. (10.8)) confirming that the model displays positive degree correlations.

In this case, the degree distribution (see Fig. 10.1) in each layer $P(k)$ is given by the degree distribution of the BA network model, i.e.

$$P(k) = \frac{2m(m+1)}{k(k+1)(k+2)}. \tag{10.11}$$

In Ref. [173] a calculation of the Pearson correlation coefficient r (defined in Sec. 7.2) has been carried out for the parameter values $c_{1,1} = c_{2,2} = 1 - \epsilon$, finding in the limit $t \to \infty$

$$r = \frac{\langle(k - \langle k\rangle)(q - \langle q\rangle)\rangle}{\sigma_k \sigma_q} = \begin{cases} \frac{1}{2} & \text{for } \epsilon = 0 \\ 1 & \text{for } \epsilon > 0, \end{cases} \tag{10.12}$$

where $k = k^{[1]}$ indicates the degree in layer 1 and $q = k^{[2]}$ indicates the degree in layer 2. Therefore, for $\epsilon > 0$ the correlations are more significant than in the case $\epsilon = 0$.

For a treatment of a growing multiplex network model with more than two layers, see Ref. [214].

10.2.3 Generalized non-linear preferential attachment

Growing multiplex network models can also generate negative degree correlations between the layers. In particular, this is the case in the non-linear attachment kernel used in Ref. [229]. The model includes growth and non-linear preferential attachment depending on the degree of the same node in the different layers. In a single layer, non-linear preferential attachment has been shown [178] to display a rich phenomenology, including a gelation phase transition in which the oldest node acquires a finite fraction of all the links (see discussion in Sec. 2.8.4).

The growing duplex network model with non-linear preferential attachment proposed in Ref. [229] is described by the following algorithm.

Starting at time $t = 1$ from a duplex network with $n_0 \geq m$ nodes (with a replica node in each of the two layers) connected by m_0 links in each layer, the model proceeds as follows:

- *Growth:* At each time $t > 1$, a node with a replica node in each of the two layers is added to the multiplex. Each newly added replica node is connected to the other nodes of the same layer by m links.
- *Generalized non-linear preferential attachment:* The new links are attached to node i with probability $\Pi_i^{[1]}$ in layer 1 and with probability $\Pi_i^{[2]}$ in layer 2 with

$$\Pi_i^{[1]} \propto \left(k_i^{[1]}\right)^\eta \left(k_i^{[2]}\right)^\theta,$$
$$\Pi_i^{[2]} \propto \left(k_i^{[2]}\right)^\eta \left(k_i^{[1]}\right)^\theta. \tag{10.13}$$

The model depends on the values of the two parameters η and θ. When η, θ are both positive, nodes with high degrees in both layers are more likely to acquire new links in both layers. However, when $\eta > 0$ but $\theta < 0$ the nodes that have high degree in a given layer but low degree in the other are more likely to acquire new links in the layer in which they are more connected. Therefore, for $\eta > 0$ we expect to observe positive correlations for $\theta > 0$ and negative correlations for $\theta < 0$.

This model displays a very interesting phenomenology, generating multiplex networks with different types of degree distributions and degree–degree correlations, and with a condensation of the links. The condensation of the links occurs when the oldest node acquires a finite fraction of all the links of the network. The model can be predicted theoretically [229] to display the condensation of the links for

$$\eta > \min(1, 1 - \theta). \tag{10.14}$$

These predictions perfectly match the simulation results and reduce to the condition $\eta > 1$ valid in single layers for $\theta = 0$. The degree distribution is predicted to be scale-free only on the transition line $\eta = \min(1, 1 - \theta)$. For $\eta < \min(1, 1 - \theta)$ the distribution is homogeneous while for $\eta > \min(1, 1 - \theta)$ it is dominated by the condensation phenomenon with all the nodes different from the oldest one acquiring only a small degree.

In order to investigate the interlayer degree correlations generated naturally by this model, let us consider the Kendall correlation coefficient $\tau(t)$ between the degree of corresponding replica nodes, calculated only over the nodes arrived at time $t' < t$ (see Fig. 10.2). For $\eta > 0, \theta > 0$ all the nodes have positively correlated degrees in the two layers. In the case $\eta > 0, \theta < 0$ only the older nodes have negatively correlated degrees in the two different layers, while the recently added nodes in the duplex network have positively correlated degrees. This is indicated by the dependence of $\tau(t)$ as a function of t which is increasing, eventually acquiring even a positive value (see Fig. 10.2). This phenomenon is due to the fact that the attachment kernel is symmetric in the two layers, and initially every node starts by having the same degree in both layers. Only with time, for $\eta > 0, \theta < 0$ each node will end up having much larger degree in one layer than in the other. Therefore, the degrees of the younger nodes will follow a stochastic dynamics where layers compete to become the most connected layer before one of the layers becomes clearly the dominating one. Note that on average half of the nodes will end up having the first as the dominating layer and half of the nodes will have the other, respecting the overall symmetry of the growing dynamics.

10.2.4   Growing multiplex networks with communities

Triadic closure is an important mechanism driving the evolution of growing multiplex networks.

Consider, for example, collaboration networks of scientists working on different topics. In this case, as in many social networks, it is possible to argue that links are not random, and that instead scientists usually exploit the neighbourhood of their collaborators in a specific field to establish new collaborations. Therefore, it is often the case that the new collaborator of a given scientist already active in a scientific topic

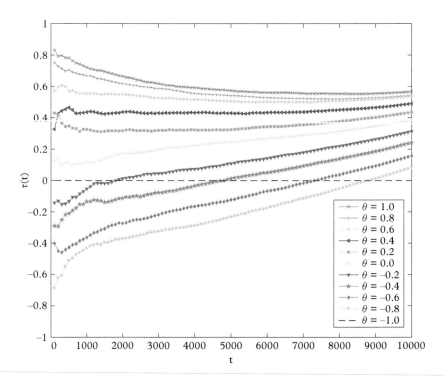

Fig. 10.2 *The Kendall coefficient $\tau(t)$, measuring the interlayer degree correlations among the nodes arrived up to time* t *is here shown for the growing duplex network model with non-linear preferential attachment proposed in Ref. [229] with $\eta = 0$. The dashed black line corresponds to $\tau = 0$ and is reported for visual reference. For $\theta > 0$ the interlayer degree correlations are always positive. For $\theta < 0$ the interlayer degree correlations for older nodes are disassortative (yielding $\tau(t) < 0$ for small* t*), while for younger nodes they are positive (yielding $\tau(t) > 0$ for large* t*).*

is already a second neighbour in the collaboration network corresponding to the given topic (*intralayer* triadic closure). Similarly, when opening himself to new scientific fields, a researcher usually takes into account the neighbourhoods of his past colleagues from previous collaborations in other fields (*interlayer* triadic closure).

Interestingly, in Ref. [29] it is found that growing multiplex network models enforcing triadic closure are able to generate multiplex networks with tunable community structure, reproducing the patterns observed in real-world collaboration networks.

Therefore, triadic closure can be considered a basic microscopic mechanism responsible for the mesoscale structure of multiplex networks.

In Ref. [29] a stylized multiplex network with $M = 2$ layers, including both *interlayer and intralayer triadic closure* has been proposed.

Initially (at time $t = 1$) each layer is formed by a clique of $n_0 \geq m$ nodes. At each time step $t > 1$ a new node is added to the multiplex network. Each of its two replica

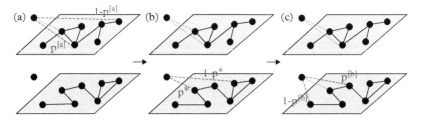

Fig. 10.3 *Schematic representation of network growth with intralayer and interlayer triadic closure. Reprinted from Ref. [29] ©2016 Battiston et al.*

nodes is connected to the other replica nodes of its own layer by m links according to the following algorithm (see Fig. 10.3):

- *Selection of the first layer.*
  One of the two layers $\alpha \in \{1, 2\}$ is chosen uniformly at random.

  (a) *Random initial attachment.*
  The new node i connects its replica node (i, α) to a random replica node of the network called node (j, α).

  (b) *Intralayer triadic closure.*
  Each of the remaining $m - 1$ edges of the replica node (i, α) is attached with probability $p^{[\alpha]}$ to a random neighbour (in layer α) of replica node (j, α) and with probability $1 - p^{[\alpha]}$ to a random replica node of layer α.

- *Selection of the second layer.*
  The links in layer $\beta \neq \alpha$ are placed according to the following algorithm.

  (a') *Interlayer triadic closure.*
  The replica node (i, β) connects to the replica node (j, β) with probability $p^\star$ and with probability $1 - p^*$ to one of the other replica nodes of layer β, chosen uniformly at random. The node to which this first link is attached is called (j', β).

  (b') *Intralayer triadic closure.*
  The remaining $m - 1$ links at layer β are attached with probability $p^{[\beta]}$ to one of the first neighbours of (j', β) chosen uniformly at random, and with probability $1 - p^{[\beta]}$ to a random replica node in layer β.

This general model has four tunable parameters, namely the number of new edges m which determines the average degree on each layer and the three probabilities $p^{[1]}, p^{[2]}$ and p^*, which are responsible for the formation of intra- and interlayer triangles. In fact, by varying the parameter $p^{[\alpha]}$ it is possible to tune the strength of the intralayer triadic

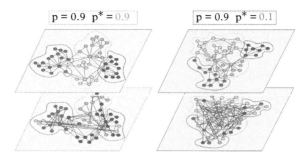

Fig. 10.4 *In the framework of the model proposed in Ref. [29] enforcing triadic closure, the effect of the value of the interlayer triadic closure parameter p^* on the multiplex community structure is displayed. The two top layers show two typical realizations of the simplest version of the network model with $N = 50$, $m = 2$ and $p^{[1]} = p^{[2]} = p = 0.9$. Nodes belonging to the same community are given the same colour and are drawn close to each other. The two layers at the bottom of each multiplex are obtained by setting $p^* = 0.9$ (left) and $p^* = 0.1$ (right) respectively. The nodes maintain the same placement in space on the second layer, but are coloured according to the community they belong to in that layer (colours are chosen in order to maximize the number of nodes that have the same colour in the two layers). It is evident that the community structures of the two layers on the left, corresponding to $p^* = 0.9$, are very similar, while the partition into communities of the upper layer on the left panel is substantially different from the one observed in the bottom layer of that multiplex. Reprinted figure from Ref. [29] ©2016 Battiston et al.*

closure mechanism in each layer α, i.e the probability of forming triangles on the given layer. In particular, larger values of $p^{[\alpha]}$ will foster the creation of a larger number of triangles within layer α. Conversely, the parameter p^* tunes the interlayer triadic closure mechanism, and in particular high values of p^* correspond to a higher probability that the neighbourhoods of node i at the two layers will exhibit a certain level of overlap. These two simple attachment rules, namely intralayer and interlayer triadic closure, aim to describe the real mechanisms characterizing the evolution of collaboration networks.

For large values of $p^{[1]}, p^{[2]}$ this model favours the establishment of communities within each layer. For large values of the parameter p^* the communities across the two layers are correlated, displaying a significant overlap, while for small values of p^* they are not correlated (see Fig. 10.4).

10.2.5   Other attachment kernels

Other attachment kernels have been considered in different papers. For instance, in Ref. [173] an initial attractiveness has been added to the generalized preferential attachment, while in Ref. [42] a model with generalized preferential attachment and an internal ability (fitness) of the nodes to acquire new links has been considered. Both models are able to reproduce scale-free multiplex networks with tunable power-law exponent γ. By modifying the attachment probability, multiplex networks with exponential layers or multiplex networks including both scale-free and exponential layers have been generated [228]. Finally, in Refs [84, 230] multiplex networks growing by the subsequent attachment of entire new layers have been considered.

10.3   Multiplex network ensembles

10.3.1   Ensembles as null models of networks

While growing non-equilibrium models aim at proposing a basic mechanism for explaining the emergence of complex features (such as scale-free degree distributions, degree correlations, communities), ensembles of multiplex networks provide the ideal scenario for constructing null models. Null models are models of multiplex networks that preserve some structural property but are otherwise totally random. Specifically, they are the least biased models of multiplex networks satisfying a set of constraints. As such, ensembles of multiplex networks are widely used in network analysis because they provide a reference model to compare real multiplex networks to. Additionally, they are used to probe the interplay between structure and dynamics when dynamical processes are defined on top of them.

Ensembles of multiplex networks can reproduce a wide class of properties of real multiplex networks. They include null models of independent layers with given degree sequence, multiplex networks with given activity of the nodes, multiplex networks with given multidegree sequence and controlled level of link overlap and spatial random multiplex networks. Therefore, this framework is very flexible and can be very useful in a wide variety of cases.

10.3.2   The theoretical framework

Consider a multiplex formed by N labelled nodes $i = 1, 2, \ldots, N$ and M layers where we indicate by $\vec{G} = (G_1, G_2, \ldots, G_M)$ the set of all the networks G_α at layer $\alpha = 1, 2, \ldots, M$ forming the multiplex. A multiplex network ensemble is specified when the probability $P(\vec{G})$ for each possible multiplex network is given. Specifically, in a multiplex network ensemble we can neglect to treat the interlinks. In fact, even when the interlinks are explicitly taken into account, since they are placed deterministically among each pair of replica nodes the probability of a multiplex network only depends on the random topology of its layers $\vec{G}$.

In a multiplex network ensemble, the entropy S is defined as

$$S = -\sum_{\vec{G}} P(\vec{G}) \log P(\vec{G}). \tag{10.15}$$

The entropy measures the logarithm of the typical number of multiplex networks in the ensemble.

The least biased method of constructing multiplex networks with given structural properties is to maximize the entropy of the ensemble, given the set of structural constraints.

One can distinguish between microcanonical and canonical multiplex network ensembles. The microcanonical ensembles enforce a given set of hard constraints, i.e. every multiplex network of the ensemble has a given structural property such as total number of links in each layer, given degree sequences within each layer, given multidegree sequences. On the contrary, the canonical ensembles (also called exponential random graphs) enforce a given set of soft constraints. This means, for example, that instead of fixing the total number of links in each layer the number of links in each layer is allowed to fluctuate as long as their average in the ensemble is fixed. Similarly, soft constraints might constrain the expected degree sequence of each layer or the expected multidegree sequences. The statistical mechanics of randomized network ensembles (*exponential random graphs*) has been extended to describe multiplex ensembles [41, 209, 304, 148, 120, 308, 245], and applied to analysing different multiplex data sets [209, 185, 154]. An alternative class of equilibrium network models with coloured nodes were proposed in 2003 by Söderberg [271–3].

In the following paragraphs we discuss multiplex network ensembles adopting the main lines of the discussion presented in Ref. [41].

10.3.3   Ensembles with independent layers

Independent layers and correlated degrees

Multiplex network ensembles can be distinguished between the ones with independent layers and the ones in which the layers are not independent. When the layers are independent, the probability $P(\vec{G})$ of the multiplex network is factorized in the probabilities $P_\alpha(G_\alpha)$ of each single layer, i.e.

$$P(\vec{G}) = \prod_{\alpha=1}^{M} P_\alpha(G_\alpha). \qquad (10.16)$$

Therefore, each layer can be drawn independently from the others. This is the simplest way to generate multiplex networks. Specifically, one can consider the multiplex in which each layer is drawn using a configuration model, i.e. with a given degree sequence, or, instead, in which each layer is drawn using an exponential random graph with given expected degree sequence. Note that in this scenario multiplex networks with any pattern of degree correlations can be constructed since this approach is applicable to any set of degree sequences (or expected degree sequences) specified for any layer.

Canonical multiplex network ensemble

Let us consider the canonical multiplex network ensemble where the expected degrees $k_i^{[\alpha]}$ of each node i in every layer α are fixed. Indicating by $P_C(\vec{G})$ the probability of the multiplex networks in the ensemble, we require that the following constraints are satisfied:

$$\sum_{\vec{G}} P_C(\vec{G}) \left[\sum_{j=1}^{N} a_{ij}^{[\alpha]} \right] = k_i^{[\alpha]}, \tag{10.17}$$

for $i = 1, 2, \ldots, N$ and $\alpha = 1, 2 \ldots, M$. The least biased ensemble that satisfies the constraints is obtained by maximizing the entropy of the ensemble, obtaining for $P_C(\vec{G})$ the log-linear expression

$$P_C(\vec{G}) = \frac{1}{Z_C} \exp \left[-\sum_{i=1}^{N} \sum_{\alpha=1}^{M} \lambda_i^{[\alpha]} \sum_{j=1}^{N} a_{ij}^{[\alpha]} \right], \tag{10.18}$$

where $\lambda_i^{[\alpha]}$ are the Lagrangian multipliers enforcing the constraints in Eq. (10.17) and Z_C is the normalization constant. The probability $P_C(\vec{G})$ can equivalently be written as

$$P_C(\vec{G}) = \prod_{i<j} \prod_{\alpha=1}^{M} \left[p_{ij}^{[\alpha]} a_{ij}^{[\alpha]} + (1 - p_{ij}^{[\alpha]})(1 - a_{ij}^{[\alpha]}) \right], \tag{10.19}$$

where $p_{ij}^{[\alpha]}$ indicates the marginal probability that node i and node j are connected in layer α. The entropy S determining the typical number of multiplex networks in this ensemble can also be expressed in terms of the marginal distributions as

$$S = -\sum_{i<j} \sum_{\alpha=1}^{M} \left[p_{ij}^{[\alpha]} \ln p_{ij}^{[\alpha]} + (1 - p_{ij}^{[\alpha]}) \ln(1 - p_{ij}^{[\alpha]}) \right]. \tag{10.20}$$

The value of the marginal probabilities $p_{ij}^{[\alpha]}$ is given in terms of the Lagrangian multipliers $\lambda_i^{[\alpha]}$ as

$$p_{ij}^{[\alpha]} = \frac{e^{-\lambda_i^{[\alpha]} - \lambda_j^{[\alpha]}}}{1 + e^{-\lambda_i^{[\alpha]} - \lambda_j^{[\alpha]}}}. \tag{10.21}$$

Here the Lagrangian multipliers are fixed by the constraint that each node i has expected degree $k_i^{[\alpha]}$ in layer α, which reads

$$k_i^{[\alpha]} = \sum_{j=1}^{N} p_{ij}^{[\alpha]}. \tag{10.22}$$

In the presence of a structural cutoff for each layer α, when the expected degrees of the nodes satisfy

$$k_i^{[\alpha]} \ll \sqrt{\langle k^{[\alpha]} \rangle N} \tag{10.23}$$

the marginal probabilities $p_{ij}^{[\alpha]}$ take the uncorrelated form

$$p_{ij}^{[\alpha]} = \frac{k_i^{[\alpha]} k_j^{[\alpha]}}{\langle k^{[\alpha]} \rangle N}. \tag{10.24}$$

Microcanonical ensemble

In the microcanonical model, in which we fix the degree $k_i^{[\alpha]}$ of each node i in every layer α (also called the configuration model of multiplex networks), the probability of the multiplex network is given by

$$P_M(\vec{G}) = \sum_{\vec{G}} \prod_{i=1}^{N} \prod_{\alpha=1}^{M} \delta\left(k_i^{[\alpha]}, \sum_{j=1}^{N} a_{ij}^{[\alpha]}\right). \tag{10.25}$$

The marginal $p_{ij}^{[\alpha]}$ expressing the probability that node i and node j are connected in layer α takes the same expression as for the canonical network model (Eq. (10.21)). In particular, if the layers of the multiplex network have the structural cutoff (i.e. they satisfy Eq. (10.23)), the probability of a link between node i and node j in layer α follows Eq. (10.24). However, the microcanonical and canonical ensembles are not equivalent, as emerges clearly from the calculation of the entropy Σ of the microcanonical network ensemble, which in the thermodynamic limit is not equal to the entropy S of the canonical ensemble given by Eq. (10.20). In fact, Σ and S are related by the equation [44,7,39]

$$\Sigma = S - \Omega \tag{10.26}$$

where

$$\Omega = -\ln\left[\sum_{\vec{G}} P_C(\vec{G}) \prod_{i=1}^{N} \prod_{\alpha=1}^{M} \delta\left(k_i^{[\alpha]}, \sum_{j=1}^{N} a_{ij}^{[\alpha]}\right)\right], \tag{10.27}$$

where $P_C(\vec{G})$ is defined in Eqs (10.18) and (10.19). The quantity Ω is extensive in the number of nodes N of the multiplex network and is a non-negligible contribution. In the presence of the structural cutoffs when the degrees $k_i^{[\alpha]}$ of the nodes in layer α satisfy Eq. (10.23), Ω has an explicit analytical expression given by

$$\Omega = -\sum_{i=1}^{N} \sum_{\alpha=1}^{M} \ln\left[\frac{1}{k_i^{[\alpha]}!} \left(k_i^{[\alpha]}\right)^{k_i^{[\alpha]}} e^{-k_i^{[\alpha]}}\right]. \tag{10.28}$$

Most notably, this configuration model for multiplex networks can be used to model networks $P(\mathbf{k})$. In these multiplex networks the degree of the nodes across different layers can be either uncorrelated or correlated. When the degrees across different layers are uncorrelated we have

$$P(\mathbf{k}) = \prod_{\alpha=1}^{M} P^{[\alpha]}\left(k^{[\alpha]}\right) \tag{10.29}$$

where $P^{[\alpha]}\left(k^{[\alpha]}\right)$ is the degree distribution in layer α. In this case, we have for each pair of distinct layers α and β with $\alpha \neq \beta$

$$\left\langle k^{[\alpha]} k^{[\beta]} \right\rangle = \left\langle k^{[\alpha]} \right\rangle \left\langle k^{[\beta]} \right\rangle, \tag{10.30}$$

where the average indicates the average over all the nodes of the network, i.e.

$$\left\langle k^{[\alpha]} k^{[\beta]} \right\rangle = \frac{1}{N} \sum_{i=1}^{N} k_i^{[\alpha]} k_i^{[\beta]}. \tag{10.31}$$

Instead if the degrees across the different layers are correlated we have

$$P(\mathbf{k}) \neq \prod_{\alpha=1}^{M} P^{[\alpha]}\left(k^{[\alpha]}\right) \tag{10.32}$$

and in this case

$$\left\langle k^{[\alpha]} k^{[\beta]} \right\rangle \neq \left\langle k^{[\alpha]} \right\rangle \left\langle k^{[\beta]} \right\rangle. \tag{10.33}$$

These latter multiplex networks include multiplex networks with Maximally Positively correlated and Maximally Negatively correlated degrees represented in Fig. 7.1.

Construction of multiplex ensembles with independent layers

Multiplex network ensembles with independent layers can be created simply by generating each layer independently. Therefore, when we want to preserve the expected degree of each node in each layer we can draw the network in each separate layer independently from the canonical network ensemble (exponential random graph). However if we aim at preserving exactly the degree of each node in each layer every network of each layer can be independently constructed using the microcanonical network ensemble (configuration model).

10.3.4   Independent layers do not display significant overlap

In multiplex networks with independent layers we can evaluate, following Ref. [41], the average global overlap $\left\langle O^{[\alpha,\alpha']} \right\rangle$ between the layers α and α' and the average local overlap

$\left\langle o_i^{[\alpha,\alpha']}\right\rangle$ between two layers α and α' where the global overlap $O^{[\alpha,\alpha']}$ and the local overlap $o_i^{[\alpha,\alpha']}$ are defined in Eq. (7.13) and Eq. (7.14) respectively. These quantities are given by

$$\left\langle O^{[\alpha,\alpha']}\right\rangle = \sum_{i<j} p_{ij}^{[\alpha]} p_{ij}^{[\alpha']},$$

$$\left\langle o_i^{[\alpha,\alpha']}\right\rangle = \sum_{j|j\neq i} p_{ij}^{[\alpha]} p_{ij}^{[\alpha']}. \tag{10.34}$$

For sparse multiplex ensembles with given expected degree of the nodes in each layer, when $p_{ij}^{[\alpha]}$ can be approximated by Eq. (10.24), these equations can be expressed as

$$\left\langle O^{[\alpha,\alpha']}\right\rangle = \frac{1}{2}\left(\frac{\left\langle k^{[\alpha]}k^{[\alpha']}\right\rangle^2}{\left\langle k^{[\alpha]}\right\rangle\left\langle k^{[\alpha']}\right\rangle}\right),$$

$$\left\langle o_i^{[\alpha,\alpha']}\right\rangle = k_i^{[\alpha]}k_i^{[\alpha']}\frac{\left\langle k^{[\alpha]}k^{[\alpha']}\right\rangle}{\left\langle k^{[\alpha]}\right\rangle\left\langle k^{[\alpha']}\right\rangle N}. \tag{10.35}$$

If the degrees in the different layers are uncorrelated, i.e. Eq. (10.31) holds, the global and local overlaps are given by

$$\left\langle O^{[\alpha,\alpha']}\right\rangle = \frac{1}{2}\left(\left\langle k^{[\alpha]}\right\rangle\left\langle k^{[\alpha']}\right\rangle\right) \ll N,$$

$$\left\langle o_i^{[\alpha,\alpha']}\right\rangle = \frac{k_i^{[\alpha]}k_i^{[\alpha']}}{N} \ll \min\left(k_i^{[\alpha]}, k_i^{[\alpha']}\right). \tag{10.36}$$

Therefore, the overlap is negligible. Degree correlation between different layers can enhance the overlap, but as long as $\left\langle k^{[\alpha]}k^{[\alpha']}\right\rangle \ll N$ the average global $\left\langle O^{[\alpha,\alpha']}\right\rangle$ and the local $\left\langle o_i^{[\alpha,\alpha']}\right\rangle$ overlap continue to remain negligible with respect to the total number of nodes in the two layers and the degrees of the node i in the two layers. In sparse multiplex network ensembles without an embedding space, if we want to obtain a significant link overlap we need to consider multiplex ensembles with dependent layers. Examples of ensembles in this category will be considered in the following paragraphs.

10.3.5   Multiplex networks with given multidegree sequence or expected multidegree sequence

Ensemble with dependent layers

One way to construct multiplex networks with the desired amount of link overlap is to consider multiplex network ensembles which enforce a given multidegree sequence or, alternatively, an expected multidegree sequence. The multidegree $k_i^{\vec{m}}$, defined in Sec. 7.3, determines the number of multilinks $\vec{m}$ connected to node i. Every different multilink determines in which layers the two connected nodes are linked. Therefore, by fixing the multidegree sequence it is possible to generate multiplex networks with the desired level of link overlap. In these cases the layers are not any more independent, and as a consequence of this the probability $P(\vec{G})$ of a multiplex network cannot be expressed as a product of the probabilities $P_\alpha(G_\alpha)$ of the single layers taken in isolation

$$P(\vec{G}) \neq \prod_{\alpha=1}^{M} P_\alpha(G_\alpha). \tag{10.37}$$

Canonical ensemble

Let us first consider the canonical network ensemble when we fix the expected non-trivial multidegrees $k_i^{\vec{m}}$ of each node. Therefore, by indicating with $P_C(\vec{G})$ the probability of a multiplex network of the ensemble we impose the constraints

$$\sum_{\vec{G}} P_C(\vec{G}) \left[\sum_{j=1}^{N} A_{ij}^{\vec{m}} \right] = k_i^{\vec{m}} \tag{10.38}$$

for all $i = 1, 2 \ldots, N$ and $\vec{m} \neq \vec{0}$. Here $A_{ij}^{\vec{m}}$ indicates the elements of the multiadjacency matrices defined in Sec. 7.3. Again, the probability $P_C(\vec{G})$ of the least biased ensemble satisfying these constraints can be obtained by maximizing the entropy getting

$$P_C(\vec{G}) = \frac{1}{Z_C} \exp \left[-\sum_{i=1}^{N} \sum_{\vec{m} \neq \vec{0}} \lambda_i^{\vec{m}} \sum_{j=1}^{N} A_{ij}^{\vec{m}} \right], \tag{10.39}$$

where $\lambda_i^{\vec{m}}$ are the Lagrangian multipliers enforcing the constraints in Eq. (10.38) and Z_C is the normalization constant. An alternative expression of the probability $P_C(\vec{G})$ is

$$P_C(\vec{G}) = \prod_{i<j} \left[\sum_{\vec{m}} p_{ij}^{\vec{m}} A_{ij}^{\vec{m}} \right], \tag{10.40}$$

where $p_{ij}^{\vec{m}}$ indicates the marginal probability that a pair of nodes (i,j) is connected by a multilink $\vec{m}$. The entropy of these canonical multiplex network ensembles can also be expressed in terms of the marginal probabilities $p_{ij}^{\vec{m}}$ as

$$S = -\sum_{i<j} \sum_{\vec{m}} p_{ij}^{\vec{m}} \ln p_{ij}^{\vec{m}}. \tag{10.41}$$

The marginals $p_{ij}^{\vec{m}}$ are given in terms of the Lagrangian multipliers by

$$p_{ij}^{\vec{m}} = \frac{e^{-\lambda_i^{\vec{m}} - \lambda_j^{\vec{m}}}}{1 + \sum_{\vec{m} \neq \vec{0}} e^{-\lambda_i^{\vec{m}} - \lambda_j^{\vec{m}}}} \tag{10.42}$$

for $\vec{m} \neq \vec{0}$ and for $\vec{m} = \vec{0}$ by

$$p_{ij}^{\vec{0}} = 1 - \sum_{\vec{m} \neq \vec{0}} p_{ij}^{\vec{m}}. \tag{10.43}$$

In Eq. (10.42), $\lambda_i^{\vec{m}}$ indicate the Lagrangian multipliers which are fixed by the constraints

$$k_i^{\vec{m}} = \sum_{j=1}^{N} p_{ij}^{\vec{m}} \tag{10.44}$$

with $\vec{m} \neq \vec{0}$. In presence of a suitably defined structural cutoff, when the non-trivial multidegrees $\vec{m} \neq \vec{0}$ satisfy

$$k_i^{\vec{m}} \ll \sqrt{\langle k^{\vec{m}} \rangle N} \tag{10.45}$$

the marginal probabilities of multilinks $p_{ij}^{\vec{m}}$ for non-trivial multilinks $\vec{m} \neq \vec{0}$ are given by the simple expression [41]

$$p_{ij}^{\vec{m}} = \frac{k_i^{\vec{m}} k_j^{\vec{m}}}{\langle k^{\vec{m}} \rangle N}. \tag{10.46}$$

Construction of the canonical ensemble with expected multidegree sequence

A multiplex network with expected multidegree sequence can easily be constructed by following the steps below:

- Calculate the probability $p_{ij}^{\vec{m}}$ to have a multilink $\vec{m}$ between nodes i and j.
- For every pair of nodes i and j, draw a multilink $\vec{m}$ with probability $p_{ij}^{\vec{m}}$ and consequently put a link in every layer α where $m^{[\alpha]} = 1$ and put no link in every layer α where $m^{[\alpha]} = 0$.

Microcanonical ensemble

In the microcanonical model, when we fix the multidegrees $k_i^{\vec{m}}$ of each node i, the probability of the multiplex network is given by

$$P_M(\vec{G}) = \sum_{\vec{G}} \prod_{i=1}^{N} \prod_{\vec{m}\neq\vec{0}} \delta\left(k_i^{\vec{m}}, \sum_{j=1}^{N} A_{ij}^{\vec{m}}\right). \tag{10.47}$$

The marginal $p_{ij}^{\vec{m}}$ expressing the probability that node i and node j are connected by a multilink $\vec{m}$ takes the same form as for the canonical network model (Eqs (10.42) and (10.43)). In particular, when Eq. (10.45) is satisfied $p_{ij}^{\vec{m}}$ is given by Eq. (10.46). However, the microcanonical ensemble is not equivalent to the corresponding canonical ensemble, as it emerges clearly from the calculation of the entropy Σ of the microcanonical network ensemble, that it is not equal to the entropy of the corresponding canonical ensemble given by Eq. (13.47) in the thermodynamic limit. Specifically, Σ is smaller than S as it satisfies [39]

$$\Sigma = S - \Omega \tag{10.48}$$

where

$$\Omega = -\ln\left[\sum_{\vec{G}} P_C(\vec{G}) \prod_{i=1}^{N} \prod_{\vec{m}\neq\vec{0}} \delta\left(k_i^{[\alpha]}, \sum_{j=1}^{N} A_{ij}^{\vec{m}}\right)\right], \tag{10.49}$$

with $P_C(\vec{G})$ given by Eqs (10.39) and (10.40). The quantity Ω is extensive in the number of nodes N of the multiplex network and is a non-negligible contribution. In the presence of the structural cutoffs when the multidegrees $k_i^{\vec{m}}$ of non-trivial multilinks $\vec{m} \neq \vec{0}$ satisfy Eq. (10.45), Ω has an explicit analytical expression given by

$$\Omega = -\sum_{i=1}^{N} \sum_{\vec{m}\neq\vec{0}} \ln\left[\frac{1}{k_i^{\vec{m}}!} \left(k_i^{\vec{m}}\right)^{k_i^{\vec{m}}} e^{-k_i^{\vec{m}}}\right]. \tag{10.50}$$

This configuration model for multiplex networks includes multiplex networks with given multidegree distributions, where each multidegree is drawn independently, i.e.

$$P(\mathbf{k}^{\vec{m}}) = \prod_{\vec{m}\neq\vec{0}} P^{\vec{m}}(k^{\vec{m}}). \tag{10.51}$$

For instance, it is possible to consider duplex networks in which each different type of multilink is Poisson-distributed with average degree $\langle k^{\vec{m}} \rangle = c^{\vec{m}}$, i.e.

$$P^{\vec{m}}(k^{\vec{m}}) = \frac{\left(c^{\vec{m}}\right)^{k^{\vec{m}}}}{k^{\vec{m}}!} e^{-c^{\vec{m}}}. \tag{10.52}$$

Interestingly, this ensemble can also capture more complex scenarios where the multidegree distributions are not independent, i.e.

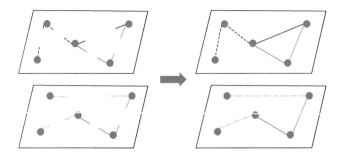

Fig. 10.5 *Schematic representation of the construction of a multiplex network with given multidegree sequence. Stubs of different types are associated with nodes. Solid line, dashed line and dot-dashed line stubs refer to multilinks (1,1), multilinks (1,0) and multilinks (0,1) respectively. Stubs of the same type are randomly matched, forming the multiplex network.*

$$P(\mathbf{k}^{\vec{m}}) \neq \prod_{\vec{m} \neq \vec{0}} P^{\vec{m}}(k^{\vec{m}}). \tag{10.53}$$

For instance, this ensemble can capture the topology of duplex networks with nodes with higher multidegree $k^{(1,1)}$, having also high multidegrees $k^{(1,0)}, k^{(0,1)}$, or, on the contrary, duplex networks in which nodes with high multidegree $k^{(1,1)}$ tend to have low multidegrees $k^{(1,0)}, k^{(0,1)}$.

Construction of microcanonical multiplex ensemble with given multidegree sequence

A multiplex network with given multidegree sequence can be constructed following these steps (see Fig. 10.5 for a schematic representation of the algorithm):

- Assign to each node i, $k_i^{\vec{m}}$ stubs of type $\vec{m} \neq \vec{0}$.
- Randomly pair stubs belonging to different nodes, matching exclusively stubs of the same type $\vec{m}$ and never matching more than two stubs for any pair of nodes.
- If two nodes have matched stubs of type $\vec{m}$ connect them by a multilink $\vec{m}$.

10.3.6   Spatial multiplex network ensembles

When the multiplex networks are embedded in a real (as in the case of the airport multiplex network in Ref. [71]) or a hidden space, the overlap of the links can emerge naturally from the correlations induced by the distance in the embedding space [148]. Intuitively, if in every layer links are more likely between nodes that are closer in the embedding space, then one can observe non-negligible overlap of the links because two nodes that are close in the embedding space are more likely to be connected to each other in every layer of the multiplex. Let us consider, for instance, the case in which the probability $P(\vec{G}|\{\vec{r}_i\})$ of a network $\vec{G}$ with N nodes at positions $\{\vec{r}_i\}$ is given by

$$P(\vec{G}|\{\vec{r}_i\}) = \prod_{\alpha=1}^{M} P_\alpha(G_\alpha|\{\vec{r}_i\}), \tag{10.54}$$

where $P_\alpha(G_\alpha|\{\vec{r}_i\})$ is the probability of a network in layer α given by

$$P_\alpha(G_\alpha|\{\vec{r}_i\}) = \prod_{i<j} \left[p_{ij}^{[\alpha]} a_{ij}^{[\alpha]} + (1 - p_{ij}^{[\alpha]})(1 - a_{ij}^{[\alpha]}) \right]. \tag{10.55}$$

The probability $p_{ij}^{[\alpha]}$ of a link between node i and node j in layer α can be modulated either by an exponential function of the distance d_{ij} between the nodes in the embedding space (when for each link (i,j) there is an associated 'cost' increasing linearly with the distance d_{ij}),

$$p_{ij}^{[\alpha]} = \frac{e^{-\lambda_i^{[\alpha]} - \lambda_j^{[\alpha]} - d_{ij}/d_0^{[\alpha]}}}{1 + e^{-\lambda_i^{[\alpha]} - \lambda_j^{[\alpha]} - d_{ij}/d_0^{[\alpha]}}} \tag{10.56}$$

or by a power law of the distance, (when for each link (i,j) there is an associated 'cost' increasing linearly with the order of magnitude of the distance d_{ij}),

$$p_{ij}^{[\alpha]} = \frac{\dfrac{e^{-\lambda_i^{[\alpha]} - \lambda_j^{[\alpha]}}}{d_{ij}^{\theta^{[\alpha]}}}}{1 + \dfrac{e^{-\lambda_i^{[\alpha]} - \lambda_j^{[\alpha]}}}{d_{ij}^{\theta^{[\alpha]}}}}. \tag{10.57}$$

Note that here $\lambda_i^{[\alpha]}, d_0^{[\alpha]}, \theta^{[\alpha]}$ are Lagrangian multipliers. In both cases it is possible to observe a significant overlap of the links [148].

Recently also a multiplex network model in hyperbolic hidden geometry has been proposed in Ref. [175]. In this model each layer is a network in the hyperbolic hidden geometry. In each layer α, every node i is assigned a radial $r_i^{[\alpha]}$ and an angular $\theta_i^{[\alpha]}$ position in a $(d = 2)$-dimensional Poincaré disk. The radial coordinate is approximately determined by the node degree in layer α (satisfying approximately $r_i^{[\alpha]} = \ln N - \ln k_i^{[\alpha]}$), while the angular coordinate allows for the identification of node 'similarities'. In this model pairs of nodes (i,j) that have a small hyperbolic distance

$$x_{ij}^{[\alpha]} \sim r_i^{[\alpha]} + r_j^{[\alpha]} + 2 \ln \sin(\Delta\theta_{ij}^{[\alpha]}/2), \tag{10.58}$$

with

$$\Delta\theta_{ij}^{[\alpha]} = \pi - |\pi - |\theta_i^{[\alpha]} - \theta_j^{[\alpha]}|| \tag{10.59}$$

indicating the angular distance between the nodes, are more likely to be connected. This model can be used in inference problems: for any given layer of a multiplex network the angular coordinates in the hidden hyperbolic space can be estimated. In this way, clusters of soft communities revealed by the position of closed sets of nodes in the hyperbolic space can be detected. Interestingly, when applying these techniques on real multiplex networks important correlations are observed between the hidden geometry of different layers. For instance, the probability $P(\beta|\alpha)$ that a random pair of nodes is connected in layer β given their hyperbolic distance in layer α reveals that nodes that are connected in layer β tend to be close in space also in layer α.

10.3.7 Multiplex networks with heterogeneous activity of the nodes

As many real networks display a heterogeneous activity of the nodes (defined in Sec. 7.5) it is important to generate ensembles displaying this important structural property. Since a node is active in a layer only if it is connected at least to one other node, the first and most direct way to generate networks with heterogeneous activity of the nodes is to consider degree sequences $\{k_i^{[\alpha]}\}$ or multidegree sequences $\{k_i^{\vec{m}}\}$. When these degree sequences allow a fraction of nodes to be disconnected in one or more layers it is possible to generate multiplex networks with heterogeneous activity of the nodes.

However, this procedure implies that we know exactly which nodes are active in which layers.

In a number of cases, we do not have this information, but we need to construct multiplex networks with given activity distribution. In this scenario the following constructive procedure proposed in Ref. [75] can be used:

(a) Construct the bipartite network between nodes and layers where each node is linked to a layer if it is active on it (see Fig. 7.5). For instance, this bipartite network can be constructed using the configuration model where each node i has given activity B_i (its degree in the bipartite network) and each layer α has its own layer activity N^α (its degree in the bipartite network).

(b) Assign to each layer α of the multiplex network the set of nodes i that are active on it, as indicated in the bipartite network constructed in point (a).

(c) Assign to the active nodes in each layer α a degree greater than or equal to one according to a desired degree distribution $P^{[\alpha]}(k)$ and generate the network with the resulting degree sequence.

10.4 Randomization algorithms

Randomization algorithms are very useful for generating null models starting from real network data. Here we provide an overview of three different randomization algorithms. The first keeps the same network in each layer but randomizes the one-to-one mapping between the nodes and removes the effect of interlayer degree correlations. The second algorithm keeps the same interlayer degree correlations but randomizes each layer independently, therefore removing the effect of link overlap among the layers and the effect of intralayer degree correlations. Finally, the third algorithm keeps the multidegree sequence and therefore keeps at the same time the same link overlap and the same interlayer degree sequences as the original network.

Randomization of the replica nodes

A given null model of a real multiplex network can be obtained by randomizing the one-to-one mapping between the replica nodes of the network while keeping the same layers. To this end, for every layer α we can consider a random permutation of the labels i of the nodes. In this way, while the networks in each layer remain unchanged the interlayer degree correlations are removed.

Independent randomization of each layer

In this case, each single network can be randomized separately while keeping the degree sequences and the interlayer degree correlations. To this end it is possible to use the swap algorithm [200] extensively used in the context of single layers to randomize a network while preserving the degree sequence. The algorithm applied to each single layer α proceeds as follows

- Consider two random links of layer α connected to four distinct nodes. We assume that the first link is connected to nodes i_1 and j_1 and that the second link is connected to nodes i_2, j_2.

- Swap the two links substituting them with other two links connecting node i_1 to node j_2 and node i_2 to node j_1 only if the move is allowed. The move is allowed if none of the links (i_1, j_2) and (i_2, j_1) already exists in layer α.

These swaps proceed until the network is completely randomized and basic structural properties of the network do not change by increasing the number of iterations.

Randomization algorithm preserving the multidegree sequence

In this case the existing multiplex network is randomized by keeping the same multidegree sequence. To this end, it is possible to generalize the widely used swap algorithm of single networks [201] for randomizing a multiplex network keeping the multidegree sequence and therefore the same overlap of the links between any two layers. This generalized swap algorithm is defined as follows:

- Consider two random multilinks of the same type $\vec{m} \neq \vec{0}$ connected to four distinct nodes of the network. Let us assume that the first multilink is connected to nodes i_1 and j_1 and the second multilink is connected to nodes i_2 and j_2.

- Swap the two multilinks substituting the original multilinks $\vec{m} \neq \vec{0}$ with two multilinks $\vec{m}$ connecting nodes i_1 and j_2 and nodes i_2 and j_1 if and only if the move is allowed. The move is allowed if the nodes i_1 and j_2 and the nodes i_2 and j_1 are not yet connected by any type of non-trivial multilink.

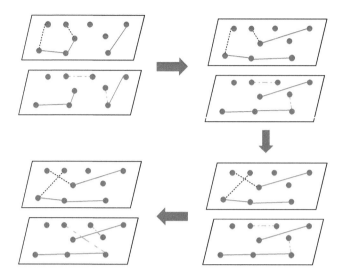

Fig. 10.6 *Schematic representation of the swap algorithm that randomizes multiplex networks preserving their multidegree sequence. Here solid line, dashed line and dot-dashed line connections refer to multilinks (1,1), multilinks (1,0) and multilinks (0,1) respectively.*

The swap algorithm proceeds until the network is completely randomized and the multiplex network measures do not change significantly if the number of iterations is increased. The swap algorithm is schematically represented in Fig. 10.6.

10.5   Models of multi-slice temporal networks

10.5.1   Modelling temporal networks

Modelling temporal networks represents a major challenge of network theory. In fact, networks that are time-varying are evolving through a non-equilibrium dynamics and often it is not at all clear if they even have a stationary state. Additionally, social temporal networks can be affected by periodic modulations in time reflecting, for instance, daily and weekly habits.

Nevertheless, several statistical common features of temporal networks can be captured by stochastic models with steady states. These models represent an ideal platform for simulating artificial datasets with controlled structural and temporal properties.

10.5.2   Models with temporal activity of the nodes

Temporal activity model: the motivation

In temporal networks some nodes might be more inclined than others to establish connections at any given time-window. This tendency of the nodes to be more active or less active in temporal networks can have profound consequences for the temporal network topology and for the dynamics defined on it. In the modelling framework proposed in Ref.[249] this heterogeneity of the nodes of the network is modelled by assuming that each node is assigned a *temporal activity* drawn randomly from a given distribution. The temporal activity of the nodes is kept fixed during the entire dynamical evolution of the network.

At each time-slice the nodes establish connections with a probability proportional to their temporal activity. Different time-slices are drawn independently, but all of them are conditioned on the temporal activity of the nodes.

Definition

Given a temporal network of N nodes, every node i is assigned a *temporal activity* $a_i = \eta x_i$ where x_i is drawn randomly from a distribution $F(x)$ and η is a parameter that can be used to tune the properties (the average number of active nodes at time t) of the temporal network.

The probability distribution $F(x)$ is arbitrary with support $[\epsilon, 1]$ with $\epsilon > 0$. However, for any given application its functional form can be dictated by the data under consideration.

The model generates a multi-slice temporal network in which every slice corresponds to a temporal window δt.

Every temporal slice is drawn independently of the previous one, but is conditioned on the temporal activities of the nodes that are quenched and do not change over time.

The algorithm generating the temporal network is simply stated. At each time t, with probability $a_i \delta t$ each node becomes active and it is connected to m other randomly chosen nodes. Note that in this way inactive nodes can also receive connections. Every slice of the temporal network is drawn independently using the same algorithm.

Derivation of the aggregated degree distribution

The degree K_i in the aggregated network of the above model is given by the sum of the connections $\tilde{K}_i^{out}$ that are drawn by node i during the time-slices when it is active and the number of connections $\tilde{K}_i^{in}$, not included in the previous set, that the node receives from other nodes during the time-slices when it is inactive, i.e.

$$K_i = \tilde{K}_i^{out} + \tilde{K}_i^{in}. \tag{10.60}$$

Since when node i is active it is connected to any node of the network randomly, the probability that in the aggregated network of $M = T$ time-slices node i is connected to any other node is given by

$$p = 1 - \left(1 - \frac{1}{N}\right)^{ma_i T} \simeq 1 - e^{-ma_i T/N}, \tag{10.61}$$

where the last expression is valid for $T/N \ll 1$. Therefore, it follows that

$$\tilde{K}_i^{out} \simeq N\left[1 - e^{-ma_i T/N}\right]. \tag{10.62}$$

The number of connections $\tilde{K}_i^{out}$ instead is given by

$$\tilde{K}_i^{in} = m\langle a \rangle \, T e^{-ma_i T/N}.$$ (10.63)

In fact, the average number of links drawn by nodes different from node i over the entire time-window T is $m\langle a \rangle NT$. Each of these links is connected to node i with probability $1/N$ giving an average number of connections $m\langle a \rangle T$. These connections should be counted in $\tilde{K}_i^{in}$ only if they are established between nodes that are not otherwise connected. Therefore, $\tilde{K}_i^{in}$ is given by Eq. (10.63) where the factor $e^{-ma_i T/N}$ accounts for the probability that the link is not also counted in $\tilde{K}_i^{out}$. Finally, using Eqs (10.62), (10.63) we obtain

$$K_i = N \left[1 - \left(1 + m\langle a \rangle \frac{T}{N} \right) e^{-ma_i T/N} \right] \simeq N \left[1 - e^{-ma_i T/N} \right],$$ (10.64)

valid for $T/N \ll 1$. Therefore, the aggregated degree is a one-to-one function of the temporal activity of the node. In this approximation, the cumulative distribution of the aggregated network $P(K_i \geq K)$ is given by

$$P(K_i \geq K) = P \left(x_i = \frac{a_i}{\eta} \leq -\frac{N}{\eta m T} \ln \left[1 - \frac{k}{N} \right] \right),$$ (10.65)

yielding the degree distribution $P(K)$ of the aggregated network

$$P(K) = \frac{1}{Tm\eta} \frac{1}{1 - \frac{k}{N}} F \left[-\frac{N}{\eta m T} \ln \left(1 - \frac{k}{N} \right) \right] \simeq \frac{1}{Tm\eta} F \left[\frac{k}{Tm\eta} \right].$$ (10.66)

Therefore, the degree distribution of the aggregated network is determined by the distribution of temporal activities. This theoretical model can be simulated starting from empirically measured activity parameters and can be show to capture important properties of temporal networks (see Fig. 10.7).

Variations of the model

Different variations of this model including memory effects in the network temporal dynamics have been considered in subsequent publications. In particular, a different attachment probability which goes beyond the random attachment of the nodes has been considered in Ref. [169] and in Ref. [295] a stochastic variable is attributed to each link modulating its duration.

10.5.3   Exponential random multi-slice networks

The exponential random graph framework defining the canonical ensembles of networks can be extended to temporal multi-slice networks [204]. In particular, within this

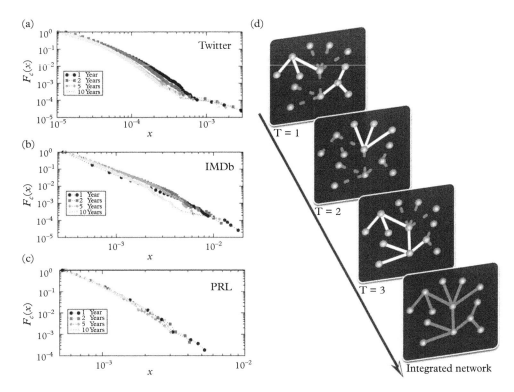

Fig. 10.7 *Cumulative distribution* $F_C(x)$ *of the activity potential* x, *empirically measured by using four different time-windows and a schematic representation of the proposed network model. In particular, the cumulative distributions of the observables for Twitter (panel (a)), for the IMDb actor collaboration network (panel (b)) and for the PRL scientific collaboration network (panel (c)) are shown. In panel (d) a schematic representation of the model is represented. Considering just 13 nodes and* m = 3, *a visualization of the resulting networks is plotted for three different time steps. The final visualization represents the network after integration over all time steps. Reprinted by permission from Macmillan Publishers Ltd: [249].*

framework it is possible to include in multi-slice networks memory effects that are not present in the original version of the model with activities of the nodes.

Let us consider a multi-slice network described by adjacency matrices $\mathbf{a}^{[\alpha]}$ with $\alpha = 1, 2 \ldots M$ each indicating the interactions occurring in the temporal slice α of the network. It is assumed that each temporal slice only depends on the previous temporal slice. It follows that the probability $P(\{\mathbf{a}^{[\alpha]}\}_{\alpha=1,2\ldots,M})$ of the entire multi-slice network follows a Markovian process

$$P(\{\mathbf{a}^{[\alpha]}\}_{\alpha=1,2\ldots,M}) = \prod_{\alpha=2}^{M} \tilde{P}(\mathbf{a}^{[\alpha]}|\mathbf{a}^{[\alpha-1]})\hat{P}(\mathbf{a}^{[1]}) \tag{10.67}$$

where $\hat{P}(\mathbf{a}^{[1]})$ is the probability that the first temporal slice $\alpha = 1$ has adjacency matrix $\mathbf{a}^{[1]}$ and $\tilde{P}(\mathbf{a}^{[\alpha]}|\mathbf{a}^{[\alpha-1]})$ is the probability that the temporal slice α has adjacency matrix $\mathbf{a}^{[\alpha]}$, given that the previous temporal slice had adjacency matrix $\mathbf{a}^{[\alpha-1]}$.

In Refs [264, 151, 150] this Markovian assumption on the probability of the multi-slice network is combined with the framework of canonical network ensembles. In particular, it is assumed that every temporal slice is a maximum entropy network satisfying a number of constraints depending also on the structural properties of the preceding temporal slice. By indicating with $F_\mu\left(\mathbf{a}^{[\alpha]}, \mathbf{a}^{[\alpha-1]}\right)$ the different structural properties of the network under consideration, the structural constraints enforced on average on the conditional probability $\tilde{P}\left(\mathbf{a}^{[\alpha]}|\mathbf{a}^{[\alpha-1]}\right)$ read

$$\sum_{\mathbf{a}^{[\alpha]}} \tilde{P}\left(\mathbf{a}^{[\alpha]}|\mathbf{a}^{[\alpha-1]}\right) F_\mu\left(\mathbf{a}^{[\alpha]}, \mathbf{a}^{[\alpha-1]}\right) = C_\mu, \tag{10.68}$$

where C_μ are constant values. The conditional probability $\tilde{P}\left(\mathbf{a}^{[\alpha]}|\mathbf{a}^{[\alpha-1]}\right)$ of the lead biased ensemble of networks satisfying these constraints is given by the maximum entropy ensemble with

$$\tilde{P}\left(\mathbf{a}^{[\alpha]}|\mathbf{a}^{[\alpha-1]}\right) = \frac{1}{Z} \exp\left(-\sum_\mu \lambda_\mu F_\mu\left(\mathbf{a}^{[\alpha]}, \mathbf{a}^{[\alpha-1]}\right)\right) \tag{10.69}$$

where λ_μ are the Lagrangian multipliers enforcing the different types of constraints. Typically, here the label μ indicates the different types of constraints. These are usually taken to be global network properties determining respectively the density, the stability, the reciprocity and the transitivity of the multi-slice network [204].

These most notable examples of constraints are described in the following:

- *Density.* This constraint fixes the expected total number of links in the temporal slice α

$$F_{density}\left(\mathbf{a}^{[\alpha]}, \mathbf{a}^{[\alpha-1]}\right) = \sum_{i,j} a_{ij}^{[\alpha]}. \tag{10.70}$$

It is an example of constraint that acts on each layer independently of the previous layer.

- *Stability.* This constraint fixes on average the total number of pairs of nodes that are either interacting or non-interacting in both layer α and the preceding layer $\alpha - 1$,

$$F_{stability}\left(\mathbf{a}^{[\alpha]}, \mathbf{a}^{[\alpha-1]}\right) = \sum_{i,j} a_{ij}^{[\alpha]} a_{ij}^{[\alpha-1]} + (1 - a_{ij}^{[\alpha]})(1 - a_{ij}^{[\alpha-1]}). \tag{10.71}$$

- *Reciprocity.* This constraint fixes on average the fraction of nodes that reciprocate a direct interaction existing in the previous time-slice,

$$F_{reciprocity}\left(\mathbf{a}^{[\alpha]}, \mathbf{a}^{[\alpha-1]}\right) = \frac{\sum_{i,j} a_{ji}^{[\alpha]} a_{ij}^{[\alpha-1]}}{\sum_{i,j} a_{ij}^{[\alpha-1]}}. \tag{10.72}$$

- *Transitivity.* This constraint fixes on average the propensity of the multi-slice network to have links at time α that close triads between triple nodes forming a wedge at the preceding time-slice,

$$F_{transitivity}\left(\mathbf{a}^{[\alpha]}|\mathbf{a}^{[\alpha-1]}\right) = \frac{\sum_{i,j,r} a_{ij}^{[\alpha]} a_{ir}^{[\alpha-1]} a_{rj}^{[\alpha-1]}}{\sum_{i,j,r} a_{ir}^{[\alpha-1]} a_{rj}^{[\alpha-1]}}. \tag{10.73}$$

These types of maximum entropy network ensembles typically are too complicated to allow an analytical determination of their entropy. As a result, these models cannot be used to estimate real temporal network parameters through maximum-likelihood estimation. Nevertheless, Markov Chain Monte Carlo methods can be used to infer the parameters of the model from real data.

10.5.4   Models with burstiness

Burstiness in temporal networks

Many temporal networks, including face-to-face interactions and mobile-phone calls, but also email correspondence and traditional exchange of mail, are bursty [13, 74, 317]. This term indicates that the interval between subsequent interactions and the duration of the interactions are power-law distributed. This implies that the establishment of new interactions does not occur at a constant rate at any given interval of time. Instead, the burstiness implies that the temporal evolution of the network includes significant memory effects.

Models of bursty temporal networks include notably queue models with priority of tasks [13] that characterize the dynamics from a single-node perspective. In this case it is possible to model the time series indicating when a single agent sends emails or letters. In this framework the main idea is that agents act with some sort of bounded rationality by performing tasks of some priority with an assigned probability distribution. This probability can range from a random uniform distribution (yielding a non-bursty temporal network) to an extremal dynamics indicating that the task with highest priority is the first to be accomplished (yielding a bursty temporal network).

Another class of models assumes that the cause of burstiness is instead a temporal reinforcement dynamics, extending to the temporal realm a sort of preferential attachment rule [288, 318, 317]. In the context of face-to-face interations this framework assumes that an interacting agent has a smaller probability of ending a conversation the longer the conversation lasts. This type of modelling has been shown to be very appropriate for describing face-to-face interactions and mobile-phone communication as well, revealing the different statistical properties of the two types of communication.

Bursty model for face-to-face interactions

In this model each node $i = 1, 2, \ldots, N$ forms or removes its links over a very fine time-window called a micro-time step.

The dynamics of the stochastic model is extremely simple. Let us indicate with $n_i \in \{0, 1, 2 \ldots, N-1\}$ the number of connections of node i and with t_i the last step at which node i has acquired n_i connections. Starting from a given random initial condition, at each micro-time step t the following algorithm is simulated:

(1) A node i is chosen randomly.

(2) The node i updates its number of connections $n_i = n$ with probability $p_n(t, t_i)$. If n_i is updated, the changes in the network are determined by the rules:

   (i) If node i is isolated, i.e. $n_i = 0$, it is connected to another isolated node j chosen with probability proportional to $p_0(t, t_j)$.

   (ii) If node i is interacting in a group, i.e. $n_i = n > 0$, with probability λ the node leaves the group and with probability $(1 - \lambda)$ an isolated node j chosen with probability proportional to $p_0(t, t_j)$ joins its group.

In this model, if the probabilities $p_n(t, t')$ are chosen to be constant in time, every node has a fixed probability of forming a group or leave from a group. In this case the formation of the groups is a Poisson process and the distribution of the duration of contact is an exponential. Therefore, the duration of contacts has a characteristic time.

The experimental data on face-to-face interactions [74] strongly deviates from this traditional framework, showing a power-law distribution of the duration of contacts without a characteristic timescale. This implies that the decision of the agents to form or leave a group is driven by memory effects dictated by a reinforcement dynamics. This reinforcement dynamics can be summarized in the following statement: *the longer an agent is interacting in a group the smaller the probability that he will leave the group, the longer an agent is isolated the smaller the probability that he will form a new group.* In particular, this reinforcement principle implies that the probabilities $p_n(t, t')$ that a node degree n changes the number of its connections depends on the last time t' the node has updated its number of connections, i.e. $p_n(t, t') = p_n(t - t')$. Any function $p_n(t - t')$ taken to be a decreasing function of its argument describes a reinforcement dynamics, but the face-to-face interaction data are reproduced only for a particular choice $p_n(t - t')$. In particular, the only choice of the kernel $p_n(t - t')$ that can reproduce the power-law distribution of duration of contacts is given by

$$p_n(t - t') = \frac{b_n}{1 + (t - t')/N}. \tag{10.74}$$

In Eq. (10.74) the choice $b_n = b_1$ for every $n \geq 1$ indicates the fact that the interacting agents change their state independently of the number of other agents n with whom they are interacting, provided that $n \geq 0$. In the case in which $\lambda = 0$, only groups of size at most two are formed. At the stationary state, the number of isolated/interacting agents $N_0(t, t')/N_1(t, t')$ at time t that have been isolated/interacting since time t' decays like a power law of time, given by

$$N_0(t, t') \propto \left(1 + \frac{t - t'}{N}\right)^{-2b_0},$$

$$N_1(t, t') \propto \left(1 + \frac{t - t'}{N}\right)^{-2b_1}. \qquad (10.75)$$

In the case $\lambda \in [0, 0.5]$ a stationary state is reached where groups of different sizes are formed. The number of nodes $N_n(t, t')$ that at time t have degree n and have not changed their degree since time t' decays with time as a power law with

$$N_0(t, t') \propto \left(1 + \frac{t - t'}{N}\right)^{-b_0[2 + (1 - \lambda)\hat{\epsilon}]},$$

$$N_n(t, t')\big|_{n>0} \propto \left(1 + \frac{t - t'}{N}\right)^{-(n+1)b_1}, \qquad (10.76)$$

where $\hat{\epsilon} > 0$ can be predicted by the analytical solution of the model (see Refs [288, 318]). Since $N_n(t, t')$ decays as a power law with exponent $(n + 1)b_1$ for $n > 0$, this solution implies that larger groups are more unstable than smaller groups. In fact, in this model a large group changes size whenever any of its members decide to leave the group, leading to a higher instability of larger groups. This effect reproduces very well the behaviour observed in real datasets of face-to-face interactions coming from Sociopattern experiments (see Fig. 10.8). We note here that this model can be turned into a multi-slice network model by taking snapshots of the network after N micro-time steps.

Model for mobile-phone interactions

Mobile-phone networks have different statistical properties from face-to-face interaction networks. In fact, the duration of contacts is not power-law distributed but is distributed according to a Weibull distribution [317]. Despite this difference, since Weibull distributions cannot be obtained from Poisson processes mobile-phone interactions also cannot be described by models in which nodes change connectivity at a constant rate. It turns out that a reinforcement dynamics also has the ability to reproduce quite accurately the mobile-phone interactions. However, instead of taking a probability $p_n(t - t')$ of changing the connectivity state (from interacting to non-interacting for $n = 1$ or vice versa for $n = 0$) given by Eq. (10.74) it is opportune to take [317]

$$p_n(t - t') = \frac{b_n}{(t - t')^\beta}, \qquad (10.77)$$

with $\beta \in (0, 1)$ and $b_n > 0$. Notice that this probability again satisfies the reinforcement mechanism of the network dynamics.

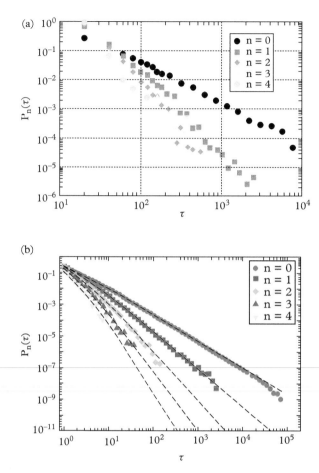

Fig. 10.8 *Probability of duration $P_n(\tau)$ of interactions with other n nodes (groups of size $n+1$). Panel (a) shows the distribution found in the Sociopattern experiment conducted between the 175 voluntary participants of the sixth European Semantic Web Conference in 2009, panel (b) shows a typical outcome of the model for bursting face-to-face interactions proposed in Refs [288, 318]. Reprinted figure from Ref. [318].*

10.6   Ensembles of more general multilayer networks

General multilayer networks can be modelled using different network ensembles depending on how interlinks connect nodes of different layers. Since in multilayer networks there is a large degree of freedom in the way the interlinks can be placed, the variety of multilayer network ensembles is remarkable. Here we will discuss three major classes of multilayer network ensembles.

Networks of networks with replica nodes

In these networks of networks, if layer α is connected with layer β in the supernetwork (i.e. $A_{\alpha\beta} = 1$), each replica node (i, α) of network α is connected to its replica node (i, β) in layer β (see Fig. 10.9 and definition in Sec. 5.5.3). Therefore, once the supernetwork is given the interlinks between different layers are fixed but the networks within any given layer can be random. Every layer can be given by a suitable single-layer network model. For instance, the layers can be given by the microcanonical network ensemble with given degree sequence or by the canonical network ensemble with expected degree sequence. Assuming that each layer is drawn from the configuration model the probability $P(\mathcal{M})$ of a network of networks $\mathcal{M}$ in this ensemble is given by

$$P(\mathcal{M}) = \frac{1}{Z} \prod_{i=1}^{N} \prod_{\alpha=1}^{M} \left[\delta \left(k_i^{[\alpha]}, \sum_{j=1}^{N} \mathcal{A}_{i\alpha,j\alpha} \right) \prod_{\beta\neq\alpha} \delta \left(A_{\alpha\beta}, \mathcal{A}_{i\alpha,i\beta} \right) \right], \qquad (10.78)$$

where here and in the following $\mathbf{A}$ is the adjacency matrix of the supernetwork, $\mathcal{A}$ is the supra-adjacency matrix of the network of networks, $\delta(x, y)$ is the Kronecker delta and Z is a normalization factor. In Sec. 11.6 we will study percolation in the presence of interdependencies for this class of networks of networks.

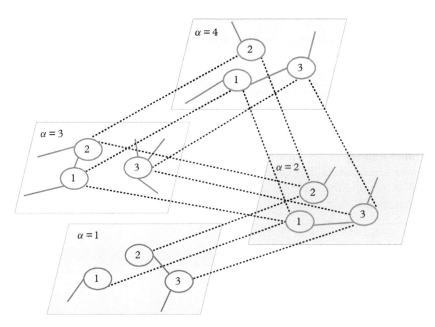

Fig. 10.9 *Schematic view of a typical network of networks with replica nodes. Interdependencies (interlinks between nodes from different levels) are shown by the black dashed lines. Intralinks between nodes within layers are shown as solid red lines. Reprinted figure from Ref. [49].*

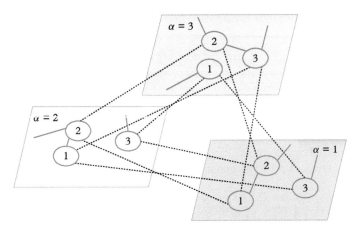

Fig. 10.10 *Schematic representation of a network of networks with fixed supernetwork and random permutations of the labels of the nodes. The supernetwork can be arbitrary but in the drawing we show a specific realization in which the supernetwork is a loop formed by M = 3 layers. Reprinted from [54] ©2014 with permission from Elsevier.*

Networks of networks without replica nodes

In this case the networks of networks do not have replica nodes but have a supernetwork. If network α is connected with network β in the supernetwork, each node (i, α) of network α is connected to a single node (j, β) of network β randomly chosen from the nodes of layer β (see Fig. 10.10). The one-to-one mapping between the nodes of two different layers is performed by defining a permutation $\pi_{\alpha,\beta}$ of the indices $\{i\}$ such that $j = \pi_{\alpha,\beta}(i)$ if and only if node (i, α) is linked to node (j, β). In order to define an undirected network of networks, we need to impose that the permutation $\pi_{\beta,\alpha}$ is the inverse permutation of $\pi_{\alpha,\beta}$, enforcing that if $j = \pi_{\alpha,\beta}(i)$ then $i = \pi_{\beta,\alpha}(j)$.

Ensembles of networks of networks of this type can be defined for any given supernetwork and set of permutations $\pi_{\alpha,\beta}(i)$ by assuming that every network in each distinct layer is random (either generated by a microcanonical or the canonical network ensemble). Assuming that each layer has the same number of nodes (i.e. $N_{\alpha} = N \; \forall \alpha$) and is drawn from a configuration model, the probability $P(\mathcal{M})$ of a network of networks $\mathcal{M}$ in this ensemble is given by

$$P(\mathcal{M}) = \frac{1}{Z} \prod_{i=1}^{N} \prod_{\alpha=1}^{M} \left[\delta\left(k_i^{[\alpha]}, \sum_{j=1}^{N} A_{i\alpha,j\alpha} \right) \delta\left(A_{i\alpha,\pi_{\alpha\beta}(i)\beta}, A_{\alpha\beta} \right) \right]. \qquad (10.79)$$

The present case differs substantially from the first discussed type of network of networks, having both a supernetwork and replica nodes. For instance, consider the case of a supernetwork forming a loop of size M between the different layers. In the previous case, starting from each node (i, α) and following only interlinks between different layers we can reach only M other nodes, while in the present scenario (as each permutation

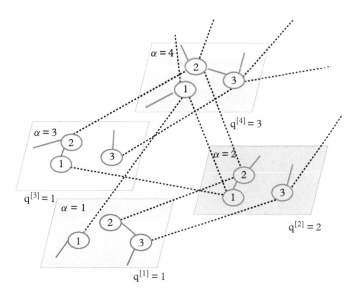

Fig. 10.11 *Schematic representation of a network of networks with given superdegree distribution. Each layer* α *has given superdegree* q_α. *However each node* (i, α) *can be linked to any other replica nodes. Reprinted figure from Ref. [48].*

$\pi_{\alpha,\beta}(i)$ is random) the number of different nodes that can be reached following interlinks can be significantly higher than M (see Fig. 10.10).

In Sec. 11.6 we will study percolation in the presence of interdependencies for this class of network of networks.

Multilayer networks with replica nodes and with given supradegree distribution

In this case, each node (i, α) can be linked only to its replica nodes (i, β), but there is no fixed supernetwork between the layers. Therefore, each set of replica nodes (i, α) with $\alpha = 1, 2, \ldots, M$ is connected by a network formed exclusively by interlinks. In order to have a simple model we assume this network is a random uncorrelated network where the replica node (i, α) in layer α is linked to $q = q_\alpha$ replica nodes (i, β) of randomly chosen layers β [48] (see Fig. 10.11). The value q_α is called the supradegree of layer α. Assuming that each layer is drawn from the configuration model, the probability $P(\mathcal{M})$ of a multilayer network $\mathcal{M}$ in this ensemble is given by

$$P(\mathcal{M}) = \frac{1}{Z} \prod_{i=1}^{N} \prod_{\alpha=1}^{M} \left[\delta \left(k_i^{[\alpha]}, \sum_{j=1}^{N} \mathcal{A}_{i\alpha,j\alpha} \right) \delta \left(q_\alpha, \sum_{\beta \neq \alpha} \mathcal{A}_{i\alpha,i\beta} \right) \right]. \tag{10.80}$$

In this ensemble of network of networks every layer is drawn randomly from a suitable network ensemble but also the interlinks are drawn randomly. Therefore, for a sufficiently

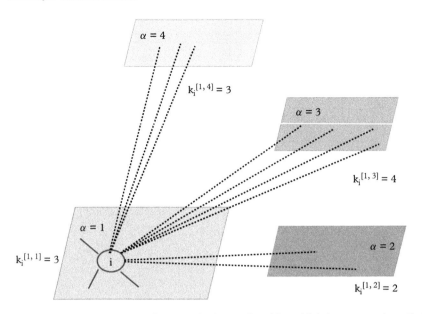

Fig. 10.12 *Schematic representation of a network of networks with multiple interconnections. Each node* (i, α) *has* $k_i^{[\alpha, \beta]}$ *links with nodes in layer* β. *Reprinted from [54] ©2014 with permission from Elsevier.*

sparse distribution of supradegree the network formed by the interlinks is uncorrelated and any interlink between node (i, α) and node (i, β) with $\alpha \neq \beta$ has probability $\pi_{i\alpha; i\beta}$ given by

$$\pi_{i\alpha; i\beta} = \frac{q_\alpha q_\beta}{\langle q \rangle M}. \tag{10.81}$$

In Sec. 11.6 we will study percolation in the presence of interdependencies for this class of network of networks and we will indicate with $P_S(q)$ the supradegree distribution, while we will indicate with $P_L(k)$ the degree distribution of the intralink within each layer, assumed for simplicity to be the same for every layer.

Multilayer networks with multiple interconnections

The last example of multilayer networks assumes that the nodes of a layer can be linked to any other node in other layers, and that the number of interlinks of each node is arbitrary. In this scenario it is possible to consider the ensemble in which every node (i, α) has $k_i^{[\alpha, \beta]}$ connections with nodes in layer $\beta = 1, 2, \ldots, M$. Therefore, assuming that all the layers have the same number of nodes (i.e. $N_\alpha = N \ \forall \alpha$), the probability of a generic multilayer network $\mathcal{M}$ is given by

$$P(\mathcal{M}) = \frac{1}{Z} \prod_{i=1}^{N} \prod_{\alpha=1}^{M} \prod_{\beta=1}^{M} \delta\left(k_i^{[\alpha,\beta]}, \sum_{j=1}^{N} A_{i\alpha,j\beta}\right).$$ (10.82)

A random uncorrelated multilayer network of this ensemble has a link between node (i, α) and node (j, β) with probability

$$\pi_{i\alpha;j\beta} = \frac{k_i^{[\alpha,\beta]} k_j^{[\beta,\alpha]}}{\langle k^{[\alpha,\beta]}\rangle N}.$$ (10.83)

The typical structure of a network of networks in this ensemble is shown in Fig. 10.12. In Sec. 12.2.3 we will study classical percolation in this class of multilayer networks.

11

Interdependent Multilayer Networks

11.1 Interdependencies in multilayer networks

Multilayer networks are usually formed by interdependent layers. While in classical percolation the failure of a node does not necessarily induce the failure of its neighbouring nodes, when two nodes are interdependent it is impossible for a node to be functional if the other node is not also functional. Therefore, the failure of one of the two interdependent nodes causes the failure of the other one. Global infrastructures are dependent on each other as a node in one layer can control or regulate other nodes in other layers. A similar scenario holds for multilayer financial networks where different financial institutions are related to each other by financial contracts. Biological networks in the cell formed by the combination of transcription networks, signalling networks, protein interaction networks and the metabolic network are strongly interdependent and, as a result, the cell is not alive if any of these networks is not functional.

Having a reliable theory for characterizing the robustness of multilayer networks in the presence of interdependencies is of fundamental importance for policy makers, and in general for a comprehensive understanding of the response of complex systems to damage.

In a seminal paper by Buldyrev et al. [66] it has been shown that in the presence of interdependencies multilayer networks can be much more fragile than single networks. In particular, as an increasing fraction f of nodes is damaged interdependent multilayer networks are dismantled even if their single layers would be functional if in isolation. Additionally, the disruption of the network occurs at a discontinuous transition characterized by large avalanches of failure events that propagate back and forth across the different layers of the network.

Significant progress has been made in characterizing this transition, and variants of the models (including partial [319] and redundant interdependencies [257]) have been proposed as possible scenarios for improving the robustness of interdependent multilayer networks.

In this chapter we aim at providing a comprehensive overview of the rich literature covering these important problems, highlighting the main results and their relevance to understanding the robustness of multilayer networks.

Multilayer Networks. Ginestra Bianconi, Oxford University Press (2018).
© Ginestra Bianconi. DOI: 10.1093/oso/9780198753919.001.0001

Additionally, from the theoretical perspective, the characterization of the robustness of interdependent multilayer networks allows us to formulate a generalized percolation problem that opens an entirely new scenario in the percolation theory.

11.2   Percolation of interdependent networks

11.2.1   The Mutually Connected Giant Component of a multiplex network

A multilayer network is *interdependent* if all the interlinks imply the interdependence of the connected nodes. Two nodes are interdependent if the damage of one node implies the damage of the other interdependent node, independently of the rest of the network. In presence of interdependencies, the robustness of a multilayer network when a random damage affects a fraction of its nodes can be evaluated by calculating the size of its *Mutually Connected Giant Component* (MCGC) [66]. In Fig. 11.1, it is shown how to construct the MCGC for the case of a multiplex formed by two layers where the replica nodes are interdependent. Starting from a random initial damage in a given layer, the damage will affect all the nodes of the same layer that are not part of its giant component. The damage caused in one layer propagates to the other layer through the interlinks because the interlinks imply interdependencies. In each of these layers all the nodes that are not in the giant component are damaged. The MCGC of a multiplex interdependent network is the giant component that remains after the random damage propagates back and forth in the different layers. Therefore, initial damage to the nodes of the multiplex network can cause an avalanche of failure events.

The algorithm that defines the MCGC in a general multilayer network formed by M layers is as follows [66]. Given an initial damage of the nodes

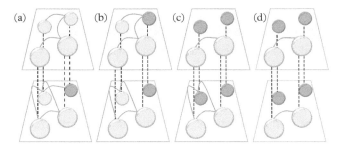

Fig. 11.1 *The algorithm for extracting the MCGC from a multiplex interdependent network formed by two layers is here shown after the damage (attack) of a single initial node in the bottom layer. The damage propagates from one layer to the other until the algorithm stops and no more nodes can be damaged. The remaining nodes belong to the MCGC. Here damaged nodes are indicated as darker nodes.*

   (i) the giant component of each layer $\alpha = 1, 2, \ldots, M$ is determined, evaluating the effect of the damaged nodes in each single layer;

  (ii) each node that has at least an interdependency with a node that does not belong to the giant component of its proper layer is damaged;

(iii) if there are no new damaged nodes the algorithm stops; otherwise it proceeds, starting again from step (i).

At the end of the iteration the nodes that are not damaged by the iterative process form the MCGC.

The MCGC in a multiplex network admits also another equivalent definition [33]. In fact, it is the giant subgraph of the multiplex network induced by the set of nodes such that every pair of nodes in the set is connected at least by one path in each of the M layers of the multiplex network where these paths need to traverse only nodes belonging to the MCGC.

11.2.2 Hybrid phase transition and avalanches of failure events

In the case in which the nodes are randomly damaged, it is possible to characterize the non-linear response of the system by determining the fraction S of nodes that remain in the MCGC. If the nodes of the multilayer networks are damaged with increasing probability $f = 1 - p$ it is therefore possible to characterize the entire function $S = S(p)$, determining the robustness of the system. Interestingly, by studying this phase transition in different multilayer network topologies it is observed that multilayer networks are more fragile than single layers taken in isolation, and that the initial damage of an interdependent network can generate avalanches of cascading failures.

This important result provides a clear theoretical proof that, including interdependencies in multilayer infrastructures or transportation networks, can greatly reduce the robustness properties of multilayer networks.

Commenting on the specific scenario of a multiplex interdependent network with Poisson layers can be particularly instructive. Assuming that each of the M layers of the multiplex network has a Poisson degree distribution with average degree c, the fraction S of nodes in the MCGC when nodes are damaged with probability $f = 1 - p$ satisfies

$$S = p\left(1 - e^{-cS}\right)^{M}. \tag{11.1}$$

This equation can be compared with the equation determining the fraction S of nodes in the giant component of a single Poisson network with average degree c, when a fraction $f = 1 - p$ of its nodes are randomly damaged, which is given by

$$S = p\left(1 - e^{-cS}\right). \tag{11.2}$$

The only difference between these equations is the exponent $M > 1$ on the right hand side of Eq. (11.1). Nevertheless, this apparently small change is responsible for a dramatic change in the nature of the percolation-phase transition observed in interdependent multiplex networks. While on a single layer the percolation transition occurs continuously at a second-order transition, in interdependent multiplex networks the MCGC emerges discontinuously [66] (see Fig. 11.2). If one approaches the percolation threshold $p = p_c$, from $p > p_c$ the interdependent networks are affected by cascading failures that propagate throughout the structure, rapidly dismantling the entire multiplex network. Moreover, the percolation transition occurs for a critical value $p = p_c$ which is increasing with the number of layers M, indicating an increased fragility of the system. In fact, the multiplex network with larger number of layers M is destroyed when a smaller fraction of nodes $f_c = 1 - p_c$ is initially removed.

Finally, it has been shown that the critical behaviour of the model is characteristic of a *hybrid* phase transition [33] (see Sec. 3.2). In other words, the MCGC emerges at a discontinuous phase transition but is also characterized by a singular behaviour as the probability p that a node is damaged approaches its critical value p_c from above, i.e. for $p \to p_c^+$.

On a side note, we mention here that it is also possible to generalize the MCGC by defining the K-core percolation of interdependent multiplex networks. In this context the K-core is the giant component formed by a set of nodes such that every pair of nodes is connected in every layer at least by K paths. In Ref. [11] it has been shown that the K-core of interdependent multiplex networks emerges at the discontinuous hybrid phase transition for every $K \geq 1$ as long as the multiplex is non-trivial, i.e. it includes $M > 1$ layer. Given the space limitation here we will focus our attention on the MCGC.

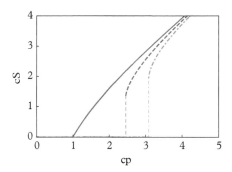

Fig. 11.2 *Percolation transition in single and multiplex networks. The solid line indicates the fraction S of nodes in the giant component of a single network (M = 1) with Poisson degree distribution with average c as a function of the probability p that a random node is not initially damaged. The dashed and the dot-dashed lines indicate the fraction of nodes in the MCGC of a multiplex network with Poisson layers with the same average degree c formed respectively by M = 2 and M = 3 layers.*

11.3   Interdependent multiplex networks without link overlap

11.3.1   The message-passing algorithm

On a locally tree-like multiplex network without link overlap, the MCGC can be found by using a suitable message-passing algorithm. This algorithm was first proposed by Son et al. in Ref. [281]. According to this algorithm, nodes send messages along their links to neighbouring nodes. These messages can be used to determine whether a node is in the MCGC or not. We distinguish three versions of the algorithm valid respectively when the network and the initially damaged nodes are known, when the network is known but the identities of the initially damaged nodes are not known and when we do not know the exact topology of the network.

Given network and initially damaged nodes

Assuming that the variable s_i assigned to each node i indicates whether a node is initially damaged ($s_i = 0$) or not ($s_i = 1$), each message sent by a node i to a node j indicates whether node i connects node j to nodes in the MCGC. Specifically, the message $\sigma^\alpha_{i \to j}$ that node i sends to a neighbouring node j in layer α is equal to one $\left(\sigma^\alpha_{i \to j} = 1\right)$ if the following conditions are met:

(a) node i is not initially damaged, i.e. $s_i = 1$;
(b) node i belongs to the MCGC even if the link between node j and node i is removed from the multiplex, i.e. for every layer $\beta = 1, 2 \ldots, M$ node i receives at least one positive message $\sigma^\beta_{\ell \to i} = 1$ from a node $\ell \neq j$ that is its neighbour in layer β.

If these conditions are not met, then $\sigma^\alpha_{i \to j} = 0$. These messages determine whether a node i belongs ($\sigma_i = 1$) or not ($\sigma_i = 0$) to the MCGC. In fact, node i belongs to the MCGC if and only if

(a) node i is not initially damaged, i.e. $s_i = 1$;
(b) node i receives at least one positive message $\sigma^\alpha_{\ell \to i} = 1$ from a neighbour ℓ of node i in every layer α.

These two algorithms directly translate to the message-passing equations

$$\sigma^\alpha_{i \to j} = s_i \left[1 - \prod_{\ell \in N_\alpha(i) \backslash j} (1 - \sigma^\alpha_{\ell \to i}) \right] \prod_{\beta \neq \alpha} \left[1 - \prod_{\ell \in N_\beta(i)} \left(1 - \sigma^\beta_{\ell \to i}\right) \right].$$

$$\sigma_i = s_i \prod_{\alpha=1}^{M} \left[1 - \prod_{\ell \in N_\alpha(i)} (1 - \sigma^\alpha_{\ell \to i}) \right], \tag{11.3}$$

where $N_\alpha(i)$ indicates the set of neighbours of node i in layer α. The fraction of nodes S in the MCGC is given by

$$S = \frac{1}{N} \sum_{i=1}^{N} \sigma_i. \tag{11.4}$$

This algorithm allows us to determine for every real network, as long as it is locally tree-like, which nodes are in the MCGC once the initially damaged nodes are given.

Note that the message $\sigma_{i \to j}^{\alpha}$ is one only if node i receives at least one positive message in each layer from node other than j. Therefore, this algorithm is not reducible to the algorithm determining the giant component in each single layer. Interestingly, this algorithm admits an epidemic-spreading interpretation [281] according to which each node i can spread the epidemics (send a positive message) to a given neighbour j only if in each layer of the multiplex network it is connected to at least one other neighbour $\ell \neq j$ which is spreading the infection to node i.

Given network and random damage of the nodes

Let us consider a random initial damage of the nodes $\{s_i\}_{i=1,2,\dots,N}$ where each node is damaged with probability $1-p$. The initially damaged nodes are indicated by the variables $\{s_i\}_{i=1,2,\dots,N}$ drawn from the distribution

$$\hat{P}(\{s_i\}) = \prod_{i=1}^{N} p^{s_i} (1-p)^{1-s_i}. \tag{11.5}$$

In this case we can formulate a message-passing algorithm which is able to determine with which probability a node is in the MCGC. The messages $\hat{\sigma}_{i \to j}^{\alpha}$ of this algorithm are sent between pairs of connected nodes and are given by the average of the messages $\sigma_{i \to j}^{\alpha}$ over the distribution $\hat{P}(\{s_i\})$ defined in Eq. (11.5). The messages $\hat{\sigma}_{i \to j}^{\alpha}$ determine the probability $\hat{\sigma}_i$ that the generic node i is in the MCGC. Here $\hat{\sigma}_i$ is the average of σ_i over the $\hat{P}(\{s_i\})$ distribution. The equations determining the messages $\hat{\sigma}_{i \to j}^{\alpha}$ and the probabilities $\hat{\sigma}_i$ are given by

$$\hat{\sigma}_{i \to j}^{\alpha} = p \left[1 - \prod_{\ell \in N_\alpha(i) \backslash j} \left(1 - \hat{\sigma}_{\ell \to i}^{\alpha} \right) \right] \prod_{\beta \neq \alpha} \left[1 - \prod_{\ell \in N_\beta(i)} \left(1 - \hat{\sigma}_{\ell \to i}^{\beta} \right) \right],$$

$$\hat{\sigma}_i = s_i \prod_{\alpha=1}^{M} \left[1 - \prod_{\ell \in N_\alpha(i)} \left(1 - \hat{\sigma}_{\ell \to i}^{\alpha} \right) \right]. \tag{11.6}$$

This algorithm allows us to establish with which probability a given node of the multiplex network is in the MCGC when only the probability of the initial damage of the nodes is given. Therefore, the expected fraction of nodes in the MCGC is given by

$$S = \frac{1}{N} \sum_{i=1}^{N} \hat{\sigma}_i. \tag{11.7}$$

Random network and random damage of the nodes

Let us consider a random multiplex network taken with probability given by Eq. (10.25) and a random realization of the initial damage described by the probability given by Eq. (11.5). The probability S_α' that by following a link in layer α we reach a node in the MCGC can be calculated by averaging the messages $\hat{\sigma}_{i \to j}^\alpha$ and the fraction of nodes in the MCGC, S can be obtained by averaging the probabilities $\hat{\sigma}_i$. In this way we obtain that S and S_α' satisfy

$$S_\alpha' = p \sum_{\mathbf{k}} \frac{k^{[\alpha]}}{\langle k^{[\alpha]} \rangle} P(\mathbf{k}) \left[1 - \left(1 - S_\alpha'\right)^{k^{[\alpha]}-1} \right] \prod_{\beta \neq \alpha} \left[1 - \left(1 - S_\beta'\right)^{k^{[\beta]}} \right],$$

$$S = p \sum_{\mathbf{k}} P(\mathbf{k}) \prod_{\alpha=1}^{M} \left[1 - \left(1 - S_\alpha'\right)^{k^{[\alpha]}} \right], \tag{11.8}$$

where $\mathbf{k}$ indicates the multiplex degree and $P(\mathbf{k})$ indicates the multiplex degree distribution.

Let us consider the case in which there are no correlations between the degrees of a node in different layers, and the degree distribution $P(\mathbf{k})$ follows

$$P(\mathbf{k}) = \prod_{\alpha=1}^{M} P^{[\alpha]}\left(k^{[\alpha]}\right), \tag{11.9}$$

where $P^{[\alpha]}(k^{[\alpha]})$ indicates the degree distribution of layer α. In this case Eqs (11.8) reduce to

$$S_\alpha' = p \left[1 - G_1^{[\alpha]}\left(1 - S_\alpha'\right) \right] \prod_{\beta \neq \alpha} \left[1 - G_0^{[\beta]}\left(1 - S_\beta'\right) \right],$$

$$S = p \prod_{\alpha=1}^{M} \left[1 - G_0^{[\alpha]}\left(1 - S_\alpha'\right) \right]. \tag{11.10}$$

Here the generating functions $G_0^{[\alpha]}(z)$ and $G_1^{[\alpha]}(z)$ of the degree distribution $P^{[\alpha]}\left(k^{[\alpha]}\right)$ of layer α are given by

$$G_0^{[\alpha]}(z) = \sum_{k^{[\alpha]}} P^{[\alpha]}\left(k^{[\alpha]}\right) z^{k^{[\alpha]}},$$

$$G_1^{[\alpha]}(z) = \sum_{k^{[\alpha]}} \frac{k^{[\alpha]}}{\langle k^{[\alpha]} \rangle} P^{[\alpha]}\left(k^{[\alpha]}\right) z^{k^{[\alpha]}-1}. \tag{11.11}$$

This algorithm allows us to determine which is the expected fraction of nodes in the MCGC in a random multiplex network. This algorithm has the advantage that by treating ensembles of networks with given multilayer degree distribution it is possible to explore the large size limit $N \to \infty$, characterizing in this way the properties of the phase transition.

11.3.2   Cascading failures

In an interdependent multiplex network, after the initial damage of the nodes failures propagate back and forth from one layer to the others according to the algorithmic definition of the MCGC given in Sec. 11.2.1. Here we discuss how cascades of failure events [66] take place on multiplex networks of M layers where every layer α is an uncorrelated network with degree distribution $P^{[\alpha]}\left(k^{[\alpha]}\right)$.

Initially, the nodes are independently damaged with probability $f = 1 - p$. Therefore the probability $\hat{P}(\{s_i\})$ follows Eq. (11.5).

Subsequently, for every layer $\alpha = 1, 2, \ldots, M$ only a fraction $S_\alpha(1)$ of replica nodes belonging to layer α are left in the giant component. The fractions of nodes $S_\alpha(1)$ are determined by the probabilities $S'_\alpha(1)$ that by following a random link in layer α we reach a replica node in the giant component of the same layer. Therefore, $S_\alpha(1)$ and $S'_\alpha(1)$ satisfy the equations valid for the percolation in each single layer (see Sec. 3.3.2)

$$S'_\alpha(1) = p\left[1 - G_1^{[\alpha]}(1 - S'_\alpha(1))\right],$$

$$S_\alpha(1) = p\left[1 - G_0^{[\alpha]}(1 - S'_\alpha(1))\right]. \tag{11.12}$$

After this first step, the damage propagates from one layer to the other through the interlinks and every node that has at least one replica node not in the giant component of its own layer is considered damaged in all the layers. This coupling between the layers, caused by the built-in interdependencies, induces avalanches of failure events propagating among the layers of the multiplex network. At every step $n > 1$ of the cascade dynamics, only a fraction $S_\alpha(n)$ of replica nodes are left in the giant component of layer α. These quantities depend on the probabilities $S'_\alpha(n)$ and $S'_\alpha(n-1)$ that by following a link of layer α we reach a replica node in the giant component of the same layer at iteration n and at iteration $n - 1$ respectively. Specifically, $S'_\alpha(n)$ and $S_\alpha(n)$ are determined by the equations

$$S'_\alpha(n) = p\prod_{\beta \neq \alpha}\left[1 - G_0^{[\beta]}\left(1 - S'_\beta(n-1)\right)\right]\left[1 - G_1^{[\alpha]}\left(1 - S'_\alpha(n)\right)\right],$$

$$S_\alpha(n) = p\prod_{\beta \neq \alpha}\left[1 - G_0^{[\beta]}\left(1 - S'_\beta(n-1)\right)\right]\left[1 - G_0^{[\alpha]}\left(1 - S'_\alpha(n)\right)\right]. \tag{11.13}$$

The first equation indicates that at iteration n, by following a link of layer α we reach a replica node (i, α) in the giant component of the same layer if and only if:

(a) the node i is not initially damaged, which occurs with probability p;

(b) all the replica nodes (i, β) belong to the giant component of their own layer β at iteration step $n - 1$, which occurs with probability

$$\prod_{\beta \neq \alpha} \left[1 - G_0^{[\beta]} \left(1 - S'_\beta(n - 1) \right) \right]$$

as long as we assume that node i is not initially damaged;

(c) among all the remaining links of replica node (i, α) in layer α, at least one reaches a node in the giant component of the same layer at iteration step n, which occurs with probability

$$\left[1 - G_1^{[\alpha]} \left(1 - S'_\alpha(n) \right) \right]$$

assuming that node i is not initially damaged.

Similarly, the second equation indicates that at iteration n the replica node (i, α) is in the giant component of the same layer if and only if:

(a) node i is not initially damaged, which occurs with probability p;

(b) all the replica nodes (i, β) belong to the giant component of their own layer at iteration step $n - 1$, which occurs with probability

$$\prod_{\beta \neq \alpha} \left[1 - G_0^{[\beta]} \left(1 - S'_\beta(n - 1) \right) \right]$$

as long as we assume that node i is not initially damaged;

(c) among all the links of replica node (i, α) in layer α, at least one reaches a node in the giant component of the same layer at iteration step n which occurs with probability

$$\left[1 - G_0^{[\alpha]} \left(1 - S'_\alpha(n) \right) \right]$$

assuming that node i is not initially damaged.

Equations (11.13) describe the cascade of failure events propagating back and forth between the interdependent layers of the multiplex network. Eventually these cascading events stop when the damage does not propagate further in the duplex network. The remaining nodes are the nodes in the MCGC. In fact, by putting

$$S'_\alpha(n) = S'_\alpha(n - 1) = S'_\alpha,$$
$$S_\alpha(n) = S_\alpha(n - 1) = S_\alpha \tag{11.14}$$

in Eqs (11.13) we observe these equations reduce to Eqs (11.10) with

$$S_\alpha = S \tag{11.15}$$

for every layer α where S indicates the fraction of nodes in the MCGC.

In Ref. [66] it was shown that the cascade of failure events close to the percolation threshold $p \simeq p_c$ are characterized by an average value of iterations $\langle n \rangle$ which scales as

$$\langle n \rangle \simeq \begin{cases} (p_c - p)^{-1/2} & \text{for } p < p_c, \\ N^{1/4} & \text{for } p = p_c, \\ [\ln N] (p - p_c)^{-1/2} & \text{for } p > p_c. \end{cases} \tag{11.16}$$

Therefore, close to the phase transition $p \simeq p_c$ the cascades become systemic events and propagate until the mutually connected component completely disappears from the network (see Fig. 11.3). At the critical point $p = p_c$ the cascades typically involve a diverging number of iterations $\langle n \rangle \to \infty$ in the limit $N \to \infty$. This phenomenon reveals that the interdependent networks can be much more fragile than single networks and are prone to abrupt cascading failures.

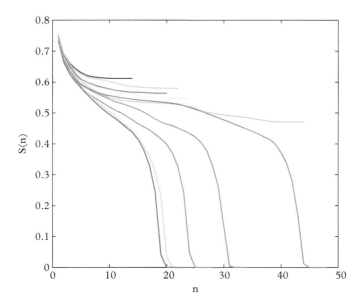

Fig. 11.3 *The fraction of nodes S(n) in the MCGC of a duplex network after n steps of the avalanche dynamics for several instances of networks from the same ensemble. The layers of the duplex network have Poisson degree distribution with the same average degree c = 2.455 and a number of nodes N = 10⁴. The probability p is set to p = 1.*

11.3.3   Case of a multiplex formed by M Poisson networks with the same average degree

In this paragraph we consider the case of a multiplex network formed by M Poisson layers with the same average degree, showing that the percolation transition for interdependent multiplex networks is discontinuous and hybrid.

In a multiplex formed by M Poisson networks with the same average degree $\langle k^{[\alpha]} \rangle = c$, $\forall \alpha$, where the degree distribution $P^{[\alpha]}\left(k^{[\alpha]}\right)$ in layer α is given by

$$P^{[\alpha]}\left(k^{[\alpha]}\right) = \frac{1}{k^{[\alpha]}!} c^{k^{[\alpha]}} e^{-c}, \tag{11.17}$$

for every $\alpha = 1, 2, \ldots, M$ the generating functions $G_0^{[\alpha]}(z)$ and $G_1^{[\alpha]}(z)$ reduce to

$$G_0^{[\alpha]}(z) = G_1^{[\alpha]}(z) = e^{-c(1-z)}. \tag{11.18}$$

Therefore, Eqs (11.10) reduce to

$$S' = S = p\left(1 - e^{-cS}\right)^M. \tag{11.19}$$

This equation can be written as

$$g_y(x) = x - y\left(1 - e^{-x}\right)^M = 0, \tag{11.20}$$

where $y = cp$ and $x = cS$. In Fig. 11.4 we plot the function $g_y(x)$ for $M = 2$. The equation $g_y(x) = 0$ always has a solution $S = 0$, but at $y = cp = y_c$ another solution appears discontinuously when the function $g_y(x)$ is tangent to the x-axis at the point

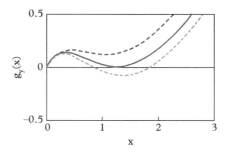

Fig. 11.4 *The function* $g_y(x)$ *for* $M = 2$ *is here plotted versus* x *for* y $= 2.2$ *(blue, dashed line)* y $= y_c =$ *2.455 (solid line) and* y $= 2.6$ *(yellow, dot-dashed line). The function* $g_y(x)$ *is zero for* x $= 0$ *and has a minimum. This minimum is tangent to the x-axis for* y $= y_c$. *Therefore, for* y $<$ y_c *the unique solution to the equation* $g_y(x) = 0$ *is* x $= 0$. *Instead for* y $=$ y_c *a non-zero solution emerges discontinuously at* x $=$ x_c.

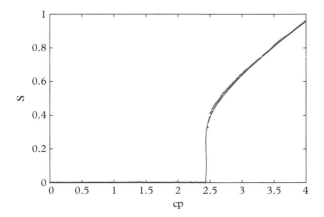

Fig. 11.5 *The fraction S of nodes in the MCGC of a Poisson multiplex network with M = 2 layers with the same average degree c and a number of nodes N = 10⁴ (symbol) is compared to the theoretical prediction (solid line). Here p indicates the probability that a random node is not initially damaged. At p = p_c = 2.455/c the size S of the MCGC has a discontinuous jump. For p → p_c⁺, S shows a square-root singularity.*

$x = x_c, y = y_c$. Therefore, in order to find the critical values $x = x_c$ and $y = y_c$, we can impose the set of equations [66, 281]

$$g_y(x) = 0$$
$$\frac{dg_y(x)}{dx} = 0. \qquad (11.21)$$

Solving numerically this system of equations for $M = 2$ we get $y = y_c = cp_c = 2.45541\ldots$ and $cS_c = x_c = 1.25643\ldots$ Therefore, the size of the MCGC for $M = 2$ jumps from a value $S_c = 1.25643/c$ to $S = 0$ at $p_c = 2.45541/c$ (see Fig. 11.5).

Comparing this result with the result obtained for a single Poisson network we observe [66]:

(a) for $M = 2$ the transition is discontinuous, while for $M = 1$ it is continuous;

(b) for $M = 2$ the system is much more fragile because we observe the transition already when a fraction $f_c = 1 - p_c = 1 - 2.45541/c$ of nodes are initially removed, while for $M = 1$ the transition only occurs for $f_c = 1 - 1/c$.

For general $M > 1$, it can be shown that we always get a minimum of $g_y(x)$, yielding a discontinuous percolation transition in interdependent multiplex networks. Interestingly, f_c decreases for a larger value of M, showing that a multiplex network with more layers is more fragile than a multiplex network with fewer layers.

In a multiplex network with M layers formed by Poisson networks with the same average degree c, the MCGC emerges at a discontinuous *hybrid transition* [33] characterized by a square-root singularity; in other words the order parameter, close to the transition, scales as

$$S - S_c \propto (p - p_c)^{1/2} \tag{11.22}$$

for $cp - y_c = \epsilon$ and $0 < \epsilon \ll 1$.

Let us discuss in detail the simple derivation of this result. For $p = p_c + \delta p$ with $\delta p > 0$ (or equivalently $y \simeq y_c$ and $y = y_c + \delta y$ with $\delta y > 0$) we can expand Eq. (11.20) around the point $y = y_c$, $x = x_c$, obtaining

$$0 = g_y(x) = x_c + \delta x - (y_c + \delta y)(1 - e^{-x_c})^M - (y_c + \delta y) \left[\frac{d(1 - e^{-x})^M}{dx} \bigg|_{x=x_c} \delta x \right.$$

$$+ \frac{1}{2} \frac{d^2 (1 - e^{-x})^M}{dx^2} \bigg|_{x=x_c} (\delta x)^2 \Big] + o\left((\delta x)^2 \right) \bigg].$$

Since at $x = x_c$ and $y = y_c$ we have $g_y(x) = 0$ and $\frac{dg_y(x)}{dx} = 0$, it follows that

$$x_c = y_c \left(1 - e^{-x_c} \right)^M,$$

$$1 = y_c \frac{d \left(1 - e^{-x} \right)^M}{dx} \bigg|_{x=x_c}. \tag{11.23}$$

Using these two equations, neglecting higher-order terms we get

$$0 = \delta y \left(1 - e^{-x_c} \right)^M + y_c \frac{1}{2} \frac{d^2 \left(1 - e^{-x} \right)^M}{dx^2} \bigg|_{x=x_c} (\delta x)^2. \tag{11.24}$$

Since it can be shown that at $x = x_c$ we have $\frac{d^2 (1 - e^{-x})^M}{dx^2} \bigg|_{x=x_c} < 0$, it follows that

$$\delta y \propto (\delta x)^2, \tag{11.25}$$

which for $p \to p_c^+$ (by putting $y = cp$ and $x = cS$) yields the singularity

$$S - S_c \propto (p - p_c)^{\hat{\beta}}, \tag{11.26}$$

with the dynamical critical exponent $\hat{\beta}$ given by

$$\hat{\beta} = \frac{1}{2}. \tag{11.27}$$

11.3.4 Case of a duplex network formed by two Poisson networks with different average degree

In a network formed by $M = 2$ Poisson networks (layer A and layer B) with respectively average degrees c_A and c_B, Eqs (11.10) determining the emergence of the MCGC reduce to a single equation for $S = S'_{[1]} = S'_{[2]}$ given by [281]

$$S = p \left(1 - e^{-c_A S}\right) \left(1 - e^{-c_B S}\right). \tag{11.28}$$

The emergence of the mutually connected component is described by a discontinuous hybrid transition at the points of the phase diagram satisfying simultaneously the following set of equations

$$h(S) = S - p \left(1 - e^{-c_A S}\right) \left(1 - e^{-c_B S}\right) = 0$$
$$h'(S) = 0. \tag{11.29}$$

The phase diagram of the model in the plane (pc_A, pc_B) is shown in Fig. 11.6.

11.3.5 Multiplex networks formed by layers of scale-free networks

For interdependent multiplex networks formed by scale-free layers, the MCGC emerges at a discontinuous hybrid transition at a *finite percolation threshold* $p = p_c$ [66, 33]. As for Poisson networks, the transition is characterized by a square-root singularity for $p \to p_c^+$. In Fig. 11.7 from Ref. [33] we show the size of the MCGC for two scale-free networks with the same degree distribution given by

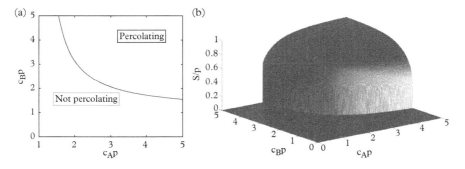

Fig. 11.6 *The phase diagram of percolation with interdependencies for a duplex network formed by two layers with Poisson degree distribution and average degree respectively* c_A *and* c_B *(panel (a)). Fraction* S *of nodes in the MCGC of the same network as a function of the average degrees* c_A *and* c_B *and the probability that a node is not randomly damaged is* p *(panel (b)).*

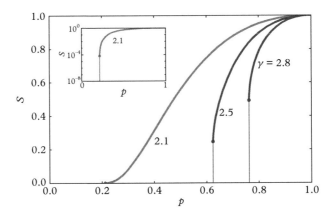

Fig. 11.7 *Fraction of nodes S in the MCGC as a function of p for two symmetric power-law-distributed networks with, from right to left, power-law exponent γ = 2.8, 2.5 and γ = 2.1 and same minimal degree. The height of the jump becomes very small as the power-law exponent γ approaches two, but is not zero, as seen in the inset, which shows S vs p on a logarithmic vertical scale for γ = 2.1. Reprinted figure with permission from Ref. [33] ©2012 by the American Physical Society.*

$$P^{[\alpha]}\left(k^{[\alpha]}\right) = C\frac{1}{\left(k^{[\alpha]}\right)^{\gamma}}. \tag{11.30}$$

Keeping the minimum degree constant, as $\gamma \to 2$ the discontinuity of S becomes smaller and smaller but is not vanishing as long as $\gamma > 2$. In Ref. [66] it was shown that a major difference exists between scale-free single-layer and multiplex networks. In fact, if we consider a multiplex network formed by two scale-free networks with the same degree distribution and compare their percolation threshold p_c keeping the average degree $\langle k^{[\alpha]}\rangle$ constant and changing only the power-law exponent γ, we obtain that the percolation threshold p_c increases as γ is decreased. This implies that multiplex networks formed by broader scale-free distributions are more fragile than multiplex networks with a steeper degree distribution. This surprising result shows that the robustness of multiplex networks has very new and unexpected features. Therefore, for interdependent multiplex networks, having scale-free layers does not provide an advantage with respect to their robustness, as happens for single networks. This is due to the fact that high-degree nodes in one layer might be interdependent with low-degree nodes of the other layer, greatly increasing the fragility of the entire interdependent multiplex network.

11.3.6   Percolation in networks with degree correlations

The presence of degree correlations can modulate significantly the robustness properties of interdependent multiplex networks by moving the position of the percolation threshold p_c[67, 240, 33, 213]. The effect of different types of correlation between the degree of the nodes in different layers has been investigated in several papers including Refs [33, 213].

For Maximally Negative (MN) correlated multiplex networks, the hub nodes are interdependent on low-degree nodes. Therefore, the hub nodes that play an important role in keeping each layer connected, can be easily damaged if their interdependent low-degree nodes are damaged. It follows that Maximally Negative degree correlations yield more fragile multiplex networks (larger value of p_c). Conversely, for Maximally Positive (MP) correlated multiplex networks, where the hub nodes are interdependent on hub nodes instead, more robust multiplex networks are generated (smaller value of p_c). The networks with uncorrelated degrees in the different layers have a percolation threshold p_c that is in between the previous two. In Fig. 11.8, the dependence of the fraction of nodes in the MCGC of correlated Poisson networks is shown as a function of the average degree of the nodes.

In Ref. [213], the role of targeted attack of the multiplex networks is discussed. Other papers have investigated the effect of the degree correlations within a layer on the percolation properties of the network, and in particular assortativity [320] and the effect of clustering [162].

11.3.7 Percolation in spatial multiplex networks

The vast majority of infrastructures, from the power grid to the water supply network, are strongly affected by their embedding space that influences the robustness properties

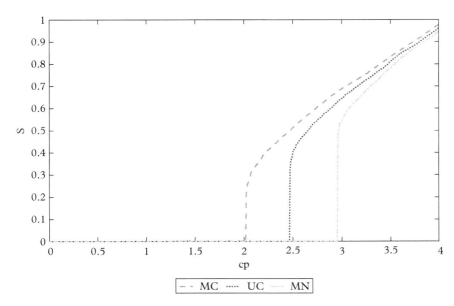

Fig. 11.8 *The size S of the MCGC for duplex networks with different levels of degree correlation. The duplex networks have two Poisson layers with the same average degree c = 4 and with N = 10^4 nodes. Here MP indicates Maximally Positive degree correlations, MN indicates Maximally Negative degree correlations, UC indicates no degree correlations.*

of the network very significantly [25]. In most cases the robustness of globally inter-connected infrastructures cannot be fully captured if we do not take into consideration the effect of interdependencies. However, when multilayer networks describing inter-connected infrastructures have a non-trivial network geometry surprising phenomena can be found. For instance, when studying two-dimensional interdependent lattices the transition can become continuous also in presence of interdependencies. Additionally, for diluted random lattices the transition can become even less sharp than classical perco-lation (displaying larger dynamical critical exponent $\hat{\beta}$) [282]. However, interdependent lattices are only apparently more robust than random or scale-free networks. In fact, it is possible to observe a continuous phase transition only if the layers have the same spatial dimension and the interlinks (and therefore the interdependencies) couple the nodes that are located at corresponding equivalent or almost equivalent positions. In fact, if we assume as in Ref. [193] to have two two-dimensional interdependent lattices, where interlinks join nodes at distance r (see Fig. 11.9) there is characteristic finite value r_{max} such that for $r < r_{max}$ the transition is continuous, while for $r > r_{max}$ the transition is discontinuous.

For these spatial interdependent lattices there is also the possibility of exploring the response of the system to particular configurations of the initial damage which are localized in some region of the two-dimensional space [38]. Interestingly, it is found that even if the multiplex network can sustain the damage of a finite fraction of randomly chosen nodes, they might get completely disrupted under localized damage including an infinitesimal fraction of the nodes as long as the damage affects a large-enough region of the lattice.

Finally, it is possible to consider interdependent, spatially embedded random networks where nodes are placed on a two-dimensional grid and pairs of nodes at distance l on the grid are connected with probability $\pi(l) \propto e^{-l/l_0}$. Therefore, for $l_0 \to \infty$ the network becomes a random network, while for small values of l_0 the connections are more likely to be established between nodes that are nearest neighbours on the grid. Also, for these

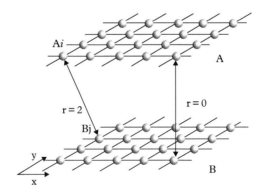

Fig. 11.9 *The nature of the percolation transition can change depending on the typical distance of interdependent nodes, from discontinuous for large typical distances to continuous for small typical distances. Reprinted figure with permission from [193] ©2012 by the American Physical Society.*

networks it is possible to observe a change in the critical properties of the percolation transition that is continuous for $l_0 < l_c$ and is discontinuous for $l_0 > l_c$ [86].

11.4   Interdependent multiplex networks with link overlap

11.4.1   The effect of link overlap on percolation

Numerous multilayer networks have a significant link overlap which explains the need to explore the percolation transition on this type of correlated multilayer structure.

The overlap of the links can change the properties of the percolation transition significantly. In fact, in the limit in which we have a multiplex network of M totally overlapping layers, the MCGC of the multiplex network reduces to the giant component of a single layer, and the percolation transition is of the second order.

For these reasons, it is important to explore whether the emergence of the mutually connected component becomes continuous when the overlap between the layers is above a threshold value.

Recently, two approaches were used to describe the transition in multiplex networks with link overlap. The first approach consists of a coarse-grained description of the multiplex network in terms of supernodes [161, 212] and is restricted to duplex networks. The second approach that we will describe in the following paragraphs is instead based on message-passing algorithms and the local tree-like approximation [76, 32] and extends to multiplex networks with an arbitrary number of layers.

Interestingly, it turns out that the message-passing algorithm valid in the absence of overlap (described in Sec.11.3.1) needs to be modified appropriately to describe the percolation of interdependent multiplex networks in realistic scenarios.

The major modifications that need to be considered are the following two:

(i)   The message sent from a node i to a node j connected to node i in at least one layer is not a scalar anymore, as in the absence of link overlap, but is instead a vector

$$\vec{n}_{i \to j} = \left(n_{i \to j}^{[1]}, n_{i \to j}^{[2]}, \ldots, n_{i \to j}^{[\alpha]}, \ldots, n_{i \to j}^{[M]} \right). \tag{11.31}$$

(ii) The message $\vec{n}_{i \to j}$ is determined under the assumption that node j belongs to the MCGC. In this hypothesis the elements $n_{i \to j}^{[\alpha]}$ indicate whether $(n_{i \to j}^{[\alpha]} = 1)$ or not $(n_{i \to j}^{[\alpha]} = 0)$ node i connects node j to the MCGC through links in layer α.

The characterization of the percolation transition in interdependent multiplex networks with link overlap reveals that the percolation remains always discontinuous, with the exception of the limiting case in which there is complete overlap of all the layers of the multiplex network.

11.4.2 Message-passing algorithms

We consider a multiplex network with M layers and adjacency matrix $\mathbf{a}^{[\alpha]}$ in each layer $\alpha = 1, 2, \ldots, M$. Initially, we assume that we know the set of nodes that are initially damaged. The configuration of the initial damage is indicated by the variables $\{s_i\}$ where $s_i = 0$ ($s_i = 1$) if node i is (is not) damaged. The message-passing algorithm for a given initial damage configuration determines whether node i does ($\sigma_i = 1$) or does not ($\sigma_i = 0$) belong to the MCGC, as long as the multiplex network is locally tree-like. The algorithm requires the determination of the message $\vec{n}_{i \to j}$ given by Eq. (11.31). The message $n_{i \to j}^{[\alpha]}$ is set to one, $n_{i \to j}^{[\alpha]} = 1$, if and only if the following four conditions are satisfied:

(i) node i is not initially damaged, i.e. $s_i = 1$;

(ii) node j is a neighbour of node i in layer α, i.e. $a_{ij}^{[\alpha]} = 1$;

(iii) assuming node j belongs to the MCGC, node i is in the MCGC, i.e. node i has in each layer at least one neighbour that connects it to nodes in the MCGC;

(iv) node i connects node j to the nodes in the MCGC through links of layer α:

These messages are determined by the recursive message-passing equations

$$n_{i \to j}^{[\alpha]} = \delta\left(v_{i \to j}, M\right) a_{ij}^{[\alpha]} s_i \left[1 - \prod_{\ell \in N(i) \setminus j} \left(1 - n_{\ell \to i}^{[\alpha]}\right) \right]. \tag{11.32}$$

Here $v_{i \to j}$ indicates in how many layers node i is connected to the MCGC assuming that node j also belongs to the MCGC, and it is given by

$$v_{i \to j} = \sum_{\alpha=1}^{M} \left\{ \left[1 - \prod_{\ell \in N(i) \setminus j} \left(1 - n_{\ell \to i}^{[\alpha]}\right) \right] + a_{ij}^{[\alpha]} \prod_{\ell \in N(i) \setminus j} \left(1 - n_{\ell \to i}^{[\alpha]}\right) \right\}. \tag{11.33}$$

Finally, the value of σ_i for any generic node i can be expressed in terms of the messages $\vec{n}_{i \to j}$ as

$$\sigma_i = s_i \prod_{\alpha=1}^{M} \left[1 - \prod_{\ell \in N(i)} \left(1 - n_{\ell \to i}^{[\alpha]}\right) \right]. \tag{11.34}$$

In Fig. 11.10 we define the complete set of non-trivial messages in a multiplex network of two layers. If node i and node j are connected only in one layer there is only one type of non-trivial multilink between them. If node i and node j are connected in both layers (i.e. they are connected by a multilink $\vec{m}_{ij} = (1, 1)$) there are three possible types of non-trivial messages that node i can send to node j, i.e. $\vec{n}_{i \to j} = (1, 0)$, $\vec{n}_{i \to j} = (0, 1)$ and $\vec{n}_{i \to j} = (1, 1)$.

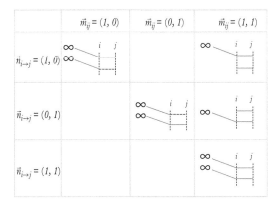

	$\vec{m}_{ij} = (1, 0)$	$\vec{m}_{ij} = (0, 1)$	$\vec{m}_{ij} = (1, 1)$
$\vec{n}_{i\to j} = (1, 0)$			
$\vec{n}_{i\to j} = (0, 1)$			
$\vec{n}_{i\to j} = (1, 1)$			

Fig. 11.10 *Possible types of non-trivial messages $\vec{n}_{i\to j}$ that can be sent from node i to node j connected by the multilink $\vec{m}_{ij} = (a^{[1]}_{ij}, a^{[2]}_{ij})$ in a duplex network. The multilink between node i and node j is represented as in Fig. 7.3; solid lines connected to the symbol of infinity indicate that through those links node i is connected to nodes in the MCGC.*

When the configuration of the initial damage is not known but only the probability $\hat{P}(\{s_i\})$ given by Eq. (11.5) is known, an alternative message-passing algorithm needs to be formulated. This novel message-passing algorithm needs to be derived from the algorithm described above by averaging over the $\hat{P}(\{s_i\})$ distribution. However, since the messages $\vec{n}_{i\to j}$ in presence of link overlap are correlated, care should be put in to performing this average. It turns out that the useful set of messages that should be used when the actual configuration of the initial damage is not known are $\hat{\sigma}^{\vec{m}_{ij}, \vec{n}}_{i\to j}$ which indicate the probability that for a random realization of the initial damage node i sends to node j the message $\vec{n}_{i\to j} = \vec{n}$.

Finally, when the topology of the network is also not known and one wishes to perform an average over a multiplex network ensemble, the relevant set of variables are: S, the expected fraction of nodes in the MCGC and the average messages $S_{\vec{m}, \vec{n}}$, indicating the probability that, starting from a given node and following a multilink $\vec{m}$, we reach a node that connects the original node to the MCGC in the layers where $n^{[\alpha]} = 1$ exclusively. In other words, $S_{\vec{m}, \vec{n}}$ is the average of $\hat{\sigma}^{\vec{m}_{ij}, \vec{n}}_{i\to j}$ over a pair of nodes (i, j) connected by a multilink $\vec{m}_{ij} = \vec{m}$.

Here we leave the somewhat technical derivation of these algorithms to Appendix D and we discuss exclusively the phase diagram of a multiplex network with Poisson multidegree distributions revealing that the continuous transition is only observed for multiplex networks with complete overlap.

We consider a duplex network having two layers with Poisson multidegree distribution and $\langle k^{(1,0)} \rangle = \langle k^{(0,1)} \rangle = c_1$ with $\langle k^{(1,1)} \rangle = c_2$. The full phase diagram of the model is displayed in Fig. 11.11. The MCGC component emerges as a continuous phase transition only when all links overlap, i.e. $pc_2 = 1, pc_1 = 0$ when we recover the case of percolation in a single Poisson network whereas if $pc_1 \neq 0$ the transition is

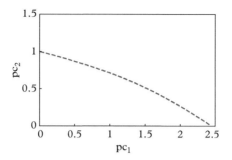

Fig. 11.11 *The critical line of a discontinuous hybrid phase transition is shown for multiplex networks with two layers and Poisson multidegree distribution with* $\langle \mathrm{k}^{(1,0)} \rangle = \langle \mathrm{k}^{(0,1)} \rangle = c_1$ *with* $\langle \mathrm{k}^{(1,1)} \rangle = c_2$. *Reprinted figure from Ref. [76].*

discontinuous. Moreover, we observe that for $c_2 = 0$ we recover the known results of percolation transition in interdependent duplex Poisson networks with no link overlap, i.e. $c_1 p_c = 2.455 \ldots$.

More complex phase diagrams, including multiple discontinuous phase transitions, can be obtained for multiplex networks having correlated multidegree distributions. For instance, in Ref. [32] very interesting phase diagrams have been observed for duplex networks in which multidegrees $k_i^{(1,1)}$ are either positively or negatively correlated with multidegrees $k_i^{(1,0)}$ and $k_i^{(0,1)}$.

11.5   Partial and redundant interdependencies

11.5.1   Partial interdependence

It is rather interesting to consider the effect of *partial interdependence* in the robustness of interdependent multiplex networks. By partial interdependence we indicate that this interdependence is not always present between the replica nodes. As a function of the probability r that a replica node is interdependent on the other replica nodes, the transition can turn from discontinuous (for $r > r_T$) to continuous (for $r = r_T$) [238, 281]. Here we show how a passage between a discontinuous and a continuous transition can occur in the simplest setting of a multiplex network, where every layer α is an uncorrelated network with degree distribution $P^{[\alpha]}\left(k^{[\alpha]}\right)$.

The equations satisfied by the fraction S of nodes in the MCGC in this ensemble of multiplex networks are here derived starting from the message-passing algorithm that determines which nodes are in the MCGC in a sparse locally tree-like multiplex network.

Using the same notation as in Sec. 11.3, the initial damage configuration is determined by the variables $\{s_i\}$ indicating whether node i is damaged ($s_i = 0$) or undamaged ($s_i = 0$). Additionally, the variables $\{q_i\}$ indicate whether that node i is interdependent ($q_i = 1$) or not interdependent ($q_i = 0$) with its replica nodes. Therefore, if $q_i = 0$ the messages $\sigma_{i \to j}$ sent by node i follow the Eqs (3.1) valid on a single layer. If, on the contrary, $q_i = 1$, the messages follow the Eqs 11.3 valid for interdependent layers. A similar argument can be made for the probability $\sigma_{i\alpha}$ that a replica node (i,α) belongs to the MCGC. These considerations lead to the message-passing equations

$$
\sigma_{i\to j}^{\alpha} = s_i \left[1 - \prod_{\ell\in N_\alpha(i)\backslash j} \left(1 - \sigma_{\ell\to i}^{\alpha}\right) \right] \prod_{\beta\neq\alpha} \left[1 - q_i \prod_{\ell\in N_\beta(i)} \left(1 - \sigma_{\ell\to i}^{\beta}\right) \right],
$$

$$
\sigma_{i\alpha} = s_i \left[1 - \prod_{\ell\in N_\alpha(i)} \left(1 - \sigma_{\ell\to i}^{\alpha}\right) \right] \prod_{\beta\neq\alpha} \left[1 - q_i \prod_{\ell\in N_\beta(i)} \left(1 - \sigma_{\ell\to i}^{\beta}\right) \right], \tag{11.35}
$$

valid in the locally tree-like multiplex networks without link overlap. When every node is damaged with probability $f = 1 - p$, and we assume that every replica node is interdependent on its other replica nodes with probability r, the novel set of message-passing equations reads

$$
\hat{\sigma}_{i\to j}^{\alpha} = p \left[1 - \prod_{\ell\in N_\alpha(i)\backslash j} \left(1 - \hat{\sigma}_{\ell\to i}^{\alpha}\right) \right] \prod_{\beta\neq\alpha} \left[1 - r \prod_{\ell\in N_\beta(i)} \left(1 - \hat{\sigma}_{\ell\to i}^{\beta}\right) \right],
$$

$$
\hat{\sigma}_{i\alpha} = p \left[1 - \prod_{\ell\in N_\alpha(i)} \left(1 - \hat{\sigma}_{\ell\to i}^{\alpha}\right) \right] \prod_{\beta\neq\alpha} \left[1 - r \prod_{\ell\in N_\beta(i)} \left(1 - \hat{\sigma}_{\ell\to i}^{\beta}\right) \right]. \tag{11.36}
$$

Here $\hat{\sigma}_{i\alpha}$ indicates the probability that node (i,α) is in the MCGC and $\hat{\sigma}_{i\to j}$ indicates the probability that node (i,α) connects node (j,α) to other nodes belonging to the MCGC.

Finally, by averaging once more this last set of equations over the ensemble of multiplex networks in which each layer is an uncorrelated network with degree distribution $P^{[\alpha]}\left(k^{[\alpha]}\right)$, we obtain that the fraction S_α of replica nodes of layer α that is in the MCGC is given by

$$
S_\alpha = p \left[1 - G_0^{[\alpha]}\left(1 - S_\alpha'\right) \right] \left\{ \prod_{\beta\neq\alpha} \left[1 - rG_0^{[\beta]}\left(1 - S_\beta'\right) \right] \right\}, \tag{11.37}
$$

where the average messages S_α' are given by

$$
S_\alpha' = p \left[1 - G_1^{[\alpha]}\left(1 - S_\alpha'\right) \right] \left\{ \prod_{\beta\neq\alpha} \left[1 - rG_0^{[\beta]}\left(1 - S_\beta'\right) \right] \right\}. \tag{11.38}
$$

In particular, if all layers α have the same degree distribution, then $G_0^{[\alpha]}(z) = G_0(z)$ and $G_1^{[\alpha]}(z) = G_1(z)$ and $S_\alpha' = S'$ is given by

$$S' = p[1 - G_1(1 - S')] \left[1 - rG_0(1 - S')\right]^{M-1}. \tag{11.39}$$

Furthermore, the probability $S = S_\alpha$ that a node in a given layer is in the mutually connected component is given by

$$S = p\left[1 - G_0(1 - S')\right]\left[1 - rG_0(1 - S')\right]^{M-1}. \tag{11.40}$$

For $r = 0$ the nodes of each layer have no interdependencies with the nodes of the other layers; therefore each layer will percolate independently, and the percolation transition will be a second-order continuous phase transition. On the contrary, when $r = 1$ we have complete interdependency of each node of a given layer on all its replica nodes in the other $M - 1$ layers, and the transition will be a discontinuous hybrid phase transition. The process for general values of r follows two different regimes separated by a tricritical point $(r = r^T, p = p^T)$ at which the transition changes its nature from discontinuous (for $r > r^T$) to continuous (for $r < r^T$). Figure 11.12 reports simulation results of these two regimes, in duplex networks of different degree distributions.

We consider a multiplex network formed by M Poisson networks with the same average degree $\langle k \rangle = c$. The order parameter $S' = S$ is the solution of the equation

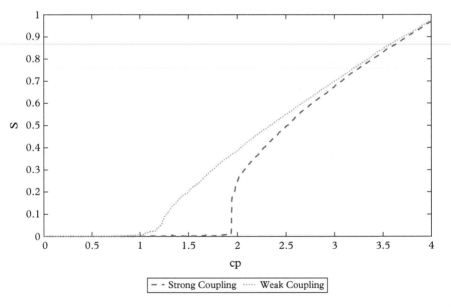

Fig. 11.12 *The fraction of nodes S in the MCGC of a duplex network with partially interdependent layers displays a continuous phase transition for weak coupling (r = 0.05) and a discontinuous transition for strong coupling (r = 0.8). Here the duplex network is formed by two Poisson layers with the same average degree c = 4 and N = 10^4 nodes.*

$$S = p \left(1 - re^{-cS}\right)^{M-1} \left(1 - e^{-cS}\right).$$
(11.41)

The equation (11.41) determining the size of the MCGC can be written as

$$h(S) = S - p \left(1 - re^{-cS}\right)^{M-1} \left(1 - e^{-cS}\right) = 0.$$
(11.42)

The points of the hybrid phase transition can be found by imposing

$$h(S) = 0,$$
$$h'(S) = 0$$
(11.43)

and by looking for a non-trivial solution $S > 0$. This solution can be found for any $r > r^T$ characterizing the level of partial interdependence of the tricritical point of the model. For $r < r^T$, the emergence of the mutually connected component is dictated by a second-order phase transition at a critical value of p that can be obtained by imposing the conditions

$$h(0) = 0,$$
$$h'(0) = 0.$$
(11.44)

The tricritical point separates these two regimes, and results from imposing

$$h(0) = 0,$$
$$h'(0) = 0,$$
$$h''(0) = 0.$$
(11.45)

This system of equations yields the parameters of the tricritical point:

$$r^T = \frac{1}{(2M-1)},$$
$$cp^T = \left(\frac{2M-1}{2M-2}\right)^{M-1}.$$
(11.46)

Therefore, the transition is continuous for $r < r^T$ and discontinuous for $r > r^T$ as can be shown by numerical simulations of the model (see Fig. 11.12).

Different papers have proposed modifications of this model and studied, for instance, the effect of different degree distributions and degree correlations [239, 27, 160, 319]. Finally, in Ref. [297], the effect of degree correlations has been studied on a duplex network with partial interdependence.

11.5.2　Redundant interdependencies

Redundant interdependencies: the motivation

This section directly addresses the robustness of multiplex networks with more than two layers.

Until now, we have always assumed that in the case of a multiplex network with more than two layers every node must be interdependent on each of its linked nodes in the other layers. This assumption predicts that by increasing the number of layers the multiplex network becomes increasingly fragile (see Fig. 11.2).

In natural and man-made systems, interdependencies do not follow this rule, and the actual rules determining the function of any given node are likely to be in general a very complex set of interdependencies, determined by complex Boolean rules.

In Ref. [257] it has been shown that in the presence of redundant interdependencies, increasing the number of layers of a multiplex can actually boost the robustness of the network.

This result opens a completely new perspective both for policy makers and practitioners that need to design interdependent infrastructures in a robust and resilient way.

Redundant Mutually Connected Giant Components

In the presence of redundant interdependencies the robustness of the network can be evaluated by considering the Redundant Mutually Connected Giant Component (RMCGC). In a multiplex network of M layers, the nodes that belong to the RMCGC can be found by following the algorithm [257]. After the initial damage to the nodes,

- (i) the giant component of each layer $\alpha = 1, 2 \dots, M$ is determined, evaluating the effect of the damaged nodes in each single layer;
- (ii) every replica node that has no other replica node in the giant component of its proper layer is removed from the network and considered as damaged;
- (iii) if no new damaged nodes are found at step (ii), then the algorithm stops; otherwise it proceeds, starting again from step (i).

The set of replica nodes that are not damaged when the algorithm stops belongs to the RMCGC.

The main difference between the percolation model discussed in Sec. 11.2.1 and the consequent definition of the MCGC is that for obtaining the MCGC step (ii) must be substituted with 'every replica node that has at least a single replica node not in the giant component of its proper layer is removed from the network and considered as damaged, i.e., if a replica node is damaged all its interdependent replica nodes are damaged'. In particular, the RMCGC and the MCGC are the same for $M = 2$ layers, but they differ as long as the number of layers is above two, i.e. $M > 2$. In the latter case, the RMCGC naturally introduces the notion of redundancy or complementarity to interdependent nodes. In fact, for a node to be in the RMCGC it does not need all of its interdependent nodes to be functional, but at least one must be.

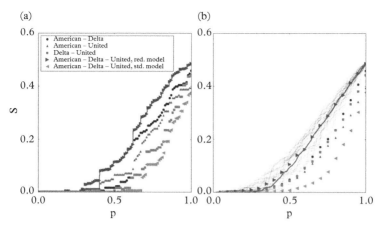

Fig. 11.13 *Percolation transition in the US air transportation network (only US domestic flights operated in January 2014 by the three major carriers in the US (American Airlines, Delta and United) are considered). The number of nodes in the network is* N = 183. *Panel (a) displays the results of a single realization of the random damage in the three possible multiplex networks composed of only two layers: American–Delta (black circles) American–United (blue triangles) and Delta–United (red squares). The size of the RMCGC and the MCGC of the multiplex network composed of all three layers are represented respectively as purple triangles and grey triangles. Panel (b) displays the same quantities as panel (a), but for average values over 100 independent realizations of the random damage. Purple thin lines stand for the 100 independent realizations of random damage considered in the case of the redundant model. As is apparent from the figure, the average doesn't capture the behaviour of single instances of disorder well and fluctuations are rather large for a wide range of possible values of p. Reprinted figure from Ref. [257].*

In Fig. 11.13 the size of the RMCGC of the airport network of major US airlines with $M = 3$ layers is compared with the MCGC of the same multiplex network, revealing that in the presence of interdependencies the percolation threshold is reduced. Therefore, the multiplex network displays an increased robustness to random damage. Interestingly, when comparing the RMCGC of the multiplex with $M = 3$ layers with the RMCGC of the multiplex networks formed by just two of its layers, an improvement on the robustness properties (signalled by a decrease in value of the percolation threshold) is also observed. Therefore, while for the MCGC increasing the number of layers is deleterious, in the presence of redundant interdependencies the robustness properties of the multiplex network will improve with the addition of new layers.

Message-passing equations determining the RMCGC

The initially inflicted random damage is defined by the variables $s_{i\alpha}$. In particular, if $s_{i\alpha} = 0$ the node (i, α) is initially damaged; otherwise $s_{i\alpha} = 1$. In this case, the indicator function $\sigma_{i\alpha}$ determines whether or not the node (i, α) is in the RMCGC and depends on the messages $\sigma_{i \to j}^{\alpha}$. In a locally tree-like multiplex network without link overlap the messages $\sigma_{i \to j}^{\alpha}$ and the indicator function $\sigma_{i\alpha}$ satisfy

$$\sigma_{i \to j}^{\alpha} = s_{i\alpha} \left[1 - \prod_{\ell \in N_\alpha(i) \setminus j} \left(1 - \sigma_{\ell \to i}^{\alpha}\right) \right] \left\{ 1 - \prod_{\beta \neq \alpha} \left[1 - s_{i\beta} + s_{i\beta} \prod_{\ell \in N_\beta(i)} \left(1 - \sigma_{\ell \to i}^{\beta}\right) \right] \right\},$$

$$\sigma_{i\alpha} = s_{i\alpha} \left[1 - \prod_{\ell \in N_\alpha(i)} \left(1 - \sigma_{\ell \to i}^{\alpha}\right) \right] \left\{ 1 - \prod_{\beta \neq \alpha} \left[1 - s_{i\beta} + s_{i\beta} \prod_{\ell \in N_\beta(i)} \left(1 - \sigma_{\ell \to i}^{\beta}\right) \right] \right\}.$$

Here we assume that the initial damage is inflicted on each replica node independently with probability $1 - p$; i.e. the probability $\hat{P}(\{s_{i\alpha}\})$ is given by

$$\hat{P}(\{s_{i\alpha}\}) = \prod_{\alpha=1}^{M} \prod_{i=1}^{N} p^{s_{i\alpha}} (1 - p)^{1 - s_{i\alpha}}. \tag{11.47}$$

When the initial configuration of the damage is not known we can gain insight into the robustness properties of the multiplex network by averaging the above message-passing equations over the distribution $\hat{P}(\{s_{i\alpha}\})$. The equations determining the probability $\hat{\sigma}_i$ that a node is in the RMCGC depend on the average messages $\hat{\sigma}_{i \to j}$ given by

$$\hat{\sigma}_{i \to j}^{\alpha} = p \left[1 - \prod_{\ell \in N_\alpha(i) \setminus j} \left(1 - \hat{\sigma}_{\ell \to i}^{\alpha}\right) \right] \left\{ 1 - \prod_{\beta \neq \alpha} \left[1 - p + p \prod_{\ell \in N_\beta(i)} \left(1 - \hat{\sigma}_{\ell \to i}^{\beta}\right) \right] \right\},$$

$$\hat{\sigma}_{i\alpha} = s_{i\alpha} \left[1 - \prod_{\ell \in N_\alpha(i)} \left(1 - \hat{\sigma}_{\ell \to i}^{\alpha}\right) \right] \left\{ 1 - \prod_{\beta \neq \alpha} \left[1 - p + p \prod_{\ell \in N_\beta(i)} \left(1 - \hat{\sigma}_{\ell \to i}^{\beta}\right) \right] \right\}.$$

Let us consider the case in which each layer α is drawn independently from the set of networks with degree distribution $P^{[\alpha]}(k^{[\alpha]})$. In this case, the equations for the average messages S_α' and for the probabilities S_α that a replica node in layer α is in the RMCGC are given by

$$S_\alpha' = p \left[1 - G_1^{[\alpha]} \left(1 - S_\alpha'\right) \right] \left\{ 1 - \prod_{\beta \neq \alpha} \left[1 - p + p \, G_0^{[\beta]} \left(1 - S_\beta'\right) \right] \right\},$$

$$S_\alpha = p \left[1 - G_0^{[\alpha]} \left(1 - S_\alpha'\right) \right] \left\{ 1 - \prod_{\beta \neq \alpha} \left[1 - p + p \, G_0^{[\beta]} \left(1 - S_\beta'\right) \right] \right\}. \tag{11.48}$$

Finally, the average number S of replica nodes in the RMCGC is given by

$$S = \frac{1}{M} \sum_{\alpha=1}^{M} S_\alpha. \tag{11.49}$$

If we consider the case of equally distributed Poisson layers with average degree c, using Eqs (11.48), one can show that $S'_\alpha = S_\alpha = S, \forall \alpha$, and S is determined by the equation

$$S = p\left(1 - e^{-cS}\right)\left\{1 - \left[1 - p + pe^{-cS}\right]^{M-1}\right\}. \tag{11.50}$$

This equation always has the trivial solution $S = 0$. In addition, a non-trivial solution $S > 0$ indicating the presence of the RMCGC emerges at a hybrid and discontinuous phase transition characterized by a square-root singularity on a line of points $p = p^\star$ corresponding to the average degree c, determined by the equations

$$h_{c,p}(S^\star) = 0,$$
$$\left.\frac{dh_{c,p}(S)}{dS}\right|_{S=S^\star} = 0, \tag{11.51}$$

where

$$h_{c,p}(S) = S - p\left(1 - e^{-cS}\right)\left\{1 - \left[1 - p + pe^{-cS}\right]^{M-1}\right\} = 0. \tag{11.52}$$

For $p > p^\star$ there is an RMCGC, for $p \leq p^\star$ there is no RMCGC. The size of the discontinuous jump at $p = p^\star$ in the fraction S of replica nodes in the RMCGC is given by $S = S^\star$.

It is found that as the number of layers M increases the percolation threshold $p^\star$ decreases for every value of the average degree c, indicating that the network becomes more robust thanks to the addition of new layers. Also, the discontinuous jump $S^\star$ decreases as the number M of layers increases for every given average degree c. Therefore, as the number of layers increases the avalanches of failure events become smaller.

Comparison between the RMCGC and the MCGC

The robustness of multiplex networks in the presence of ordinary interdependencies can be compared with the robustness of the same multiplex networks in the presence of redundant interdependencies.

To take a concrete example, here we make this comparison for a multiplex network with M Poisson layers with the same average degree c. Since for redundant interdependencies we have assumed that the damage affects each replica node independently, in order to make a meaningful comparison of the robustness of a multiplex network in the presence and the absence of redundant interdependencies we take $p = 1$ and we compare the critical value of the average degree $c = c^\star$ at which the percolation transition occurs for the RMCGC and the MCGC respectively. A smaller value of $c^\star$ for the RMCGC indicates that the multiplex network is more robust as smaller interlayer connectivities are requested to have an RMCGC.

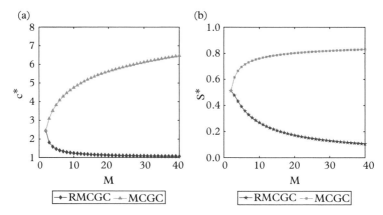

Fig. 11.14 *The critical average degree c* where the RMCGC and the MCGC emerge (panel (b)) and the corresponding size of the jump S* (panel (b)) are shown as a function of the number of layers M for Poisson multiplex networks, with layers having the same average degree. Results for the RMCGC model are displayed as diamonds. Results for the MCGC model are denoted by triangles. Reprinted figure from Ref. [257].*

The jump $S = S^*$ in the size of the RMCGC and the MCGC at the percolation threshold is the other parameter that we use to compare the two phase transitions. The smaller the value of S^* the less dramatic the avalanches of failure propagating in the multiplex network at the percolation transition. In Fig. 11.14 we display the values of c^* and S^* as a function of the number of layers M for the RMCGC and the MCGC. For $M = 2$, the two models give the same results as they are identical. For $M > 2$, differences arise. In the presence of redundant interdependencies, multiplex networks become increasingly more robust as the number M of layers increases. This phenomenon is apparent from the fact that the RMCGC emerges for multiplex networks with an average degree of their layers c^* which decreases as the number of layers M increases. On the contrary, in ordinary percolation the value of c^* for the emergence of the MCGC is an increasing function of M. Additionally, the size of the discontinuous jumps S^* at the transition point decreases with M for the RMCGC, while its increases with M for the MCGC show that the avalanches of failures have a reduced size for the RMCGC.

11.6 Percolation on interdependent multilayer networks

Since in general multilayer network nodes of different layers usually indicate different node entities, it is natural to assume that the initial damage is inflicted independently on each node (i, α). The initial damage is determined by the variables $\{s_{i\alpha}\}$ with $s_{i\alpha} = 0$ if node (i, α) is initially damaged, while $s_{i\alpha} = 1$ if it is not initially damaged. Assuming that each node is damaged with probability $f = 1 - p$, we have that the probability $P(\{s_{i\alpha}\})$ of the entire configuration of the initial damage $\{s_{i,\alpha}\}$ is given by

$$\hat{P}(\{s_{i\alpha}\}) = \prod_{\alpha=1}^{M} \prod_{i=1}^{N} p^{s_{i\alpha}} (1-p)^{1-s_{i\alpha}}.$$ (11.53)

The robustness of the multilayer network can be studied by considering the fraction S of nodes in the MCGC as a function of p.

In a locally tree-like multilayer network it is possible to study this transition using a message-passing algorithm that can be derived in the following way.

A node (i, α) is in the MCGC if and only if:

(a) it is not initially damaged, i.e. $s_{i\alpha} = 1$;

(b) it has at least one neighbour node (j, α) in layer α that connects node (i, α) to nodes in the MCGC;

(c) all the nodes (ℓ, β) that can be reached from node (i, α) by interdependent links have at least one neighbour node of their same layer β that connects them to the MCGC.

In this paragraph we will characterize the percolation transition of interdependent multilayer networks. Specifically we will focus on results applying to the following classes of networks, of networks (these ensembles of networks of networks have been described in Sec. 10.6).

- *Networks of networks with fixed supernetwork and interlinks allowed only between replica nodes.*

- *Multilayer networks with fixed superdegree distribution*, where every node (i, α) is interdependent on $q = q_\alpha$ replica nodes (i, β) of randomly chosen layers β.

- *Networks of networks without replica nodes* where there is a fixed supernetwork but there are no replica nodes.

We will show that there is a rich interplay between the structure (how interlinks are placed) and the robustness properties of the multilayer network. In fact, the properties of the percolation process change significantly in the three considered cases.

Networks of networks with replica nodes

Here we consider networks of networks in which, if network α is interdependent with network β, each node (i, α) of network α is interdependent with node (i, β) of network β, and vice versa. This ensemble has been introduced in Sec. 10.6, and it has been graphically represented in Fig. 10.9. In Ref. [49] it has been shown that for this type of network of networks, where each node can be interdependent only with its replica nodes in the other layers, the supernetwork can be a tree or can contain loops, and in both cases, as long as the supernetwork is connected, the size of the mutually connected component is determined by the same equations determining the MCGC in a multiplex network formed by the same layers.

From this result, four main conclusions can be drawn:

- there is a unique hybrid and discontinuous phase transition as soon as $M \geq 2$;
- the critical value $p = p_c$ always depends on the number of layers M;
- when the mutually connected component emerges at $p = p_c$, every layer of the network of networks contains a finite fraction of nodes that is in the mutually connected component, i.e. in each layer the percolation cluster emerges at the same time as $p = p_c$;
- if we include a partial interdependence of the nodes, the transition changes from discontinuous to continuous at a tricritical point, but also in this case either all layers percolate, or none of them do.

Multilayer networks without replica nodes and with given supradegree distribution

A much more random multilayer network topology is considered in this paragraph. In this case, the multilayer networks have replica nodes and every node of a network (layer) α is connected with $q_\alpha > 0$ randomly chosen replicas in some other networks (see Sec. 10.6, and Fig. 10.11). The supradegree distribution is given by $P_S(q)$ and for simplicity it is assumed that each layer has the same intra-layer degree distribution $P_L(k)$. The generating functions $G_0(z)$ and $G_1(z)$ of the intra-layer degree distribution $P_L(k)$ are defined as

$$G_0(z) = \sum_k P_L(k) z^k, \quad G_1(z) = \sum_k \frac{k P_L(k)}{\langle k \rangle} z^{k-1}. \tag{11.54}$$

The percolation transition for this ensemble of networks of networks has been characterized using a message-passing algorithm in the locally tree-like approximation in Ref. [48].

Each node of the layer α is in the MCGC if at least one of its neighbouring nodes in the same layer is in the MCGC and every node that can be reached by following exclusively interlinks is in the MCGC. In fact, failure of any of the nodes causes the failure of every node connected to it exclusively through interlinks because the interlinks imply interdependencies.

Therefore, assuming that the initial damage configuration is drawn randomly from the distribution $\hat{P}(\{s_{i\alpha}\})$ given by Eq. (11.53), the probability that a node i in a layer α with superdegree $q_\alpha = q$ is in the mutually connected component S_q is given by [48]

$$S_q = p \sum_s P(s|q) \left[\sum_{q'} \frac{q' P_S(q')}{\langle q \rangle} p \left[1 - G_0 \left(1 - \sigma_{q'} \right) \right] \right]^{s-1} \left[1 - G_0 \left(1 - \sigma_q \right) \right]. \tag{11.55}$$

Here $P(s|q)$ indicates the probability that a node i in layer α with $q_\alpha = q$ is connected through paths including exclusively interlinks to other s nodes of the multilayer network.

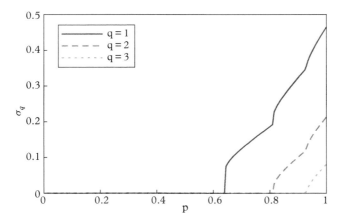

Fig. 11.15 *The fraction of nodes $\sigma_q = S_q$ of a Poisson layer with superdegree q belonging to the MCGC is shown versus the probability p that the nodes are not initially damaged for different values q = 1, 2, 3. The considered multilayer network with replica nodes has every layer with Poisson degree distribution P(k) and average degree $\langle k \rangle = c = 20$ and has a scale-free distribution $P_S(q)$ of superdegrees with power-law exponent $\gamma = 2.8$. For each value of q = 1, 2, 3, σ_q emerges discontinuously, with a jump, which becomes smaller and smaller with increasing q. The emergence of σ_2 is accompanied by a discontinuity of σ_1. The emergence of σ_3 is accompanied by discontinuities of σ_1 and σ_2. Reprinted figure from Ref. [48].*

Moreover, σ_q indicates the probability that by following a link of layer α of superdegree $q_\alpha = q$ we reach a node that is in the MCGC, which satisfies [48]

$$\sigma_q = p \sum_s P(s|q) \left[\sum_{q'} \frac{q' P_S(q')}{\langle q \rangle} p \left[1 - G_0 \left(1 - \sigma_{q'}\right)\right] \right]^{s-1} \left[1 - G_1 \left(1 - \sigma_q\right)\right]. \quad (11.56)$$

Therefore, Eq. (11.55) and Eq. (11.56) can be simplified to

$$\sigma_q = p(\Sigma)^q \left[1 - G_1 \left(1 - \sigma_q\right)\right],$$
$$S_q = p(\Sigma)^q \left[1 - G_0 \left(1 - \sigma_q\right)\right],$$
$$\Sigma = \left[\sum_{q'} \frac{q' P_S(q')}{\langle q \rangle} p \left[1 - G_0 \left(1 - \sigma_{q'}\right)\right] \right] \sum_{q'} \frac{q' P_S(q')}{\langle q \rangle} (\Sigma)^{q'-1}. \quad (11.57)$$

The parameter Σ, indicating the probability that by following an interlink we reach a node in the MCGC, determines both σ_q and S_q for any value of the superdegree q. For this reason, Σ can be considered as the order parameter. Note that for Poisson layers $G_0(z) = G_1(z)$, therefore, it follows that $\sigma_q = S_q$.

From the study of these equations, the following scenario can be drawn in the limit $M \to \infty$:

- if the degrees q_α are heterogeneous, the percolation transitions are multiple, each one corresponding to the emergence of a percolation cluster in each layer with a different value of q;
- each of these phase transitions is hybrid and discontinuous;
- the layers with a higher number of interdependencies are more fragile than those with a smaller number of interdependencies.

The increased fragility of the layers with higher superdegree is in sharp contrast with what happens in the percolation of single layers, in which nodes of high degree are more robust, revealing the important effect of interdependencies. Fig. 11.15 shows that the size of the percolation cluster $S_q = \sigma_q$ in the Poisson layers with q number of interdependencies has multiple discontinuities in correspondence with the discontinuities of the order parameter Σ. Moreover, it is shown that the percolating cluster of layers with higher superdegree q is disrupted before the percolating cluster of layers with smaller superdegree.

For partial interdependencies, when the probability that an interlink implies an interdependency is $r < 1$, Eqs (11.57) are modified as follows:

$$\sigma_q = p\,(r\Sigma + 1 - r)^q \left[1 - G_1(1 - \sigma_q)\right],$$

$$S_q = p\,(r\Sigma + 1 - r)^q \left[1 - G_0(1 - \sigma_q)\right],$$

$$\Sigma = \left[\sum_{q'} \frac{q' P_S(q')}{\langle q \rangle} p\left[1 - G_0(1 - \sigma_{q'})\right]\right] \sum_{q'} \frac{q' P_S(q')}{\langle q \rangle} (r\Sigma + 1 - r)^{q'-1}. \qquad (11.58)$$

We note here that in the case in which every node has the same number of interdependencies, i.e. $P_S(q) = \delta(q, m)$, and each layer is formed by a Poisson network with $\langle k \rangle = c$, one finds $\Sigma[r\Sigma + 1 - r] = \sigma_m = \sigma$. Therefore, σ indicating the fraction of nodes in the MCGC satisfies

$$\sigma = p\left[\frac{1}{2}\left(1 - r + \sqrt{(1 - r)^2 + 4r\sigma}\right)\right]^m (1 - e^{-c\sigma}). \qquad (11.59)$$

The emergence of the mutually connected component in the configuration model of a network of networks with $r < 1$ can always display multiple percolation transitions corresponding to the activation of layers in increasing value of q_α. Nevertheless, these transitions can be either continuous or discontinuous, depending on the value of r.

Networks of networks without replica nodes

The percolation transition in networks of networks with a supernetwork but without replica nodes, where nodes of different layer are randomly matched, has been investigated in Refs [125, 124, 123].

In this case it is found that the percolation process depends on the structure of the network of networks. If the supernetwork is a tree, i.e. does not contain any loop, the percolation transition follows the percolation transition of the networks of networks with replica nodes discussed in the previous paragraph. Therefore, in this case the MCGC is determined by the equations valid for the multiplex networks with the same layers. If, instead, the supernetwork does contain loops, the percolation transition follows a different set of equations. This difference is due to the fact that only in this case it is possible, by exclusively following interlinks, to return several times to the same layer (see Fig. 10.10).

Interestingly, when the supernetwork is a regular network with supradegree $q_\alpha = m$ for every layer α, and the layers have Poisson degree distribution with the same average degree c, the fraction of nodes in the MCGC follows Eq. (11.59) valid for multilayer networks without replica nodes.

12

Classical Percolation, Generalized Percolation and Cascades

12.1 Robustness of multilayer networks

In the previous chapter we characterized the robustness of multilayer networks with interdependencies. However, in this chapter we will show that the robustness of multilayer networks can be characterized also using different other processes.

Firstly, we will consider classical percolation in which all the links are treated on equal footing. Therefore, to a large extent in this context we consider percolation on a single large structured network. Despite the apparent absence of a multiplexity character for this process, here we will show that also in this case a rich interplay between multiplexity and robustness can be observed. In fact, the degree correlations among different layers can modulate (increase or reduce) the robustness of the entire multilayer network.

Secondly, we will present the directed percolation process on multiplex networks which describes the propagation of a disease requiring the cooperative infection from different layers of the multiplex network. We will show that this novel type of percolation transition reduces to the percolation of interdependent multiplex networks in the absence of link overlap, but characterizes a different process in the presence of link overlap and displays a rich phase diagram including both continuous and discontinuous phase transitions.

Thirdly, we will investigate a generalized percolation transition on a duplex network with antagonistic interactions where a node of one layer can be functional only if its replica node is not functional. We will show that percolation in the presence of antagonistic interactions can give rise to bistable states and hysteresis loops.

Finally, we will characterize the avalanches of failures that can affect multilayer infrastructures due to the overload of their connections. This provides an alternative scenario with respect to the avalanches caused by interdependencies discussed in the previous chapter. Relevantly also in this scenario, we observe that multiplexity has an important role in determining the distribution of avalanches. These effects can be potentially exploited to design more resilient global infrastructures.

Multilayer Networks. Ginestra Bianconi, Oxford University Press (2018).
© Ginestra Bianconi. DOI: 10.1093/oso/9780198753919.001.0001

12.2   Classical percolation

12.2.1   Classical percolation: general remarks

Classical percolation refers to the study of the robustness of multilayer networks in which all the interactions have the same valence. In this case, every link of the multilayer network is called a 'dependency link'. The response of the network to random damage of its nodes or to random damage of its links is studied by calculating the size of the giant component of the multilayer network resulting from the inflicted damage. The giant component of a multilayer network is formed by a finite fraction of nodes that are connected to each other by any type of connection. For instance, in a multilayer transportation network including bus, underground and train connections, a station is in the giant component if it is connected to a finite fraction of all the other stations by any type of transportation system. This problem can be either studied on multiplex networks or on general multilayer networks. The problem can be recast into a classical percolation model on a large structured single network in which the probability of damaging a node (or a link) depends on the layer where the node (or link) is. This problem has been studied on multilayer networks in Refs [188, 261, 43] and on multiplex networks in Refs [213, 146]. In this context it has been shown that classical percolation on multilayer networks is strongly affected by the structural correlations built in the multilayer network. These correlations determine intrinsic structural mechanisms that can change the robustness properties of the multilayer networks.

Here we will discuss the case of classical bond percolation on multilayer networks where links are randomly damaged. This choice is dictated by the fact that this model is fundamental also for understanding the property of the SIR epidemic model on multilayer networks that will be discussed in the next chapter. We will consider separately the bond percolation model defined on a multiplex network and on a general multilayer network. The node percolation problem is a straightforward generalization of the bond percolation model.

12.2.2   Classical percolation in multiplex networks

The message-passing algorithm

We consider a multiplex network formed by a set of nodes connected by different types of interactions, i.e. we disregard interlinks. The giant component of this multiplex network is formed by a pair of nodes connected to each other by at least a path composed by links belonging to any layer. Let us consider an initial damage configuration indicating whether ($s_{ij}^{\alpha} = 0$) or not ($s_{ij}^{\alpha} = 1$) the link between node i and node j in layer α is removed. Once a given configuration of the initial damage of the links is known, as long as the multiplex network is locally tree-like, it is possible to determine whether a node i is ($\sigma_i = 1$) or is not ($\sigma_i = 0$) in the giant component using a simple message-passing algorithm. This message-passing algorithm is determined by the messages $\sigma_{i \to j}$ indicating whether ($\sigma_{i \to j} = 1$) or not ($\sigma_{i \to j} = 0$) node i is in the giant component also if all the links among node i and node j are removed. A node i is in the giant component if it receives at least a positive message from a non-damaged link. A node i sends a positive message to node j if it receives at least a positive message from at least one of its remaining undamaged links, i.e.

$$\sigma_{i\to j} = 1 - \prod_{\alpha=1}^{M} \prod_{\ell\in N_\alpha(i)\backslash j} \left(1 - s_{\ell i}^{\alpha}\sigma_{\ell\to i}\right), \tag{12.1}$$

where $N_\alpha(i)$ indicates the set of nodes which are connected to node i in layer α. These messages determine the indicator σ_i as

$$\sigma_i = 1 - \prod_{\alpha=1}^{M} \prod_{\ell\in N_\alpha(i)} \left(1 - s_{\ell i}^{\alpha}\sigma_{\ell\to i}\right). \tag{12.2}$$

When the initial configuration of the damage is not known, it is important to take care that each pair of the nodes of a multiplex network can be connected in multiple layers. By assuming that every multilink $\vec{m}$ is damaged (loses all its links) with probability $f_{\vec{m}} = 1 - p_{\vec{m}}$ we can consider the average of the above message-passing algorithm, providing the probability $\hat{\sigma}_i$ that node i belongs to the giant component. The novel set of messages $\hat{\sigma}_{i\to j}$ indicating the average of $\sigma_{i\to j}$ when multilinks $\vec{m}$ are damaged with probability $f_{\vec{m}} = 1 - p_{\vec{m}}$ satisfy

$$\hat{\sigma}_{i\to j} = 1 - \prod_{\ell\in N(i)\backslash j} \left(1 - p_{\vec{m}_{\ell i}}\hat{\sigma}_{\ell\to i}\right), \tag{12.3}$$

where $N(i)$ is the set of nodes ℓ that are neighbours of node i in any layer of the multiplex network. Finally, the probability $\hat{\sigma}_i$ that node i belongs to the giant component is given by

$$\hat{\sigma}_i = 1 - \prod_{\ell\in N(i)} \left(1 - p_{\vec{m}_{\ell i}}\hat{\sigma}_{\ell\to i}\right). \tag{12.4}$$

In the case in which links are damaged independently in each layer α with probability $f_\alpha = 1 - p_\alpha$ we have

$$f_{\vec{m}} = \prod_{\alpha=1}^{M} \left(f_\alpha\right)^{m^{[\alpha]}} \tag{12.5}$$

or equivalently

$$p_{\vec{m}} = 1 - \prod_{\alpha=1}^{M} \left(1 - p_\alpha\right)^{m^{[\alpha]}}. \tag{12.6}$$

However, the above formalism allows us to also consider random damage configurations in which the links of different layers are not damaged independently.

In the case in which the multiplex network does not display link overlap the only non-trivial multilinks include a connection only in a layer. Therefore, by indicating with $p_\alpha = 1 - f_\alpha$ the probability of retaining links in layer α, the message-passing equations read

$$\hat{\sigma}_{i \to j} = 1 - \prod_{\alpha=1}^{M} \prod_{\ell \in N_\alpha(i) \backslash j} \left(1 - p_\alpha \hat{\sigma}_{\ell \to i}\right),$$

$$\hat{\sigma}_i = 1 - \prod_{\alpha=1}^{M} \prod_{\ell \in N_\alpha(i)} \left(1 - p_\alpha \hat{\sigma}_{\ell \to i}\right). \tag{12.7}$$

It follows that assuming all p_α equal, the percolation threshold p_c is equal to the inverse of the maximum eigenvalue Λ of the non-backtracking matrix, i.e. $p_c = 1/\Lambda$, where the non-backtracking matrix has elements $B_{(i,j),(r,s)} = \delta(j,r)[1-\delta(s,i)]$, where (i,j) and (r,s) are links of any layer.

Ensemble of multiplex networks without link ovelap

In order to appreciate the effect of multiplexity on classical percolation it is instructive to consider the percolation transition on random uncorrelated multiplex network ensembles without link overlap. In particular, we consider here multiplex network ensembles in which each node is assigned a multiplex degree distribution $P(\mathbf{k})$.

Let us indicate with S the fraction of nodes in the giant component of the multiplex network when each link of the generic layer α is damaged with a probability $f_\alpha = 1 - p_\alpha$. Equivalently, S can be interpreted as the average of $\hat{\sigma}_i$ or the probability that a random node belongs to the giant component of the multiplex network.

We indicate by S'_α the probability that by following a link in layer α we reach a node in the giant component. This quantity can be evaluated by averaging the messages $\hat{\sigma}_{i \to j}$ over the ensemble of multiplex networks. The probability S that a node is in the giant component is determined by the values of S'_α for $\alpha = 1, 2, \ldots, M$. Indeed, by averaging the message-passing Eqs (12.7) we have [213]

$$S = \left[1 - \sum_{\mathbf{k}} P(\mathbf{k}) \prod_{\alpha=1}^{M} \left(1 - p_\alpha S'_\alpha\right)^{k^{[\alpha]}} \right]. \tag{12.8}$$

On their turn, the probabilities S'_α satisfy the following set of recursive equations

$$S'_\alpha = \left[1 - \sum_{\mathbf{k}} \frac{k^{[\alpha]}}{\langle k^{[\alpha]} \rangle} P(\mathbf{k}) \prod_{\gamma=1}^{M} \left(1 - p_\gamma S'_\gamma\right)^{k^{[\gamma]} - \delta(\gamma,\alpha)} \right], \tag{12.9}$$

where $\delta(\gamma, \alpha)$ is the Kronecker delta. This system of equations admits a trivial solution in which $S = 0$, indicating that the giant component vanishes. This scenario is observed when the solution $S'_\alpha = 0 \ \forall \alpha$ is stable. The point at which the trivial solution becomes unstable indicates the percolation transition. In order to determine the conditions under which the multiplex network percolates, we therefore linearize the system of Eqs (12.9) for $S'_\alpha \ll 1$, finding

$$S'_\alpha = \sum_\gamma p_\gamma \frac{\left\langle k^{[\alpha]} \left[k^{[\gamma]} - \delta(\gamma, \alpha) \right] \right\rangle}{\left\langle k^{[\alpha]} \right\rangle} S'_\gamma, \tag{12.10}$$

which can be written as

$$\mathbf{S'} = \mathbf{J} \mathbf{S'}, \tag{12.11}$$

where $\mathbf{J}$ is the $M \times M$ Jacobian matrix of the system of equations (12.9) of elements

$$\mathcal{J}_{\alpha,\gamma} = p_\gamma \frac{\left\langle k^{[\alpha]} \left[k^{[\gamma]} - \delta(\gamma, \alpha) \right] \right\rangle}{\left\langle k^{[\alpha]} \right\rangle}. \tag{12.12}$$

Above the transition, this system must develop a set of non-trivial solutions. Therefore, the transition point is obtained by imposing that the maximum eigenvalue $\Lambda_{\mathcal{J}}$ of the matrix $\mathbf{J}$ satisfies

$$\Lambda_{\mathcal{J}} = 1. \tag{12.13}$$

From this derivation it follows that the multiplex networks is percolating if and only if $\Lambda_{\mathcal{J}} > 1$. Let us now consider the specific case of a multiplex network with $M = 2$ layers. By imposing $\Lambda_{\mathcal{J}} = 1$ we can show that the giant component of the multiplex network emerges for

$$\Lambda = \frac{1}{2} \left[p^{[1]} \kappa_1 + p^{[2]} \kappa_2 + \sqrt{\left(p^{[1]} \kappa_1 - p^{[2]} \kappa_2 \right)^2 + 4 p^{[1]} p^{[2]} \mathcal{K}_1 \mathcal{K}_2} \right] = 1, \tag{12.14}$$

where the variables κ_α and $\mathcal{K}_\alpha$ correspond respectively to the second moment of the degree distribution in layer α and a normalized measure of degree correlations among the two layers, i.e.

$$\kappa_\alpha = \frac{\left\langle k^{[\alpha]} (k^{[\alpha]} - 1) \right\rangle}{\left\langle k^{[\alpha]} \right\rangle}$$

$$\mathcal{K}_\alpha = \frac{\left\langle k^{[1]} k^{[2]} \right\rangle}{\left\langle k^{[\alpha]} \right\rangle}. \tag{12.15}$$

From Eq. (12.14) it is evident that positive degree correlations among the two layers (yielding large values of $\mathcal{K}_\alpha$) increase the robustness of the multiplex network, whereas negative degree correlations (yielding small values of $\mathcal{K}_\alpha$) reduce the robustness of the multiplex network.

Ensembles of multiplex networks with link overlap

Classical percolation can also be studied in multiplex network ensembles with link overlap, in which non-trivial multilinks $\vec{m}$ are damaged with probability $f_{\vec{m}} = 1 - p_{\vec{m}}$. In this scenario the percolation transition is most easily treated in the multiplex network ensemble in which the multidegree distribution $P(\{k^{\vec{m}}\})$ is fixed. The fraction of nodes S in the giant component (given by the average of the indicator functions $\hat{\sigma}_i$) is given by

$$S = \left[1 - \sum_{\{k^{\vec{m}}\}} P(\{k^{\vec{m}}\}) \prod_{\vec{m} \neq \vec{0}}^{M} (1 - p_{\vec{m}} S'_{\vec{m}})^{k^{\vec{m}}} \right]. \tag{12.16}$$

On their turn, the probabilities $S'_{\vec{m}}$ that by following a multilink $\vec{m}$ we reach a node in the giant component (equivalent to the average of the messages $\hat{\sigma}_{i \to j}$ conditioned on having $\vec{m}_{ij} = \vec{m}$) are given by

$$S'_{\vec{m}} = \left[1 - \sum_{\{k^{\vec{m}}\}} \frac{k^{\vec{m}}}{\langle k^{\vec{m}} \rangle} P\left(\{k^{\vec{m}}\}\right) \prod_{\vec{m}'} (1 - p_{\vec{m}'} S'_{\vec{m}'})^{k^{\vec{m}'} - \delta(\vec{m}, \vec{m}')} \right]. \tag{12.17}$$

In this case, linearizing the above equations for $S_{\vec{m}} \ll 1$ it can be shown that the giant component emerges when

$$\Lambda_{\hat{\mathcal{J}}} > 1. \tag{12.18}$$

Here $\Lambda_{\hat{\mathcal{J}}}$ is the maximum eigenvalue of the $(2^M - 1) \times (2^M - 1)$ Jacobian matrix $\hat{\mathbf{J}}$ with elements

$$\hat{\mathcal{J}}_{\vec{m}, \vec{m}'} = \frac{\left\langle k^{\vec{m}} \left(k^{\vec{m}'} - \delta(\vec{m}, \vec{m}') \right) \right\rangle}{\langle k^{\vec{m}} \rangle}. \tag{12.19}$$

12.2.3   Classical percolation in multilayer networks

In this paragraph we will depart form the multiplex network topology considered in the previous paragraph and we will consider general multilayer networks or also multiplex networks where we explicitly treat interlinks.

When percolation takes place on a multilayer network, multiplexity plays an important role, which is revealed by studying these processes on multilayer network ensembles including a controlled level of multilayer degree correlations [188, 261, 43].

The message-passing algorithm

Classical (bond) percolation of a single multilayer network can be studied by generalizing the message-passing algorithm used to study percolation on a single network. The giant component of the multilayer network is formed by pairs of nodes connected by at least a path formed by interlinks and intralinks of any layer. On a locally tree-like multilayer network the giant component can be determined by a message-passing algorithm. If the initial configuration of the damage is known, i.e. we know for every link connecting node (i, α) to node (j, β) whether it is damaged ($s_{i\alpha,j\beta} = 0$) or not ($s_{i\alpha,j\beta} = 1$), it is possible to determine whether node (i, α) is ($\sigma_{i\alpha} = 1$) or is not ($\sigma_{i\alpha} = 0$) in the giant component by using a suitable message-passing algorithm. Specifically, messages of this algorithm $\sigma_{i\alpha \to j\beta}$ are sent among neighbouring replica nodes and indicate whether ($\sigma_{i\alpha \to j\beta} = 1$) or not ($\sigma_{i\alpha \to j\beta} = 0$) node (i, α) is in the giant component if the link $[(i, \alpha), (j, \beta)]$ is removed. The algorithm is a direct generalization of the algorithm valid for single layers. A node (i, α) belongs to the giant component ($\sigma_{i\alpha} = 1$) if it receives at least a positive message from one of its non-damaged links. Each node (i, α) sends a positive message ($\sigma_{i\alpha \to j\beta} = 1$) to a neighbour node (j, β) if it receives at least one positive message from any of its remaining non-damaged links. Therefore, the message-passing equations read

$$\sigma_{i\alpha \to j\beta} = 1 - \prod_{(\ell,\gamma) \in N(i,\alpha) \backslash (j,\beta)} \left(1 - s_{\ell\gamma,i\alpha} \sigma_{\ell\gamma \to i\alpha}\right), \tag{12.20}$$

where $N(i, \alpha)$ indicates the set of replica nodes which are connected either by intralinks of interlinks to (i, α). These messages determine the indicator $\sigma_{i\alpha}$ as

$$\sigma_{i\alpha} = 1 - \prod_{(\ell,\gamma) \in N(i,\alpha)} \left(1 - s_{\ell\gamma,i\alpha} \sigma_{\ell\gamma \to i\alpha}\right). \tag{12.21}$$

Let us now assume that the initial configuration of the damage is not known and that only the probability $p_{\alpha\beta}$ that links connecting nodes in layer α to nodes in layer β are not damaged is available. Then, the probability $\hat{\sigma}_{i\alpha}$ that the generic node (i, α) is in the giant component is determined by the messages $\hat{\sigma}_{i\alpha \to j\beta}$ and can be found by averaging the previously discussed message-passing equation, i.e. solving the set of recursive equations

$$\hat{\sigma}_{i\alpha \to j\beta} = 1 - \prod_{(\ell,\gamma) \in N(i,\alpha) \backslash (j,\beta)} (1 - p_{\gamma\alpha} \hat{\sigma}_{\ell\gamma \to i\alpha}),$$

$$\hat{\sigma}_{i\alpha} = 1 - \prod_{(\ell,\gamma) \in N(i,\alpha)} (1 - p_{\gamma\alpha} \hat{\sigma}_{\ell\gamma \to i\alpha}). \tag{12.22}$$

It follows that assuming all $p_{\gamma\alpha}$ equal, the percolation threshold p_c is equal to the inverse of the maximum eigenvalue Λ of the non-backtracking matrix, i.e. $p_c = 1/\Lambda$, where the non-backtracking matrix has elements $B_{(i\alpha,j\beta),(r\gamma,s\xi)} = \delta(j,r)\delta(\beta,\gamma)[1 - \delta(s,i)\delta(\xi,\alpha)]$.

Ensemble of multilayer networks

Classical percolation can be studied over the multilayer network ensemble with given multilayer degree distribution $P^{[\alpha]}(\mathbf{k})$ in layer $\alpha = 1, 2, \ldots, M$ where the links among replica nodes of layer α and layer β are retained with probability $p_{\alpha\beta}$.

The probability $S'_{\alpha\beta}$ that a link going from layer α to layer β reaches a node in the giant component satisfies the set of equations [43]

$$S'_{\alpha\beta} = x_{\alpha\beta} \left[1 - \sum_{\mathbf{k}} \frac{k^{[\alpha,\beta]}}{\langle k^{[\alpha,\beta]} \rangle} P_\alpha(\mathbf{k}) \prod_{\gamma=1}^{M} \left(1 - p_{\gamma\alpha} S'_{\gamma\alpha} \right)^{k^{[\alpha,\gamma]} - \delta(\gamma,\beta)} \right], \tag{12.23}$$

where $x_{\alpha\beta} = 1$ if there is at least one connection between layer α and layer β, otherwise $x_{\alpha\beta} = 0$, i.e.

$$x_{\alpha\beta} = 1 - \delta\left(0, \langle k^{[\alpha,\beta]} \rangle\right). \tag{12.24}$$

We note here that since $\langle k^{[\alpha,\beta]} \rangle = \langle k^{[\beta,\alpha]} \rangle$, it follows that $x_{\alpha\beta} = x_{\beta\alpha}$.
The probability S_α that a generic node (i,α) of layer α is in the giant component is expressed in terms of the probabilities $S'_{\alpha\beta}$ as

$$S_\alpha = 1 - \sum_{\mathbf{k}} P_\alpha(\mathbf{k}) \prod_{\gamma=1}^{M} \left(1 - p_{\gamma\alpha} S'_{\gamma\alpha} \right)^{k^{[\alpha,\gamma]}}. \tag{12.25}$$

Finally, the fraction of nodes in the giant component is given by

$$S = \frac{1}{M} \sum_{\alpha=1}^{M} S_\alpha. \tag{12.26}$$

The percolation threshold is found by linearizing the Eqs (12.23) close to the trivial solution $S'_{\alpha\beta} = 0$, obtaining the system of equations

$$S'_{\alpha\beta} = \sum_{\gamma} p_{\gamma\alpha} x_{\alpha\beta} \frac{\langle k^{[\alpha,\beta]} [k^{[\alpha,\gamma]} - \delta(\gamma,\beta)] \rangle}{\langle k^{[\alpha,\beta]} \rangle} S'_{\gamma\alpha} \tag{12.27}$$

which can be written as

$$\mathbf{S'} = \tilde{\mathbf{J}} \mathbf{S'} \tag{12.28}$$

where $\tilde{\mathbf{J}}$ is the $M^2 \times M^2$ Jacobian matrix of the system of equations (12.23) of elements

$$\tilde{\mathcal{J}}_{\alpha\beta;\gamma\alpha} = p_{\gamma\alpha} x_{\alpha\beta} \frac{\langle k^{[\alpha,\beta]} [k^{[\alpha,\gamma]} - \delta(\gamma,\beta)] \rangle}{\langle k^{[\alpha,\beta]} \rangle}. \tag{12.29}$$

Note that in Eq. (12.29) we have adopted the following notation: whereas $\langle k^{[\alpha,\beta]} \rangle = 0$ we take $x_{\alpha\beta} \frac{\langle k^{[\alpha,\beta]} [k^{[\alpha,\gamma]} - \delta(\gamma,\beta)] \rangle}{\langle k^{[\alpha,\beta]} \rangle} = 0$. Above the transition, this system must develop a set of non-trivial solutions. Therefore, the transition point is obtained by imposing that the maximum eigenvalue $\Lambda_{\tilde{\mathcal{J}}}$ of the matrix $\tilde{\mathbf{J}}$ satisfies

$$\Lambda_{\tilde{\mathcal{J}}} = 1. \tag{12.30}$$

The case of a multilayer network ensemble with M = 2 layers

In the case of a multilayer network with $M = 2$ layers the system of Eqs (12.28) reads

$$
\begin{aligned}
S'_{11} &= \kappa_{11} p_{11} S'_{11} + \mathcal{K}_{12} p_{12} S'_{21} \\
S'_{22} &= \kappa_{22} p_{22} S'_{22} + \mathcal{K}_{21} p_{12} S'_{12} \\
S'_{12} &= \mathcal{W}_{12} p_{11} S'_{11} + \kappa_{12} p_{12} S'_{21} \\
S'_{21} &= \mathcal{W}_{21} p_{22} S'_{22} + \kappa_{21} p_{12} S'_{12}
\end{aligned}
\tag{12.31}
$$

where

$$
\kappa_{\alpha\beta} = x_{\alpha\beta} \frac{\left\langle k^{[\alpha,\beta]} (k^{[\alpha,\beta]} - 1) \right\rangle}{\left\langle k^{[\alpha,\beta]} \right\rangle}
$$

$$
\mathcal{K}_{12} = x_{11} \frac{\left\langle k^{[1,1]} k^{[1,2]} \right\rangle}{\left\langle k^{[1,1]} \right\rangle}
$$

$$
\mathcal{K}_{21} = x_{22} \frac{\left\langle k^{[2,2]} k^{[2,1]} \right\rangle}{\left\langle k^{[2,2]} \right\rangle}
$$

$$
\mathcal{W}_{12} = x_{12} \frac{\left\langle k^{[1,2]} k^{[1,1]} \right\rangle}{\left\langle k^{[1,2]} \right\rangle}
$$

$$
\mathcal{W}_{21} = x_{12} \frac{\left\langle k^{[2,1]} k^{[2,2]} \right\rangle}{\left\langle k^{[2,1]} \right\rangle}.
\tag{12.32}
$$

The transition is therefore obtained when the following condition is satisfied:

$$
0 = (1 - p_{11}\kappa_{11})(1 - p_{22}\kappa_{22}) - p_{12}^2 \mathcal{R}_{12}
\tag{12.33}
$$

where

$$
\mathcal{R}_{12} = (p_{11}\mathcal{W}_{12}\mathcal{K}_{12} + \kappa_{12} - p_{11}\kappa_{12}\kappa_{11})(p_{22}\mathcal{W}_{21}\mathcal{K}_{21} + \kappa_{21} - p_{22}\kappa_{21}\kappa_{22}).
\tag{12.34}
$$

For the case in which only the links within each layer exist, i.e. $x_{12} = 0$ or $p_{12} = 0$, we recover the transitions in the single layers [78, 227]

$$
p_{\alpha,\alpha}\kappa_{\alpha\alpha} = 1,
\tag{12.35}
$$

or equivalently,

$$
p_{\alpha\alpha} = \frac{\left\langle k^{[\alpha,\alpha]} \right\rangle}{\left\langle k^{[\alpha,\alpha]} \left(k^{[\alpha,\alpha]} - 1 \right) \right\rangle}.
\tag{12.36}
$$

For the case in which only the interlinks exist $x_{11} = x_{22} = 0$ or $p_{11} = p_{22} = 0$ we obtain the result valid for bipartite networks [227]

$$p_{12}^2 \kappa_{12} \kappa_{21} = 1. \tag{12.37}$$

Therefore,

$$p_{12} = \sqrt{\frac{1}{\kappa_{12} \kappa_{21}}} = \sqrt{\frac{\langle k^{[1,2]} \rangle}{\langle k^{[1,2]} (k^{[1,2]} - 1) \rangle} \frac{\langle k^{[2,1]} \rangle}{\langle k^{[2,1]} (k^{[2,1]} - 1) \rangle}} \tag{12.38}$$

which can be vanishingly small for a heterogeneous distribution of either $k^{[1,2]}$ or $k^{[2,1]}$.

In general, we have that the relation between p_{12} at the transition point depends on the values of p_{11} and p_{22} according to

$$p_{12} = \sqrt{\frac{(1 - p_{11}\kappa_{11})(1 - p_{22}\kappa_{22})}{\mathcal{R}_{12}}}, \tag{12.39}$$

and it is therefore strongly dependent on the correlations between the degrees of the nodes within and across the layers.

Effect of interdegree and intradegree correlations

To study the effect of degree correlations let us assume we have a multilayer network formed by two layers with tunable correlations between interlayer and intralayer degrees.

To be concrete, we consider a network with identical interlayer degree sequences and intralayer degree sequences, in which the intralayer degree and the interlayer degree of each node are either Maximally Positively correlated (MC), Maximally Anti-correlated (MA) or Uncorrelated (UC). These correlated multilayer networks are constructed starting from a multilayer network with the same interlayer network structure by changing the way the intralinks are placed. For Maximally Positively correlated (MC) multilayer networks the intradegree and interdegree sequences are first sorted in descending order. To each node of rank r in the intradegree sequence the interdegree with the same rank r is assigned. Subsequently, the bipartite network between the two layers is randomly drawn in such a way that the interdegree of each node is preserved. For Maximally Anti-correlated (MA) networks we proceed as in the previous case with the exception that the intradegree and interdegree distributions are sorted in opposite order (one sequence in increasing order and the other sequence in decreasing order). Finally, for the case of Uncorrelated (UC) multilayer networks the interdegree is assigned randomly to any node of a given layer of the multilayer network by performing a random permutation of the corresponding interdegree sequence.

In Fig. 12.1 we show the percolation threshold points for Poisson multilayer networks indicated by the line $p = p(q)$ where $q = p_{12} = p_{21}$ and $p = p_{11} = p_{22}$ are respectively the probabilities for retaining the interlinks and the intralinks. For high positive correlations

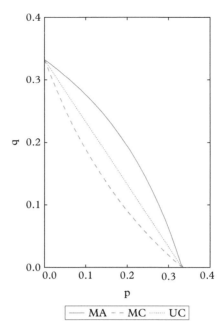

Fig. 12.1 *Percolation threshold for Maximally Correlated (MC), Maximally Anti-correlated (MA) or Uncorrelated (UC) interdegree and intradegree sequences. The multilayer networks have two layers (M = 2) with identical interlayer degree sequences and intralayer degree sequences drawn from a Poisson distribution with average degree* c = 3. *The network size is* N = 10^4 *nodes. Adapted from Ref. [43].*

of interlayer and intralayer degrees (MC multilayer network) $\mathcal{R}_{12}$ is higher, the multilayer network is more robust and the percolation threshold is smaller. For large anti-correlations of interlayer and intralayer degrees (MA multilayer network) $\mathcal{R}_{12}$ is smaller, the multilayer network is more fragile and the percolation threshold is larger.

Moreover, in Ref. [261] it has been shown that if in addition to positive interlayer and intralayer degree correlations the multilayer network is also formed by positively correlated layers where hub nodes tend to be connected to hub nodes, the robustness of the multilayer network is further increased.

12.3   Directed percolation

12.3.1   Directed mutually connected components

Directed percolation is a generalized percolation problem that can be used to probe the large-scale properties of multiplex networks. In particular, it characterizes the response of a multiplex network to random damage of its nodes by evaluating the Directed Mutually Connected Giant Component (DMCGC).

The algorithm [77, 76] that allows us to find the nodes in the DMCGC has an epidemic-spreading interpretation. In this epidemic-spreading interpretation, we assume that a different disease propagates in each layer of the multiplex network and that a node is infected only if it is in contact with at least one infected neighbour node in each layer $\alpha = 1, 2, \ldots, M$. The set of nodes that become infected are the nodes in the DMCGC.

The DMCGC reduces to the Mutually Connected Giant Component (MCGC) in the absence of link overlap and also for complete link overlap. However, the DMCGC is distinct from the MCGC in the presence of a non-negligible but partial link overlap. In particular, the algorithm determining the DMCGC has an inherent directed character due to its epidemic-spreading interpretation, while the algorithm used for detecting the MCGC in the presence of link overlap does not have this directed nature.

The MCGC can be determined by a message-passing algorithm that in the absence of overlap and for complete overlap reduces to the algorithm discussed in Sec. 11.3, but most relevantly differs from the algorithm discussed in Sec. 11.4.2 for determining the MCGC in the presence of a non-vanishing overlap [76, 212]. For example, for the duplex network in Fig. 12.2, all the nodes belong to the MCGC. In fact, every pair of nodes is connected by at least one path in each layer of the duplex network. However, according to the message-passing algorithm with the epidemic-spreading interpretation, once the

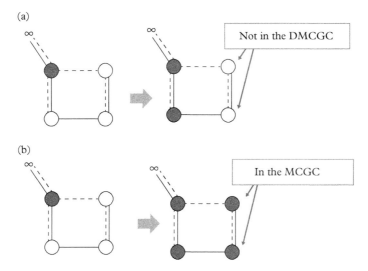

Fig. 12.2 *A multiplex network with link overlap demonstrating that the DMCGC is not equivalent to the MCGC. Here the multiplex network has* M = 2 *layers corresponding to the networks formed by links indicated respectively with solid and dashed lines. In panel (a) we assume that one node is connected to the DMCGC. By applying the message-passing algorithm for the DMCGC, we observe that two nodes of the network do not belong to the DMCGC. In panel (b) we consider the same multiplex network configuration but this time we assume that a single node is connected to the MCGC. By applying the message-passing algorithm for the MCGC with link overlap we observe that all the nodes of this network belong to the MCGC. Reprinted figure from Ref. [76].*

first two nodes have been infected neither of the two remaining nodes can receive the infection coming from both layers at the same time. As a consequence, these two nodes remain uninfected. That is, while these two nodes belong to the MCGC, they do not belong to the DMCGC.

Finally, we note that the directed percolation can be also interpreted as the limit of the multiplex network viability problem when the fraction of source nodes goes to zero [212, 211]. In the epidemic interpretation the viability cluster is the cluster obtained by assuming that some nodes are the source of the epidemics and the other nodes get the infection and can further propagate if they are receiving the infection from links coming from each layer. When the fraction of source nodes goes to zero the viable cluster reduces to the DMCGC.

12.3.2   The message-passing algorithm

In a locally tree-like multiplex network the DMCGC can be determined by the following message-passing equation [77, 76]. Let $s_i = 0, 1$ indicate if a node i is removed or not from the network and let $\sigma_i = 0, 1$ indicate whether the node i is in the DMCGC. The value of σ_i is determined by the messages $\sigma_{j \to i}$ that each of the neighbouring nodes j send to node i. The value of the generic message $\sigma_{i \to j}$ sent from node i to the neighbouring node j is set to one, $\sigma_{i \to j} = 1$, if and only if the following two conditions are satisfied:

(a) node i is not initially damaged, i.e. $s_i = 1$;

(b) node i belongs to the DMCGC even if the multilink $\vec{m}^{ij}$ between node i and node j is removed from the multiplex, i.e. node i receives at least one positive message $\sigma_{\ell \to i} = 1$ from a nearest neighbour $\ell \neq j$ in every layer α of the multiplex network.

If any of these conditions are not satisfied then the message is zero, i.e. $\sigma_{i \to j} = 0$.

Additionally, node i is in the DMCGC ($\sigma_i = 1$) if the following conditions are satisfied:

(a) node i is not initially damaged;

(b) for every layer α node i receives at least one positive message $\sigma_{\ell \to i} = 1$ from a neighbour ℓ in layer α.

This algorithm directly translates into the following message-passing equations for σ_i and $\sigma_{i \to j}$:

$$\sigma_i = s_i \prod_{\alpha=1}^{M} \left[1 - \prod_{j \in N_\alpha(i)} \left(1 - \sigma_{j \to i} \right) \right], \tag{12.40}$$

$$\sigma_{i \to j} = s_i \prod_{\alpha=1}^{M} \left[1 - \prod_{\ell \in N_\alpha(i) \backslash j} \left(1 - \sigma_{\ell \to i} \right) \right], \tag{12.41}$$

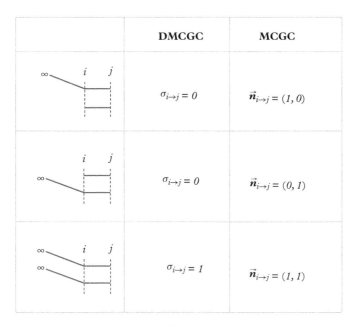

	DMCGC	MCGC
i j	$\sigma_{i \to j} = 0$	$\vec{n}_{i \to j} = (1, 0)$
i j	$\sigma_{i \to j} = 0$	$\vec{n}_{i \to j} = (0, 1)$
i j	$\sigma_{i \to j} = 1$	$\vec{n}_{i \to j} = (1, 1)$

Fig. 12.3 *Possible types of non-trivial messages $\vec{n}_{i \to j}$ that can be sent from node i to node j connected by the multilink $\vec{m}_{ij} = (a_{ij}^{[1]}, a_{ij}^{[2]})$ in a duplex network. The multilink between node i and node j is represented as in Fig. 7.3; solid lines connected to the symbol of infinity indicate that through those links node i is connected to nodes in the MCGC.*

where $N_\alpha(i)$ indicates the set of neighbouring nodes of node i in layer α. This algorithm reduces to the one defining the MCGC in the absence of link overlap (see Sec. 11.3) but it is distinct from the one determining the MCGC in the presence of link overlap (see Sec. 11.4.2) In Fig. 12.3 we emphasize the difference between this algorithm and the one used to detect the MCGC in multiplex networks with link overlap. It is particularly clear that the directed nature of the algorithm defining the DMCGC does not allow the non-trivial messages to propagate in situations where instead the messages determining the MCGC are non-trivial.

Let us consider a random realization of the initial damage in which each node is damaged, with probability $1 - p$ and a random realization of the multiplex network with given multidegree distribution chosen with probability $P(\vec{G})$ given by Eq. (10.47). The average $S_{\vec{m}}$ of the messages $\hat{\sigma}_{i \to j}$ along a generic non-trivial multilink $\vec{m}^{ij} = \vec{m}$ determines the average number of nodes in the DMCGC S (see Appendix D for the details of the derivation).

We consider a duplex network with Poisson multidegree distribution and average multidegrees $\langle k^{(1,1)} \rangle = c_2$, $\langle k^{(1,0)} \rangle = \langle k^{(0,1)} \rangle = c_1$. The full phase diagram of the model is displayed in Fig. 12.4. We note that for $c_2 = 0$ the phase transition is hybrid and discontinuous and reduces to the known transition in duplex networks with no link overlap, while for $c_1 = 0$, indicating complete overlap of the layers, the transition is continuous and reduces to the percolation transition on a single Poisson network. For

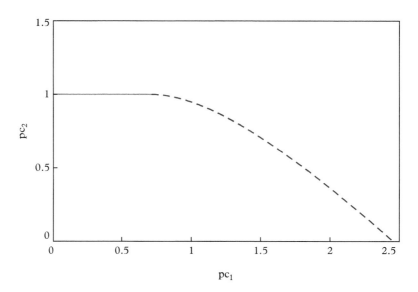

Fig. 12.4 *The critical lines of discontinuous hybrid phase transition (dashed line) and continuous phase transition (solid line) describing the emergence of the DMCGC are shown for the case of a multiplex network with two layers and the Poisson multidegree distribution with* $\langle k^{(1,0)} \rangle = \langle k^{(0,1)} \rangle = c_1$ *and* $\langle k^{(1,1)} \rangle = c_2$*. Reprinted figure from Ref. [76].*

$c_2 < c_1\sqrt{2}$ the DMCGC emerges at a discontinuous phase transition while for $c_2 > c_1\sqrt{2}$ the transition is continuous. The point $c_2 = c_1\sqrt{2}$ separating the line of continuous from the line of discontinuous phase transition if called tricritical point.

When this phase diagram is compared with the corresponding phase diagram determining the emergence of the MCGC (see Fig. 11.11), one observes firstly that the transition points at $c_2 = 0$ (absence of overlap) and at $c_2 = 1$ (complete overlap) are the same; secondly one notices that while for the DMCGC there is a tricritical point at a non-zero value of the link overlap $c_2/c_1 = \sqrt{2}$, in the phase diagram for the MCGC the only tricritical point is the trivial one for $c_2 = 1$. This is clear evidence that although the MCGC and the DMCGC are the same in the absence of link overlap and in the presence of complete link overlap, they describe a very distinct set of nodes in the presence of a non-negligible but non-complete link overlap. In particular, the DMCGC is a proper subset of the MCGC as long as the overlap between the layers is not vanishing or complete, in which cases the DMCGC coincides with the MCGC.

12.4   Antagonist percolation

The treatment of interdependencies in multilayer networks has introduced the effect of cooperative interactions in percolation problems. Nevertheless, in a variety of systems, including most notably biological networks, it is possible to assume that actual resilience and percolation properties of the network can be affected both by cooperative interactions (interdependencies) and antagonistic interactions. Possibly in the future it will be possible to formulate a well-defined percolation problem on multilayer networks that will include different types of combinatorial relations between the nodes and overall determine the stability and robustness of the multilayer structure.

In Ref. [316], a first step in this direction has been made by considering the case in which the interactions across different layers of a multiplex network have an antagonistic nature. In particular, a duplex network is considered in which the function or activity of a node is incompatible with the function or activity of the replica node in the other layer. The nodes in layer α are indicated in the following by (i, α), while the nodes in layer β are indicated by (i, β), with $i = 1, 2, \ldots, N$. The two nodes (i, α) and (i, β) are replica nodes with an antagonistic interaction. In order to determine whether a node (i, α) is in the percolation cluster of layer α, or a node (i, β) is in the percolation cluster of layer β, the following algorithm has been proposed.

A node (i, α) is part of the percolation cluster in layer α if the following recursive set of conditions is satisfied:

(i)   at least one neighbour (j, α) of node (i, α) in layer α is in the percolation cluster of layer α;

(ii)   none of the neighbours (j, β) of node (i, β) are in the percolation cluster of layer β.

Similarly, a node (i, β) is part of the percolation cluster in layer β if the following recursive set of conditions is satisfied:

(a)   at least one neighbour (j, β) of node (i, β) in layer β is in the percolation cluster of layer β;

(b)   none of the neighbours (j, α) of node (i, α), are in the percolation cluster of layer α.

For a given locally tree-like multiplex network without link overlap it is possible to construct a message-passing algorithm that determines whether node (i, α) belongs to the percolation cluster of layer α.

We denote by $\sigma^{\alpha}_{i \to j} = 1, 0$ the message within a layer α going from node (i, α) to node (j, α). The message $\sigma^{\alpha}_{i \to j} = 1$ indicates that node (i, α) is in the percolation cluster of layer α when we consider the network in which the link (i, j) in network α is removed. In addition, we indicate with $s_{i\alpha} = 0$ a node that is removed from the network as an effect of the damage inflicted on the network, otherwise $s_{i\alpha} = 1$. The message-passing equations take the form

$$\sigma^{\alpha}_{i \to j} = s_{i\alpha} \left[1 - \prod_{\ell \in N_{\alpha}(i) \setminus j} \left(1 - \sigma^{\alpha}_{\ell \to i} \right) \right] \prod_{\ell \in N_{\beta}(i)} \left(1 - \sigma^{\beta}_{\ell \to i} \right), \qquad (12.42)$$

where here and in the following $\alpha \neq \beta$. The indicator function $\sigma_{i\alpha}$ indicates that the node (i, α) is ($\sigma_{i\alpha} = 1$) or is not ($\sigma_{i\alpha} = 0$) in the percolation cluster of layer α. The variables $\sigma_{i\alpha}$ can be expressed in terms of the messages, i.e.

$$\sigma_{i\alpha} = s_{i\alpha}\left[1 - \prod_{\ell \in N_\alpha(i)} \left(1 - \sigma^\alpha_{\ell \to i}\right)\right] \prod_{\ell \in N_\beta(i)} \left(1 - \sigma^\beta_{\ell \to i}\right). \tag{12.43}$$

If we only know that the replica nodes α are damaged with probability $f_\alpha = 1 - p_\alpha$, the probability $\hat{\sigma}_{i\alpha}$ that the replica node (i, α) is in the percolation cluster after the inflicted damage is given by

$$\hat{\sigma}_{i\alpha} = p_\alpha\left[1 - \prod_{\ell \in N_\alpha(i)} \left(1 - \hat{\sigma}^\alpha_{\ell \to i}\right)\right] \prod_{\ell \in N_\beta(i)} \left(1 - \hat{\sigma}^\beta_{\ell \to i}\right), \tag{12.44}$$

where $\hat{\sigma}^\alpha_{i \to j}$ satisfies

$$\hat{\sigma}^\alpha_{i \to j} = p_\alpha\left[1 - \prod_{\ell \in N_\alpha(i)\backslash j} \left(1 - \hat{\sigma}^\alpha_{\ell \to i}\right)\right] \prod_{\ell \in N_\beta(i)} \left(1 - \hat{\sigma}^\beta_{\ell \to i}\right). \tag{12.45}$$

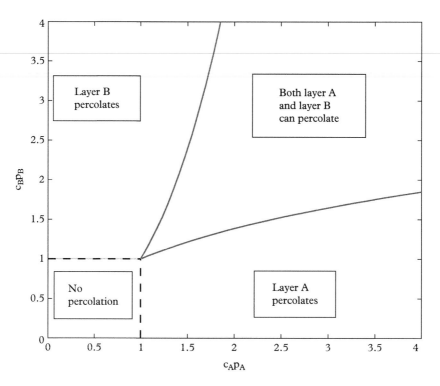

Fig. 12.5 *Phase diagram of the percolation of two antagonistic Poisson networks (layer A and layer B) with average degree given respectively by* c_A *and* c_B.

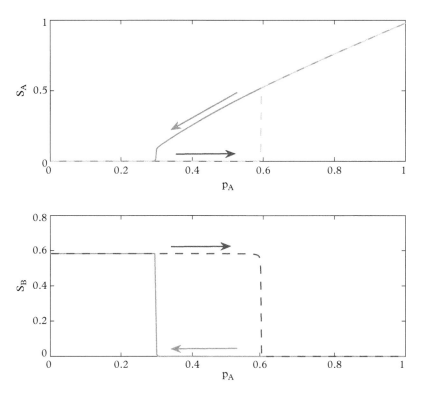

Fig. 12.6 *Panels (a) and (b) show the hysteresis loop for the percolation of a duplex formed by two antagonistic Poisson layers with average degree* $c_A = 4$, $c_B = 1.5$ *as a function of the probability* p_A *of not damaging the nodes in layer A. Here the nodes in layer B are not damaged, i.e.* $p_B = 1$.

If we consider a duplex formed by two uncorrelated networks (layer A and layer B) with degree distributions $P_A(k)$ and $P_B(k)$ respectively, we can average the messages in each layer, getting the equations for S'_A indicating the average of $\hat{\sigma}^{\alpha}_{i \to j}$ and S'_B indicating the average of $\hat{\sigma}^{\beta}_{i \to j}$, that read as

$$S'_A = p_A \left[1 - G^A_1 \left(1 - S'_A \right) \right] G^B_0 \left(1 - S'_B \right)$$
$$S'_B = p_B \left[1 - G^B_1 \left(1 - S'_B \right) \right] G^A_0 \left(1 - S'_A \right), \qquad (12.46)$$

where $G^A_0, G^B_0, G^A_1, G^B_1$ are the generating function of the degree distribution $P^A(k)$, $P^B(k)$, $\frac{k}{\langle k \rangle_A} P^A(k)$, $\frac{k}{\langle k \rangle_B} P^B(k)$ respectively. The probabilities S_A (or S_B) that a random node in layer A (or layer B) is in the percolation cluster of layer A (or layer B) are given by

$$S_A = p_A \left[1 - G^A_0 \left(1 - S'_A \right) \right] G^B_0 \left(1 - S'_B \right)$$
$$S_B = p_B \left[1 - G^B_0 \left(1 - S'_B \right) \right] G^A_0 \left(1 - S'_A \right). \qquad (12.47)$$

This novel percolation problem has surprising features [316]. For example, for a duplex network formed by two Poisson networks with average degree $\langle k \rangle_A = c_A$ and $\langle k \rangle_B = c_B$ respectively, the phase diagram in Fig. 12.5 shows that this generalized percolation problem displays a bistable phase. For $c_A p_A < 1$ and $c_B p_B < 1$ we observe a phase in which neither layer can percolate. The remaining three regions of the phase space include: one region in which only layer A contains a percolation cluster; a symmetric region in which only layer B contains a percolation cluster; a third region in which the solution of the model is bistable, and depending on the initial condition of the message-passing algorithm, either network A or network B are percolating. Due to the presence of this bistable region, this generalized percolation problem can display a hysteresis loop, as shown in Fig. 12.6. This means that the size of the giant component in layer A and in layer B depends on the previous history of the process. If, for instance, we start from a configuration in which there is a giant component in layer A and no giant component in layer B and we increase subsequently the number of damaged nodes in layer A, the size of the giant component is reduced until it reaches a tipping point where there is no more a giant component in layer A (solid line in Fig. 12.6). However, the response of the system might be significantly different if we start from an initial configuration in which layer B is percolation and layer A is not and the nodes in layer A are subsequently activated (dashed line in Fig. 12.6). Therefore, the same type of random damage can yield two different and opposite configurations as long as the system is in the bistable region.

In the following chapters we will show that antagonistic or competing networks have also been discussed in the framework of epidemic-spreading processes (Sec. 13.5), election models (Sec. 16.3.1) and models for competing resources (Sec. 16.3.2).

12.5   Cascades on multilayer networks

12.5.1   Cascades without interdependencies

Cascades in multilayer networks can occur due to interdependencies between the nodes of different layers or can be caused by alternative mechanisms including cascades of loads [63, 187] described, for instance, by the Bak–Tang–Wisenfeld sandpile model [12], by the threshold cascade model [65] proposed by Watts in Ref. [309] or by novel cascade processes [186].

12.5.2   Cascade of loads and sandpile model

Cascades of loads have been extensively studied in simplex networks, shedding light on the response of these networks to the flow of physical quantities in the network. These works are especially important for applications ranging from power grids and communication and transportation systems to financial markets [218]. In Ref. [63] cascades of loads described by the stylized Bak–Tang–Wisenfeld sandpile model [12] have been proposed to study the robustness of multilayer infrastructures and power grids.

In this model it is assumed that 'grains of sand' flow in the network according to a discrete dynamics of activations or 'toppling events'. Each node is assigned a *load capacity*, i.e. the maximum possible number of particles of sand it can support. Usually the load capacity of each node is given by its degree. When a new grain of sand arrives on a given node and the number of sand grains on the node is larger than its load capacity the node 'topples' and all the grains of sand in the node are distributed to its neighbour nodes, adding one grain of sand to each neighbour node. Eventually, this dynamics can trigger additional toppling events, generating a cascade in the system. Typically in the sandpile model, the system is subjected to the subsequent addition of single grains of sand, and the size of the resulting avalanche is monitored. In this way the number of grains of sand in the network will continuously increase, eventually saturating all the nodes. In order to avoid this undesired state it is important that the dynamics includes some mechanism for dissipating the grains of sand. In the traditional realization of the sandpile model on a two-dimensional finite-square lattice the dissipation is enforced by open boundary conditions where all the grains of sand that move out of the lattice boundaries are dissipated. In complex networks, the concept of boundary is usually lost, so in this case the dissipation mechanism is enforced by introducing a small but non-zero probability f of deleting grains of sand as they shed.

Power-grid stability and sandpile model on multilayer network

The sandpile model defined on single networks can be generalized without modifications to modular multilayer networks formed by different interconnected networks. For instance, this is the most realistic description of the US power grid which includes thousands of distinct local and regional utilities organized in a multilayer structure (see Fig. 12.7). In Ref. [63] the sandpile model is studied in random multilayer networks and on real power-grid datasets in order to predict the response of these networks to cascades of loads and the optimal multilayer network structures. The resulting sandpile model on multilayer networks extends the model defined on single networks. Specifically, the multilayer network ensembles where this model is defined take into account that links might belong to different networks and therefore the multilayer network structure might be non-trivial. Nevertheless, similarly to what happens for classic percolation, if one does not average over the multilayer network ensemble the dynamics is the same as the one defined on a large single modular network. The differences observed with respect to a single random network are therefore due exclusively to the multilayer network (modular) topology. The main result of Ref. [63] is that the probability of large avalanches can be affected by the interlinks between layers. In fact, starting from two distinct layers and increasing the probability p of having random interlinks, it is observed that the probability of large avalanches has a minimum for a non-zero value $p = p^\star$ (see Fig. 12.8). In fact, a smaller probability of interconnections leads to a higher probability of large avalanches; because the network has more bottlenecks, a larger probability of the interconnections can increase the capacity and the total possible load, eventually generating larger cascades. Therefore, moderately increasing the interlayer connectivity can reduce the size of the largest cascade in each layer, but too many interconnections can become detrimental for the stability of the multilayer system.

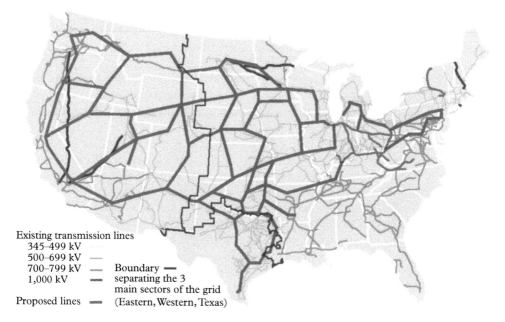

Existing transmission lines
345–499 kV
500–699 kV   —
700–799 kV   —   Boundary   —
1,000 kV   —   separating the 3
              main sectors of the grid
Proposed lines   —   (Eastern, Western, Texas)

Fig. 12.7 *The power grid of the United States is formed by three main regions—West, East and Texas—forming a multilayer network. Here the power grid is shown schematically together with the proposed new lines for transport of wind power. Reprinted figure from Ref. [63].*

Sandpile model on multiplex networks

In Ref. [187] the sandpile dynamics on multiplex networks was further explored. It has been shown that the avalanche distribution is not changed with respect to the one observed on single layers. However, the multiplexity has a role in increasing the fragility of high-degree nodes in multiplex networks.

12.5.3   Watts contagion model

Cascades of adoption of behaviour in social and financial contexts can be captured by the Watts model [309]. These cascades include propagation of social opinions, emergence of novel political or social movements and diffusion of financial strategies. The Watts model is a contagion model indicating that a node's decision to change state (opinion, behaviour) is influenced by its neighbours. However, differently from percolation, it is assumed that each individual has some resistance in changing his own state. Specifically, an individual will not change his state if only a single neighbour adopts a different behaviour, while only a larger number of neighbours will exert sufficient social pressure to induce a change of state. Therefore, the Watts model assumes that each individual will change opinion only if a given fraction (or a given number) of his neighbours acquires another opinion.

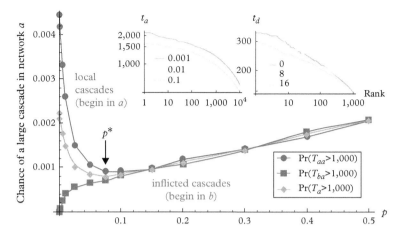

Fig. 12.8 *The main plot displays the probability of large avalanches propagating within a single layer* T_{aa}, *inflicted on one layer by the other one* T_{ba}, *and the probability of large avalanches* T_a *occurring in one layer regardless of the origin of the avalanches as a function of the probability of interconnections* p. *Here it is evident that* T_a *has a minimum at* p*, *indicating that there is an optimal value of interconnectivity that reduces the likelihood of large avalanches. The inset shows the distribution of avalanche sizes for different values of the interconnectivity* p = 0.001, 0.01, 0.1 *for a given artificial multilayer network (left inset) and for a duplex network formed by two real subsets of the US power grid (right inset) connected by* c = 0, 8, 16 *interlinks. Reprinted figure from Ref. [63].*

In Ref. [65, 186] the Watts model is extended to the multiplex networks and modified to capture the effects of multiplexity. It is shown that multiplexity can actually increase the network vulnerability to global cascades. Additionally, in some regions of the phase diagram of the model proposed in Ref. [186] it is possible to observe an abrupt emergence of cascading events. These results show that advertising strategies can be more effective in multiplex networks. However, whereas this model is appropriate for modelling financial systems, these results also indicate that multiplex financial networks have a larger systemic risk than single-layer financial networks. Additionally, the discontinuous emergence of cascading events reveals a novel mechanism for capturing the sudden adoption of a product, the uprising of a movement or the emergence of a social opinion.

13

Epidemic Spreading

13.1 Epidemics and multiplexity

Epidemic spreading is one of the most interesting and studied dynamical processes on complex networks, with relevance for the modelling of biological epidemics, social contagions, success of memes and tweets in social online networks and for designing immunization strategies and identifying influential spreaders. Therefore, extending the study of epidemic spreading to multilayer networks is natural. In fact, both biological epidemics and social contagion generally occur on multilayer networks. For example, sexually transmitted diseases can propagate across the multiplex networks formed by layers consisting on heterosexual and homosexual contacts, while the co-existence of several online social networks allows social contagion to span across the multiplex networks formed by different online platforms such as Facebook, Twitter and LinkedIn. Additionally, the multilayer set-up allows for modelling interacting contagion processes as, for instance, the simultaneous spread of influenza on the social contact network and the spread of the awareness behaviour on online social networks that do not require physical contact between the individuals.

A rich and surprising phenomenology is observed when epidemic spreading propagates in a multilayer network. It is found that in general multiplexity allows for the propagation of the epidemics even if the single layers of the network are not able to sustain the epidemics if taken in isolation. Additionally, the degree correlations existing in a multilayer network can significantly affect the properties of the epidemic spreading.

The bibliography on epidemic processes in multilayer networks is already very rich and includes several review articles [88, 305, 269]. Due to space limitation it will be impossible here to cover entirely this very active field of research, so we decided instead to focus on fewer representative results, showing the important effect of multiplexity on epidemic spreading.

Multilayer Networks. Ginestra Bianconi, Oxford University Press (2018).
© Ginestra Bianconi. DOI: 10.1093/oso/9780198753919.001.0001

13.2 SIS model

13.2.1 General remarks

The Susceptible–Infected–Susceptible (SIS) dynamics plays a fundamental role in modelling epidemic spreading in complex networks. This model assumes that nodes can be in two possible states: susceptible (S) or infected (I). Susceptible nodes can become infected if they have at least one infected neighbour, while infected nodes can spread the infection and become susceptible at a constant rate in time. The SIS dynamics can result in a dynamical state in two different regimes (or dynamical phases): the absorbing phase and the endemic phase. In the absorbing phase the epidemic does not spread fast enough and after a transient it dies out. The endemic phase instead is characterized by a steady state in which a finite fraction of the nodes of the network is infected at any given time.

13.2.2 Definition

The SIS dynamics on multiplex and general multilayer networks can be defined in two different ways. On a multiplex network it is possible to assume that each node has the same state on each of its replica nodes and that the infection proceeds at different rates depending on the pattern of connections between any two linked nodes. In this case, we can assume that each node i of the multiplex network is either susceptible (indicated with S_i) or infected (indicated with I_i). A suceptible individual that is connected to an infected individual by a multilink $\vec{m}_{ij}$ gets the infection with rate $\xi^{\vec{m}_{ij}}$, i.e.

$$S_i + I_j \xrightarrow{\xi^{\vec{m}_{ij}}} I_i + I_j, \tag{13.1}$$

while an infected individual I_i becomes susceptible with rate μ, i.e.

$$I_i \xrightarrow{\mu} S_i. \tag{13.2}$$

In the particular case in which the multiplex network has no link overlap, the infection probabilities $\xi^{\vec{m}_{ij}}$ only depend on the single layer α in which node i and node j are connected, i.e. $\xi^{\vec{m}_{ij}} = \xi^{[\alpha]}$.

Alternatively, the SIS dynamics can be defined assuming that every node (i, α) of a general multilayer network has a different dynamical state, being either susceptible (indicated with $S_{i\alpha}$) or infected (indicated with $I_{i\alpha}$). Considering the same recovery rate for all the nodes of the multilayer network given by μ, this SIS dynamics can be summarized by the following two processes. A susceptible node (i, α) connected to a node (j, β) either by intralinks ($\alpha = \beta$) or interlinks ($\alpha \neq \beta$) gets the infection at rate $\xi^{[\alpha, \beta]}$, i.e.

$$S_{i\alpha} + I_{j\beta} \xrightarrow{\xi^{[\alpha,\beta]}} I_{i\alpha} + I_{j\beta}. \tag{13.3}$$

An infected node (i, α) becomes susceptible with rate μ, i.e.

$$I_{i\alpha} \xrightarrow{\mu} S_{i\alpha}. \tag{13.4}$$

Here, given space limitations, we will focus on this latter definition of the SIS dynamics, which has been investigated extensively in the literature [271, 82, 87]. The derivation of the theoretical predictions for the SIS dynamics on multilayer networks can be directly extended to the SIS dynamics on multiplex networks where each node is characterized by the same dynamical state.

13.2.3 General formalism for the SIS dynamics on multilayer networks

Let us here derive the general formalism that describes the SIS dynamics on multilayer network in discrete time. Let us indicate with $X_{i\alpha}(t) = 0, 1$ whether node (i, α) is infected $X_{i\alpha}(t) = 1$ or susceptible $X_{i\alpha}(t) = 0$ at time t. At each given time a susceptible node (i, α) becomes infected it has been infected at least by one of its infected neighbours, while an infected node (i, α) becomes susceptible with probability μ independently of its neighbours. Therefore, the SIS dynamics on multilayer networks can be summarized by the set of dynamic equations

$$X_{i\alpha}(t+1) = \begin{cases} (1 - X_{i\alpha}(t)) & \text{with probability } \pi_{i\alpha}, \\ X_i(t) & \text{with probability } 1 - \pi_{i\alpha}, \end{cases} \tag{13.5}$$

where

$$\pi_{i\alpha} = \begin{cases} \left[1 - \prod_{\beta=1}^{M} \prod_{j=1}^{N} \left(1 - \mathcal{R}_{i\alpha,j\beta} X_{j\beta}(t)\right)\right] & \text{if } X_{i\alpha}(t) = 0 \\ \mu & \text{if } X_{i\alpha}(t) = 1. \end{cases} \tag{13.6}$$

In Eq. (13.5) the reaction supra-matrix $\mathcal{R}$ is expressed in terms of the adjacency matrices $\mathbf{a}^{[\alpha,\beta]}$ and the infection rates $\xi^{[\alpha,\beta]}$ and has a block structure of the type

$$\mathcal{R} = \begin{pmatrix} \xi^{[1,1]}\mathbf{a}^{[1,1]} & \xi^{[1,2]}\mathbf{a}^{[1,2]} & \cdots & \xi^{[1,M]}\mathbf{a}^{[1,M]} \\ \hline \xi^{[2,1]}\mathbf{a}^{[2,1]} & \xi^{[2,2]}\mathbf{a}^{[2,2]} & \cdots & \xi^{[2,M]}\mathbf{a}^{[2,M]} \\ \hline \vdots & \vdots & \ddots & \vdots \\ \hline \xi^{[M,1]}\mathbf{a}^{[M,1]} & \xi^{[M,2]}\mathbf{a}^{[M,2]} & \cdots & \xi^{[M,M]}\mathbf{a}^{[M,M]} \end{pmatrix}. \tag{13.7}$$

From the study of Eq. (13.5) we can derive the exact equation for the probability $p_{i\alpha}(t) = \langle X_{i,\alpha}(t+1) \rangle$ that a node (i, α) is infected at time t, i.e.

$$p_{i\alpha}(t) = \langle X_{i\alpha}(t+1) \rangle = \left\langle (1 - X_{i\alpha}(t)) \left[1 - \prod_{\beta=1}^{M} \prod_{j=1}^{N} \left(1 - \mathcal{R}_{i\alpha,j\beta} X_{j\beta}(t)\right) \right] \right\rangle$$
$$+ (1 - \mu) \langle X_{i\alpha}(t) \rangle. \tag{13.8}$$

13.2.4   Individual mean-field approximation

The exact numerical integration of the SIS dynamics on multilayer networks, and on single networks as well, constitutes until now one of the major challenges of network theory. In the absence of an exact solution of this model, as we have discussed in the context of single layers, two approximations for studying these equations have been proposed: the individual mean-field approximation and the annealed network approximation. These two approximations can be used to compare simulations to analytical predictions. Simulations have shown that in many cases of interest the mean-field approaches can give a good indication of the simulation results. Nevertheless, the limitations of these approaches have also raised significant attention in recent years. The individual mean-field approach consists in neglecting correlations and assuming that

$$\langle X_{i\alpha}(t) X_{j\beta}(t) \rangle \simeq \langle X_{i\alpha}(t) \rangle \langle X_{j\beta}(t) \rangle \tag{13.9}$$

and

$$\langle X_{i_1\alpha_1} X_{i_2\alpha_2} \ldots X_{i_q\alpha_q} \rangle \simeq \prod_{n=1}^{q} \langle X_{i_n\alpha_n} \rangle. \tag{13.10}$$

With these assumptions we can write the dynamical equation for $p_{i\alpha}(t) = \langle X_{i\alpha}(t) \rangle$ starting from Eq. (13.5) as

$$p_{i\alpha}(t+1) = \left(1 - p_{i\alpha}(t)\right) \left[1 - \prod_{\beta=1}^{M} \prod_{j=1}^{N} \left(1 - \mathcal{R}_{i\alpha j\beta} p_{j\beta}(t)\right) \right]$$
$$+ (1 - \mu) p_{i\alpha}(t), \tag{13.11}$$

which has the steady-state solution

$$p_{i\alpha}(t) = p_{i\alpha}(t+1) = p_{i\alpha}^\star \tag{13.12}$$

for $t \gg 1$ given by

$$0 = -\mu p_{i\alpha}^\star + \left(1 - p_{i\alpha}^\star\right) \left[1 - \prod_{\beta=1}^{M} \prod_{j=1}^{N} \left(1 - \mathcal{R}_{i\alpha j\beta} p_{j\beta}^\star\right) \right]. \tag{13.13}$$

This equation always has a solution $p_{i\alpha}^\star = 0$, $\forall (i, \alpha)$. A non-trivial solution appears for values of sufficiently high infection rates. The onset of the endemic regime can be studied by linearizing Eq. (13.13) around the solution $p_{i\alpha}^\star = 0$. In this way, by putting $p_{i,\alpha} = \epsilon_{i\alpha}$ with $\epsilon_{i\alpha} \ll 1$ we get

$$0 = \sum_{\beta=1}^{M} \sum_{j=1}^{N} [\mathcal{R}_{i\alpha j\beta} - \mu \delta(i,j)\delta(\alpha,\beta)] \epsilon_{j\beta}, \tag{13.14}$$

where $\delta(x, y)$ indicates the Kronecker delta. For simplicity we assume that the propagation of the disease has the same timescale within each layer $\xi^{[\alpha,\alpha]} = \lambda$ and that the disease propagates across each pair of layers $\alpha \neq \beta$, with another timescale set by $\xi^{[\alpha,\beta]} = \eta$. In these conditions we have that the endemic regime sets in when λ/μ is given by [82, 303]

$$\frac{\lambda}{\mu} = \frac{1}{\Lambda(\tilde{\mathcal{R}})} \tag{13.15}$$

where $\Lambda(\tilde{\mathcal{R}})$ is the maximum eigenvalue of the normalized reaction matrix $\tilde{\mathcal{R}} = \mathcal{R}/\lambda$ of the multilayer network. We note here that, as in the case of single layers, this result requires additional caution. In fact, it is only valid as long as the eigenvector associated with the maximum eigenvalue $\Lambda(\tilde{\mathcal{R}})$ is delocalized over at least a finite fraction of the nodes of the network. Under these conditions, this result gives interesting insights regarding the properties of the epidemic spreading in multilayer networks. In fact, the epidemic on a multilayer network can become endemic also for infection rates λ for which the epidemic does not spread in the individual layers if the layers are taken in isolation. For example, in the case of a multilayer network formed by two layers we have [82, 303]

$$\Lambda(\tilde{\mathcal{R}}) \geq \max\left[\Lambda(\mathbf{a}^{[1,1]}), \frac{\eta}{\lambda}\Lambda(\mathbf{a}^{[2,2]})\right]. \tag{13.16}$$

In Fig. 13.1 of Ref. [87] the fraction of infected individuals at steady state is reported against the control parameter λ in the case in which the multilayer network is a multiplex network formed by two layers with interlinks joining the replica nodes (i, α) with $\alpha = 1, 2, \ldots, M$. From this figure it is possible to see that the epidemic threshold for the onset of the endemic state in the multiplex network is smaller than the epidemic threshold for each of the two layers taken in isolation. This result shows in a concrete example that multiplexity in general will favour the spread of the epidemics. Interestingly, the multilayer network will in any case preserve some 'memory' of the transition occurring on its single layer taken in isolation. This is particularly evident when the susceptibility of the system is measured. The susceptibility χ is defined as

$$\chi = \frac{\langle(N^I)^2\rangle - \langle N^I\rangle}{\langle N^I\rangle}, \tag{13.17}$$

where N^I is the number of infected nodes and the average if performed over the distribution of N^I at steady state. On a single layer the susceptibility has a single peak that indicates the epidemic threshold [241]. The susceptibility of a multiplex network of two layers, however, presents two peaks occurring in correspondence with the two epidemic thresholds of its layers taken in isolation (see Fig. 13.2 from Ref. [87]). This implies that when the infection rate is larger than the critical infection rate of the layer which is more resistant to the spread of the epidemic, the infection will rapidly spread over a larger population (see also panel (a) of Fig. 13.1)

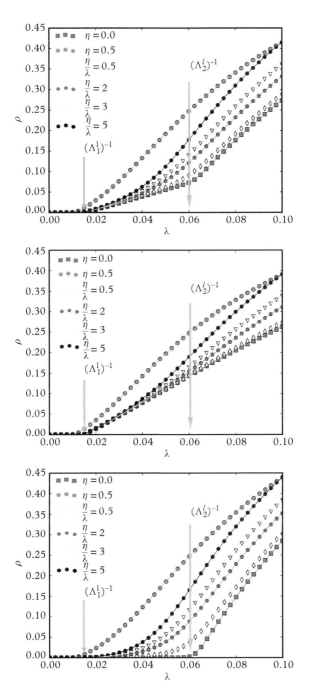

Fig. 13.1 *Individual layer behaviour of the SIS epidemic spreading over a duplex network. Each layer has* N $= 10^4$ *nodes and a fixed value of* $\mu = 1$. *The average fraction of infected nodes* ρ *at the stationary state is plotted as a function of the infectivity* λ *for the global duplex network (panel (a)) and for each single layer (panels (b) and (c)). The arrows indicate the layers' leading eigenvalues. Here the two layers have power-law degree distributions with power-law exponent* γ *given respectively by* $\gamma = 2.5$ *(First layer) and* $\gamma = 4.5$ *(Second layer). Reprinted from Ref. [87].*

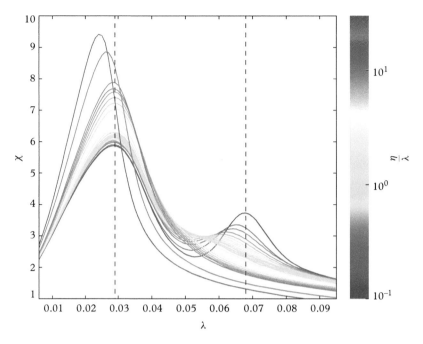

Fig. 13.2 *Susceptibility χ of the SIS dynamics as a function of the spreading rate λ for different ratios of inter- and intralayer spreading rates $\frac{\eta}{\lambda}$ for a fixed value of $\mu = 1$ over duplex networks, where each layer has $N = 10^3$ nodes, the first layer is a scale-free network with power-law exponent $\gamma \approx 2.2$ and the second layer is a scale-free network with power-law exponent $\gamma \approx 2.8$. Both layers have $\langle k \rangle \approx 8$. The simulated values are $\frac{\eta}{\lambda} = 0.1, 0.2, 0.3, 0.4, 0.5, 0.6, 0.7, 0.8, 0.9, 1.0, 1.1, 1.2, 1.3, 1.4, 1.5, 1.6, 2, 3, 4, 5, 6, 7, 8, 9, 10, 20, 30. Reprinted from Ref. [87].*

Until now the SIS epidemic has been formulated in discrete time using the mean-field Eqs (13.11). By considering a discrete time dynamics taking place at time intervals Δt, and setting

$$\lambda \to \lambda \Delta t$$
$$\delta \to \delta \Delta t$$
$$\mu \to \mu \Delta t \qquad (13.18)$$

in the limit $\Delta t \to 0$, we can recover the continuous time equation for $p_{i\alpha}(t)$ given by

$$\frac{dp_{i\alpha}}{dt} = -p_{i\alpha}(t) + (1 - p_{i\alpha}) \left[\sum_{\beta=1}^{M} \sum_{j=1}^{N} \mathcal{R}_{i\alpha,j\beta} p_{j\beta} \right], \qquad (13.19)$$

$$\frac{dp_{i\alpha}}{dt} = -\mu p_{i\alpha}(t) + (1 - p_{i\alpha}) \left[\sum_{\beta=1}^{M} \sum_{j=1}^{N} \mathcal{R}_{i\alpha,j\beta} p_{j\beta} \right] \qquad (13.20)$$

where, without loss of generality, we have set $\mu = 1$ by rescaling of time $t \to t\mu$ and $\xi^{[\alpha,\beta]} \to \lambda^{[\alpha,\beta]} = \xi^{[\alpha,\beta]}/\mu$. Therefore, in Eq. (13.19) the supramatrix $\mathcal{R}_{i\alpha,j\beta}$ is given by

$$\mathcal{R} = \begin{pmatrix} \lambda^{[1,1]}\mathbf{a}^{[1,1]} & \lambda^{[1,2]}\mathbf{a}^{[1,2]} & \cdots & \lambda^{[1,M]}\mathbf{a}^{[1,M]} \\ \lambda^{[2,1]}\mathbf{a}^{[2,1]} & \lambda^{[2,2]}\mathbf{a}^{[2,2]} & \cdots & \lambda^{[2,M]}\mathbf{a}^{[2,M]} \\ \vdots & \vdots & \ddots & \vdots \\ \lambda^{[M,1]}\mathbf{a}^{[M,1]} & \lambda^{[M,2]}\mathbf{a}^{[M,2]} & \cdots & \lambda^{[M,M]}\mathbf{a}^{[M,M]} \end{pmatrix}. \qquad (13.21)$$

Clearly, the steady-state solution of Eqs (13.19) will give results equivalent to the steady-state solution of the dynamics in discrete time analysed earlier in the paragraph.

13.2.5   Annealed network approximation

Another usually considered approximation for describing the SIS dynamics on a network is the annealed network approximation, also known as the heterogeneous mean-field approximation. This approach consists in making essentially two different approximations. First of all, one considers the mean-field approximation that is assuming a decorrelation between the dynamical states of neighbour nodes, expressed by Eqs (13.9) and (13.10). Secondly, one assumes that the network is random and substitutes the adjacency matrix elements $a_{i,j}^{[\alpha,\beta]}$ with their average value in an appropriate network ensemble such as a configuration model or a correlated network model. This second approximation, also known as the annealed network approximation, is valid in the limit in which the network is not static but instead links are continuously rewired in order to generate different network configurations in the network ensemble. For example, here we will consider the case in which links are rewired by keeping the same multilayer degree sequences.

Since in the annealed network approximation the multilayer degree of the nodes determines the role of the nodes in the network completely, the SIS equations in the heterogeneous mean-field approximation are describing the dynamical evolution of the average number $\rho_{\mathbf{k},\alpha}$ of infected nodes that are in layer α and have multilayer degree $\mathbf{k}_{i\alpha} = \mathbf{k}$, i.e.

$$\rho_{\mathbf{k},\alpha} = \langle p_{i\alpha} \rangle_{\mathbf{k}_{i\alpha}=\mathbf{k}}. \qquad (13.22)$$

In the absence of multilayer degree correlations, the annealed approximation implies substituting the elements i, j of the matrices $a^{[\alpha,\beta]}$ with their average value $\pi_{i\alpha,j\beta}$ given by Eq. (10.83) in the multilayer configuration model obtaining

$$a_{i,j}^{[\alpha,\beta]} \to \pi_{i\alpha,j\beta} = \frac{k_i^{[\alpha,\beta]} k_j^{[\beta,\alpha]}}{\langle k^{[\beta,\alpha]} \rangle}. \qquad (13.23)$$

Therefore, starting from the mean-field equation in continuous time Eq. (13.19), in the heterogeneous mean-field approximation we obtain the dynamical equations [271]

$$\frac{d\rho_{k,\alpha}}{dt} = -\rho_{k,\alpha} + (1 - \rho_{k,\alpha}) \sum_{\beta=1}^{M} \sum_{\tilde{k}_\beta} \lambda^{[\alpha,\beta]} P(\tilde{k}, \beta) \frac{k^{[\alpha,\beta]} \tilde{k}^{[\beta,\alpha]}}{\langle k^{[\beta,\alpha]} \rangle} x_{\alpha\beta} \rho_{\tilde{k},\beta}, \tag{13.24}$$

where $x_{\alpha\beta} = 0$ when $\langle k^{[\beta,\alpha]} \rangle = 0$, and otherwise $x_{\alpha\beta} = 1$. This equation is always consistent with an epidemic-free phase $\rho_{k,\alpha}(t) = 0$ for every t and every k, α. By linearizing Eq. (13.24) close to this trivial solution for $\rho_{k,\alpha} \ll 1$ the conditions for the onset of the endemic phase $\rho_{k,\alpha} \neq 0$ can be found. This equation reads

$$\frac{d\rho}{dt} = -\rho + C\rho, \tag{13.25}$$

where C is a matrix formed by $M \times M$ blocks $C^{[\alpha,\beta]}$ of elements

$$C^{[\alpha,\beta]}_{k,\tilde{k}} = \lambda^{[\alpha\beta]} \frac{k^{[\alpha,\beta]} \tilde{k}^{[\beta,\alpha]}}{\langle k^{[\alpha,\beta]} \rangle} x_{\alpha,\beta} P(\tilde{k}, \beta). \tag{13.26}$$

Eq. (13.25) allows us to study the stability of the trivial solution $\rho_{k,\alpha} = 0$ for all values of k, and α. It is found that this solution is unstable whenever the maximal eigenvalue Λ_m of C is larger than one, i.e.

$$\Lambda_m > 1. \tag{13.27}$$

Therefore, under the SIS dynamics this describes an endemic state. On the contrary, for

$$\Lambda_m \leq 1 \tag{13.28}$$

the system will be in the epidemic-free absorbing phase.

Let us now discuss in detail the case in which the multiplex network is formed by two layers, i.e. $M = 2$. In this case we have that C has the explicit form

$$C = \begin{pmatrix} C^{[1,1]} & C^{[1,2]} \\ C^{[2,1]} & C^{[2,2]} \end{pmatrix}, \tag{13.29}$$

where $C^{[\alpha,\beta]}$ are matrices of elements given by Eq. (13.26). In this case the onset of the endemic phase, when the maximal eigenvalue Λ_m of C is given by

$$\Lambda_m = 1, \tag{13.30}$$

is achieved when the infection rates $\lambda^{[\alpha,\beta]}$ satisfy [55]

$$0 = \left[1 - \lambda^{[1,1]}(\kappa_{11} + 1)\right]\left[1 - \lambda^{[2,2]}(\kappa_{22} + 1)\right] - \left(\lambda^{[1,2]}\right)^2 \hat{\mathcal{R}}_{12}, \tag{13.31}$$

where

$$\begin{aligned}
\hat{\mathcal{R}}_{12} &= \left[\lambda^{[1,1]}\mathcal{W}_{12}\mathcal{K}_{12} + \kappa_{12} + 1 - \lambda^{[1,1]}(\kappa_{12} + 1)(\kappa_{11} + 1)\right] \\
&\times \left[\lambda^{[2,2]}\mathcal{W}_{21}\mathcal{K}_{21} + \kappa_{21} + 1 - \lambda^{[2,2]}(\kappa_{21} + 1)(\kappa_{22} + 1)\right]
\end{aligned} \tag{13.32}$$

and $\kappa_{\alpha\beta}$, $\mathcal{K}_{\alpha\beta}$ and $\mathcal{W}_{\alpha,\beta}$ are given by Eqs (12.32). Indicating with $\Lambda^{[\alpha,\beta]}$ the quantities

$$\Lambda^{[\alpha,\beta]} = \begin{cases} \lambda^{[\alpha,\alpha]}(\kappa_{\alpha\alpha} + 1) & \text{for } \beta = \alpha, \\ \sqrt{\lambda^{[1,2]}\lambda^{[2,1]}(\kappa_{12} + 1)(\kappa_{21} + 1)} & \text{for } \beta \neq \alpha. \end{cases} \tag{13.33}$$

The annealed mean-field approximation of single layers predicts that the quantities $\Lambda^{[\alpha,\beta]}$ given by

$$\Lambda^{[\alpha,\beta]} = \begin{cases} \lambda^{[\alpha,\alpha]}(\kappa_{\alpha\alpha} + 1) & \text{for } \beta = \alpha, \\ \sqrt{\lambda^{[1,2]}\lambda^{[2,1]}(\kappa_{12} + 1)(\kappa_{21} + 1)} & \text{for } \beta \neq \alpha \end{cases} \tag{13.34}$$

determine the epidemic thresholds of layer 1 ($\Lambda^{[1,1]}$), layer 2 ($\Lambda^{[2,2]}$) and the bipartite network formed exclusively by interlinks ($\Lambda^{[1,2]}$). Specifically, values $\Lambda^{[1,1]} > 1$ and $\Lambda^{[2,2]} > 1$ ensure that the epidemic is in the endemic phase in the single layers taken in isolation, and values $\Lambda^{[1,2]} > 1$ ensure that the bipartite network formed exclusively by the interlinks is in the endemic phase.

It turns out that Λ_m determining the epidemic threshold of the multilayer network is always larger than the maximum of $\Lambda^{[1,1]}$, $\Lambda^{[2,2]}$ and $\Lambda^{[1,2]}$, i.e.

$$\Lambda_m \geq \max\left(\Lambda^{[1,1]}, \Lambda^{[2,2]}, \Lambda^{[1,2]}\right). \tag{13.35}$$

Therefore, also in the heterogeneous mean-field approximation we obtain that the epidemic spreading on a multilayer network might become endemic for infection rates, for which it is impossible to have an endemic state in the single layers taken in isolation. This case is captured by configurations in which $\Lambda_m > 1$, whereas $\Lambda^{[1,1]} < 1$ and $\Lambda^{[2,2]} < 1$.

The annealed approximation for studying the epidemic spreading on networks can be extended to multilayer networks with additional degree correlations [271]. To this end, we can define the probability $P_{\alpha,\beta}\left(\tilde{\mathbf{k}}, \beta | \mathbf{k}, \alpha\right)$ that a link of a node of layer α with multilayer degrees $\mathbf{k} = \left(k^{[\alpha,1]}, k^{[\alpha,2]} \dots k^{[\alpha,M]}\right)$ connects this node to a node in layer β with degrees $\tilde{\mathbf{k}} = \left(\tilde{k}^{[\beta,1]}, \tilde{k}^{[\beta,2]} \dots, \tilde{k}^{[\beta,M]}\right)$.

Close to the epidemic threshold, for $\rho_{k,\alpha} \ll 1$ the epidemic spreading is described also in this case by the linear equation

$$\frac{d\boldsymbol{\rho}}{dt} = -\boldsymbol{\rho} + \mathbf{C}\boldsymbol{\rho}, \tag{13.36}$$

whereas the expression for the elements of the matrix $C^{[\alpha,\beta]}$ is no longer given by Eq. (13.26) but instead by

$$C_{k,\tilde{k}}^{[\alpha,\beta]} = \lambda^{[\alpha,\beta]} k^{[\alpha,\beta]} P_{\alpha,\beta}(\tilde{k}, \beta | k, \alpha). \tag{13.37}$$

Therefore, in this way it is possible also to explore the epidemic spreading in multilayer networks whose single networks describing the interactions between nodes of different layers and between nodes of the same layer display degree correlations.

13.2.6   Interplay between structure and dynamics

Both the individual mean field and the annealed network approximations reveal a strong interplay between the structure of multilayer networks and the SIS dynamics which goes beyond the insights gained by characterizing the epidemic spreading in single layers.

Probably the most significant result [271, 82, 87] is that the SIS epidemic can spread in multilayer networks even if it cannot spread in the single layers that form the multilayer structures when they are taken in isolation. This is a natural effect of the increased interconnectedness of the system. Moreover, the epidemic spreading on a multilayer network is not equivalent to the epidemic spreading in a random single network because of the multilayer nature of the interconnections which impact on the dynamical behaviour of the SIS model. This effect is revealed in a duplex network by the fact that the epidemic process displays a local maximum of the susceptibility when the infectivity is given by the critical epidemic threshold of the layer that is more resistant to the spread of the epidemics when taken in isolation (see Fig. 13.2 from Ref. [87]). This implies that the epidemic, although it spreads in the entire duplex network even before reaching this value of the infectivity, when it reaches this value of the infectivity becomes suddenly much more invasive. Therefore, this is a sign that the dynamical process is somehow aware of the multilayer nature of the network.

Along a similar direction, it has also been observed in Ref. [87] that in the case of one-to-one multilayer networks with three layers it is possible to observe three peaks in the susceptibility. Interestingly, the positions of these peaks depend on how the interlinks are placed and how the corresponding network of networks is structured. For instance, in the case of multilayer networks with $M = 3$ layers the susceptibility function has a different profile if the supernetwork is a line in which the three layers are placed in a different order and/or if the supernetwork is a triangle (see Fig. 13.3).

Additionally, in general multilayer network structures where the degree distribution of interlinks is arbitrary, one observes important effects of the intralayer and interlayer

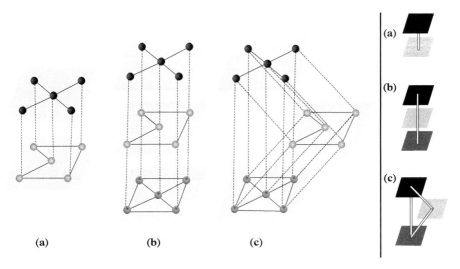

Fig. 13.3 *Schematic illustration of the three multilayer network cases considered as examples. Top panels represent the original networks which give rise to three distinct configurations for the networks of layers. Reprinted from Ref. [87].*

degree correlations [271]. These effects are made explicit by the result of the annealed mean-field calculations and are clearly apparent, for instance, from Eq. (13.31), which depends explicitly on the correlations between the degrees within each layer and across different layers for a multilayer with $M = 2$ layers.

13.3 SIR model

13.3.1 General remarks

Epidemic spreading in multilayer networks can include cases in which the infected nodes either die or become immune to the infection. This case is treated using the framework of the Susceptible-Infected-Removed (SIR) model. In this framework nodes can be in three possible dynamical states: susceptible nodes that can become infected when a neighbour node is infected, infected nodes that can spread the epidemics but become removed at a constant rate and removed nodes that cannot spread the infection but cannot be infected either. The dynamics of the SIR model is obviously different from the dynamics of the SIS model. In fact, an epidemic that obeys the SIR dynamics always dies out asymptotically in time and the number of infected nodes is zero for sufficiently long times. On the contrary, an epidemic obeying the SIS dynamics can result in an endemic stationary state with a constant average number of infected individuals. Despite this major difference, both the SIR and the SIS dynamics display a phase transition. In fact, for sufficiently small infection rates (i.e. infection rates below the epidemic threshold) both the SIS and SIR epidemics might describe an epidemic affecting only an infinitesimal fraction of the nodes of the network, while for infection rates above the epidemic threshold the epidemics will affect a finite fraction of nodes of the network.

From the analytical point of view, the SIR model not allowing infected nodes to become susceptible again is much easier to treat than the SIS model, since it can be mapped to a bond percolation problem. Specifically, while the SIS does not have an exact solution that can be expressed in a closed analytical form, the SIR dynamics is fully determined by the exact solution available for bond percolation as long as the network is locally tree-like. This technical advantage of the SIR model allows us to establish results without making use of the mean-field approximation.

The literature on the SIR model has been mostly exploring the SIR dynamics on a multiplex network without explicit use of interlinks where all the corresponding replica nodes share the same dynamical state [69, 315]. Only recently, the SIR has been studied on general multilayer networks where every replica node can be in a different dynamical state [43]. Here we will review the major results obtained so far and we will discuss the rich interplay between the multilayer network structure and the dynamical properties of the SIR epidemic spreading.

13.3.2   SIR dynamics in multiplex networks

Definition

Let us assume that every node i of a multiplex network can be in a single dynamical state, being either susceptible (indicated with S_i), infected (indicated with I_i) or removed (indicated with R_i). Additionally, we assume that the epidemic spreading can occur at different rates depending on the layers across which the epidemics spread. For instance, in the context of rumour spreading in social networks the rate at which the infection spreads on Facebook can be different from the rate at which it spreads on Twitter. Moreover, in the presence of multiplex networks with link overlap we might assume that an infected neighbour that is both neighbour in Twitter and Facebook has a higher chance of infecting a node than a neighbour connected just on one of the online social platforms. Therefore, we will assume that a node i in the susceptible state S_i connected to a node j in the infected state I_j by a non-trivial multilink $\vec{m}_{ij}$ is infected with rate $\xi^{\vec{m}_{ij}}$, i.e.

$$S_i + I_j \xrightarrow{\xi^{\vec{m}_{ij}}} I_i + I_j. \tag{13.38}$$

Moreover, each node i in the infected state I_i becomes removed with rate μ, i.e.

$$I_i \xrightarrow{\mu} R_i. \tag{13.39}$$

Theoretical predictions

The disease transmissibility $T^{\vec{m}}$ across a multilink $\vec{m}$ indicates the probability that an infected node transmits the infection to a neighbour node connected by a multilink $\vec{m}$

during its infection period. The transmissibility $T^{\vec{m}}$ is determined by the infection rate

$$\lambda^{\vec{m}} = \frac{\xi^{\vec{m}}}{\mu}.$$

The relation between $T^{\vec{m}}$ and $\lambda^{\vec{m}}$ can be derived by the stochastic properties of the epidemic spreading along the same lines as the corresponding calculations performed in a single layer (see Sec. 3.4.5) and is given by

$$T^{\vec{m}} = \frac{\lambda^{\vec{m}}}{1 + \lambda^{\vec{m}}}. \tag{13.40}$$

Alternative expressions of the transmissibility can also be considered in the case in which the lifetime of an infected individual is fixed (see Sec. 3.4.5). In a multiplex network of N nodes and M layers with given multiplex degree distribution $P(\mathbf{k})$, there is no overlap. Therefore, in this case the infection probabilities $\xi^{\vec{m}}$ only depend on the single layer α establishing the connection between two nodes, and will be indicated with $\xi^{[\alpha]}$. It follows that the transmissibilities to be considered are $T^{[\alpha]}$, expressing the probability that the infection spreads along a link of type α. The transmissibilities $T^{[\alpha]}$ are given by

$$T^{[\alpha]} = \frac{\lambda^{[\alpha]}}{1 + \lambda^{[\alpha]}} \tag{13.41}$$

where

$$\lambda^{[\alpha]} = \frac{\xi^{[\alpha]}}{\mu}.$$

These multiplex networks are also typically tree-like, and therefore in this case the SIR dynamics admits an exact solution thanks to the possibility of performing a mapping between SIR dynamics and the bond percolation model on a network.

The prevalence of the SIR epidemic S in the multiplex network is the fraction of removed individuals at the end of the epidemic and it is equal to the fraction of nodes in the giant component of the multiplex network when links in the generic layer α are randomly damaged with probability $f^{[\alpha]} = (1 - T^{[\alpha]})$. Therefore, using the techniques developed for describing percolation in a locally tree-like multiplex network without link overlap (see Sec. 12.2.2), we obtain that the SIR equation for the fraction S of nodes infected by the disease and removed at the end of the epidemics is given by

$$S = \left[1 - \sum_{\mathbf{k}} P(\mathbf{k}) \prod_{\alpha=1}^{M} \left(1 - T^{[\alpha]} S'_\alpha \right)^{k^{[\alpha]}} \right]. \tag{13.42}$$

Here S'_α indicates the probability that by following a link we reach a node that has been infected at some point in time, satisfying

$$S'_\alpha = \left[1 - \sum_{\mathbf{k}} \frac{k^{[\alpha]}}{\langle k^{[\alpha]} \rangle} P(\mathbf{k}) \prod_{\gamma=1}^{M} \left(1 - T^{[\gamma]} S'_\gamma \right)^{k^{[\gamma]} - \delta(\gamma,\alpha)} \right] \tag{13.43}$$

where $\delta(\alpha, \gamma)$ indicates the Kronecker delta. The SIR epidemic spreading describes an epidemic outbreak when the fraction of removed nodes at the end of the epidemic is not vanishing, i.e. $S > 0$. Therefore, following the same derivation as in Sec. 12.2.2 it is possible to show that the epidemic outbreak occurs for

$$\frac{1}{2} \left[T^{[1]} \kappa_1 + T^{[2]} \kappa_2 + \sqrt{\left(T^{[1]} \kappa_1 - T^{[2]} \kappa_2 \right)^2 + 4 T^{[1]} T^{[2]} \mathcal{K}_1 \mathcal{K}_2} \right] > 1, \tag{13.44}$$

where κ_α and $\mathcal{K}_\alpha$ are given by Eqs (12.15). From Eqs (13.44) it is possible to deduce the phase diagram of the SIR epidemics in multiplex networks. Interestingly, as happens for the classical percolation, the phase diagram of the SIR epidemic spreading on multiplex networks depends strongly on the degree correlations between the degree of each node in different layers.

The SIR dynamics has been extended in Ref. [69] to duplex networks having only partial coupling of the replica nodes. Additionally, the role of immunization strategies of the nodes has been addressed in Ref. [315] (see Appendix F).

We note here that the more general case of the SIR dynamics spreading on a multiplex network with link overlap can be treated along similar lines using the mapping to bond percolation in multiplex networks with link overlap as treated in Sec. 12.2.2.

13.3.3   SIR dynamics in general multilayer networks

Definition

The SIR dynamics can also be defined on general multilayer networks, including an arbitrary number of interlinks among nodes of different layers. In this case, each node (i, α) has a different dynamical state and can be either susceptible (indicated with $S_{i\alpha}$), infected (indicated with $I_{i\alpha}$) or removed (indicated with $R_{i\alpha}$). In this framework a susceptible node (i, α) connected to an infected node (j, β) gets the infection with rate $\xi^{[\alpha,\beta]}$

$$S_{i\alpha} + I_{j\beta} \xrightarrow{\xi^{[\alpha,\beta]}} I_{i\alpha} + I_{j\beta}; \tag{13.45}$$

moreover, an infected node (i, α) becomes removed with rate μ, i.e.

$$I_{i\alpha} \xrightarrow{\mu} R_{i\alpha}. \tag{13.46}$$

Theoretical predictions

The analysis of the SIR model on multilayer networks can be performed by mapping the epidemic spreading to bond percolation on multilayer networks treated in Sec. 12.2.3, and can be solved exactly as long as the network is locally tree-like. The SIR dynamics on multilayer networks reveals very strong effects of the degree correlations on the epidemic spreading.

By directly extending the results obtained in Sec. 3.4.5 for single layers it can be shown that the transmissibility $T^{[\alpha,\beta]}$ indicating the probability that the SIR epidemic spreads from an infected node in layer β to a susceptible node in layer α during the full duration of the epidemic, is given by

$$T^{[\alpha,\beta]} = \frac{\lambda^{[\alpha,\beta]}}{1 + \lambda^{[\alpha,\beta]}}, \tag{13.47}$$

where $\lambda^{[\alpha,\beta]} = \xi^{[\alpha,\beta]}/\mu$. By performing the mapping to bond percolation on multilayer networks the probability S_α that a generic node (i,α) of layer α is removed at the end of the SIR epidemics is given by

$$S_\alpha = 1 - \sum_{\mathbf{k}} P_\alpha(\mathbf{k}) \prod_{\gamma=1}^{M} \left(1 - T^{[\gamma,\alpha]} S'_{\gamma\alpha}\right)^{k^{[\alpha,\gamma]}}, \tag{13.48}$$

where $S'_{\alpha,\beta}$ satisfy the following set of equations

$$S'_{\alpha\beta} = x_{\alpha\beta} \left[1 - \sum_{\mathbf{k}} \frac{k^{[\alpha,\beta]}}{\langle k^{[\alpha,\beta]} \rangle} P_\alpha(\mathbf{k}) \prod_{\gamma=1}^{M} \left(1 - T^{[\gamma,\alpha]} S'_{\gamma\alpha}\right)^{k^{[\alpha,\gamma]}-\delta(\gamma,\beta)} \right]. \tag{13.49}$$

Here $x_{\alpha\beta}$ is given by Eq. (12.24) and the same conventions used in Sec. 12.2.3 hold.

Finally, the fraction S of nodes affected by the epidemic is given by

$$S = \frac{1}{M} \sum_{\alpha=1}^{M} S_\alpha. \tag{13.50}$$

Equations (13.48) and (13.49) are the same equations determining the classical bond percolation in multilayer networks (see Sec. 12.2.3). Therefore, under the same conditions in which the bond percolation has a giant component, the SIR dynamics give rise to an epidemic outbreak involving a finite fraction of the nodes of the multilayer network.

Let us now consider the possible scenario occurring in the case of a multilayer network formed by two layers, $M = 2$.

- *Only interlayer connectivity.* For the case in which only the links within each layer exist, the epidemic threshold $\lambda^{[\alpha,\alpha]}$ for each separate layer α is the one of the single layer taken in isolation [225, 106, 223], i.e.

$$\lambda^{[\alpha,\alpha]} = \frac{\langle k^{[\alpha,\alpha]} \rangle}{\langle (k^{[\alpha,\alpha]})^2 \rangle - 2 \langle k^{[\alpha,\alpha]} \rangle}. \tag{13.51}$$

- *Only intralayer connectivity.* For the case in which only the interlinks exist, the epidemic threshold is the one of a bipartite network and reads [223],

$$\lambda^{[1,2]} = \frac{1}{\sqrt{\kappa_{12}\kappa_{21} - 1}}, \tag{13.52}$$

$$\tag{13.53}$$

where κ_{12} and κ_{21} are given by Eq. (12.32).

- *Both interlayer and intralayer connectivity.* In general, we obtain that the SIR epidemic threshold satisfies [43]

$$\lambda^{[1,2]} = \frac{\sqrt{Q}}{\sqrt{\mathcal{R}_{12}} - \sqrt{Q}}, \tag{13.54}$$

as long as

$$\frac{\lambda^{[\alpha,\alpha]}}{1 + \lambda^{[\alpha,\alpha]}} \kappa_{\alpha\alpha} \leq 1. \tag{13.55}$$

In Eq. (13.54) the quantities Q and $\mathcal{R}_{12}$ are given

$$Q = \left(1 - \frac{\lambda^{[1,1]}}{1 + \lambda^{[1,1]}}\kappa_{11}\right)\left(1 - \frac{\lambda^{[2,2]}}{1 + \lambda^{[2,2]}}\kappa_{22}\right),$$

$$\mathcal{R}_{12} = \left(\frac{\lambda^{[1,1]}}{1 + \lambda^{[1,1]}}\mathcal{W}_{12}\mathcal{K}_{12} + \kappa_{12} - \frac{\lambda^{[1,1]}}{1 + \lambda^{[1,1]}}\kappa_{12}\kappa_{11}\right)$$
$$\times \left(\frac{\lambda^{[2,2]}}{1 + \lambda^{[2,2]}}\mathcal{W}_{21}\mathcal{K}_{21} + \kappa_{21} - \frac{\lambda^{[2,2]}}{1 + \lambda^{[2,2]}}\kappa_{21}\kappa_{22}\right), \tag{13.56}$$

with $\kappa_{\alpha\beta}$ and $\mathcal{W}_{\alpha\beta}$ given by Eq. (12.32) and indicating second moments of interdegrees and intradegrees and their correlations. The term $\mathcal{R}_{12}$ depends on $\mathcal{W}_{12}, \mathcal{W}_{21}$ and $\mathcal{K}_{12}, \mathcal{K}_{21}$ that evaluate the (normalized) correlations between the interlayer degree and the intralayer degree. As a consequence, the epidemic thresholds given by Eq. (13.54) are not only strongly dependent on the presence of broad interdegree and intradegree distributions but they are also significantly affected by the correlations between the interdegrees and the intradegrees.

In Fig. 13.4 the epidemic threshold $\eta = \eta(\lambda)$ is shown for a multilayer network with $M = 2$ layers, where λ indicates the intralayer infectivity $\lambda = \lambda^{[1,1]} = \lambda^{[2,2]}$ and η indicates the interlayer infectivity $\eta = \lambda^{[1,2]} = \lambda^{[2,1]}$. The multilayer network considered in this figure has Poisson interlayer and intralayer degree distribution but tunable correlations between interlayer and intralayer degrees. This figure provides evidence that positive correlations between interlayer and intralayer degrees favour the spread of the SIR epidemics.

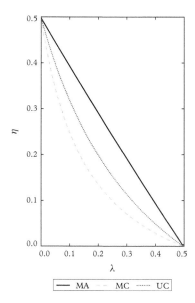

Fig. 13.4 *Epidemic threshold for Maximally Correlated (MC), Maximally Anti-correlated (MA) or Uncorrelated (UC) interdegree and intradegree sequences. The multilayer networks have two layers (i.e.* $M = 2$*) with identical interlayer degree sequences and intralayer degree sequences drawn from a Poisson distribution with average degree* $c = 3$*. The network size is* $N = 10^4$ *nodes. Adapted from Ref. [43].*

Numerical results in single multilayer networks

Multiplexity changes significantly the properties of the SIR epidemic spreading. In particular, we observe the same qualitative phenomenon observed for the SIS dynamics: mainly that the SIR epidemic can spread also if the single layers cannot sustain the epidemic when taken in isolation.

To this end, in Ref. [43] the SIR epidemic has been simulated for two different multilayer networks, with two layers ($M = 2$) obtaining very good agreement.

In the first case, (see panel (a) Fig. 13.5) the intralayer degree distribution is Poisson with average degree one, while the interlayer degree distribution is Poisson with average degree two. The intralayer infectivity $\lambda = \lambda^{[1,1]} = \lambda^{[2,2]}$ is set at a constant value $\lambda = 0.5$ and the average size of the epidemic outbreak is measured as a function of the interlayer infectivity $\eta = \lambda^{[1,2]} = \lambda^{[2,1]}$. For these parameter values the single layer cannot sustain the epidemic. Nevertheless, as η increases the multilayer network is affected by global epidemic outbreaks, i.e. $S > 0$.

In the second case (see panel (b) Fig. 13.5), the multilayer network includes only interlinks across the two layers. One layer has a very skewed scale-free interdegree distribution (power-law distribution with exponent $\gamma = 2.1$), the other layer has a Poisson interdegree distribution with an average smaller than one. Here also it is possible to observe epidemic outbreaks, i.e. $S > 0$, as a function of $\eta = \lambda^{[1,2]} = \lambda^{[2,1]}$. We notice that this occurs even if the average interdegree of one layer is Poisson with average degree

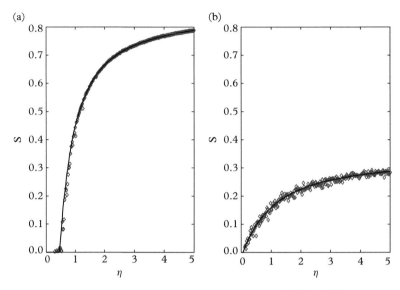

Fig. 13.5 *Size of the average epidemic outbreak* S *as function of* $\eta = \lambda^{[1,2]} = \lambda^{[2,1]}$. *In panel (a) we show the case of a multilayer with Poisson intradegree and interdegree distribution with respectively average degrees one and two. In this case* $\lambda = \lambda^{[1,1]} = \lambda^{[2,2]} = 0.5$. *In panel (b) we show the case of a multilayer network including only interlinks. The interdegree distribution is scale-free with exponent* $\gamma = 2.1$ *for one layer and Poisson for the other layer. The average interdegree is below one,* $\langle k^{[1,2]} \rangle = \langle k^{[2,1]} \rangle = 0.9$. *The number of nodes in both layers is* $N = 10^4$ *for both panel (a) and panel (b). The simulation results are averaged over 100 runs. Reprinted from Ref. [43] ©SISSA Medialab Srl. Reproduced by permission of IOP Publishing. All rights reserved.*

smaller than one, because the interdegree distribution of the other layer is sufficiently broad.

In both cases, the simulations of the SIR epidemic spreading match very well the predictions obtained by solving the corresponding bond percolation problem using the message-passing technique. Note that S indicates the average size of the epidemic outbreak, therefore events that do not span a finite fraction of the network are disregarded, as these might correspond to epidemics starting from small connected components of the multilayer network.

13.3.4   Other results on the SIR model

Among the important literature on the SIR model on multiplex and multilayer networks, here we cite some highlights which unfortunately, due to space limitations we cannot cover in more detail. In Ref. [315] (discussed in more detail in Appendix F) the impact of different immunization strategies is studied in the context of the SIR dynamics on multiplex networks. In Ref. [69] the SIR epidemic spreading on a multiplex network is studied as a function of the heterogeneous coupling between the layers. Specifically, the Authors consider a duplex network in which only a fraction q of nodes are present in both layers and determine the epidemic threshold as a function of q. In Ref. [322] the Authors provide an analytical characterization of the SIR epidemic spreading on multiplex networks with non-negligable clustering coefficient. In particular, they consider a multiplex network ensemble in which each node belongs to a given number of triangles and has a given degree. These networks are known to be locally tree-like at the level of triangles, therefore it is possible to treat them by a generalization of the generating function formalism valid for locally tree-like multiplex networks.

13.4 Interplay between awareness and epidemic spreading

The multiplex network scenario where individuals are connected by links of different types and can communicate via different types of interactions is ideal for exploring the effect that the awareness of a certain epidemic can have on the spread of the epidemic itself [305, 140, 139]. Specifically, in social networks we can distinguish between the network of physical contacts in which the epidemic spreads, and the network of virtual interactions such as the one provided by online social networks or even the network of email contacts where information about the disease spreads, changing the agent's risk perception and triggering adaptive behaviour affecting epidemic-spreading predictions. The multiplex networks allow us now to tackle, in a stylized but realistic scenario, the consequences of the coupled dynamics involving on the one side the spread of the disease itself and on the other the spreading of the awareness of the epidemics.

In Ref. [140] a duplex network formed by the physical contact layer and the virtual contact layer has been considered. On the physical contact layer the disease spreading takes place and it is modelled by a Susceptible–Infected–Susceptible (SIS) type of dynamics. On the virtual layer, the spreading of the awareness is taking place and it is modelled by an Unaware–Aware–Unaware (UAU) type of dynamics (see Fig. 13.6).

In the virtual layer, unaware (U) individuals are not informed about the disease and do not take preventing measures, while aware (A) individuals do take preventive measures, reducing their probability of becoming infected. An unaware individual becomes aware either if it is infected or if a neighbour node is aware and with rate λ informs the unaware individual. Moreover, the model assumes that the aware individuals will forget the awareness and become unaware at rate δ.

In the physical layer the individuals will be either susceptible (S) or infected (I). The diseases will spread with a dynamics of the type SIS. Susceptible individuals can become infected if they have at least one infected neighbour node. Infected individuals will become susceptible with constant rate μ. Nevertheless, the transition rate between the susceptible and the infected state will change for aware or unaware individuals. In particular, a susceptible unaware individual in contact with an infected individual will become infected at rate β, while a susceptible aware individual in a similar situation will become infected only at rate $\beta\gamma$ with $\gamma < 1$. In other words, the aware individual will have a reduced probability of being infected by a neighbour node (see Fig. 13.7).

Virtual contact UAU

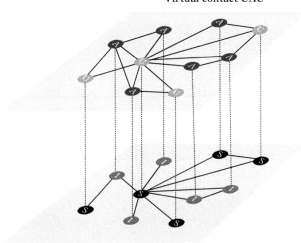

Physical contact SIS

Fig. 13.6 *Sketch of the multiplex network for the study of the simultaneous spreading of an epidemic and the awareness behaviour. The multiplex network comprises a virtual contact network where the awareness behaviour spreads according to an Unaware–Aware–Unaware (UAU) dynamics and a physical contact network where the Susceptible–Infected–Susceptible (SIS) epidemic spreading takes place. Reprinted figure with permission from [140] ©2013 by the American Physical Society.*

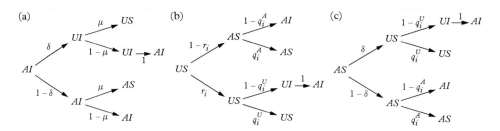

Fig. 13.7 *Transition probability trees for the states (a) AI, (b) US and (c) AS, of the UAU-SIS dynamics in the multiplex per time step. The notation is (AI) Aware–Infected, (AS) Aware–Susceptible, (UI) Unaware–Infected, (US) Unaware–Susceptible. Reprinted figure with permission from [140] ©2013 by the American Physical Society.*

In Refs [140, 139] the Authors have studied this model using a Microscopic Markov Chain Approximation (MMCA) which is essentially the individual mean-field approximation in discrete time discussed in Sec. 13.2.4. In this approximation, the dynamics is described by the temporal evolution of the probabilities $p_i^{AS}, p_i^{AI}, p_i^{US}$ that node i is respectively in the aware-susceptible, aware-infected or unaware-susceptible state.

By indicating as $r_i(t)$, $q_i^A(t)$ and $q_i^U(t)$ respectively the probability that a node i is not informed by its neighbour, the probability that an aware individual is not infected by any neighbour and the probability that an unaware individual is not infected by any neighbour, assuming the absence of dynamical correlations (mean-field assumption) we have

$$r_i(t) = \prod_j \left[1 - a_{ij}^{[1]} \left(p_j^{AS} + p_j^{AI} \right) \lambda \right],$$

$$q_i^A(t) = \prod_j \left[1 - a_{ij}^{[2]} p_j^{AI} \beta \gamma \right],$$

$$q_i^U(t) = \prod_j \left[1 - a_{ij}^{[2]} p_j^{AI} \beta \right], \tag{13.57}$$

where $\mathbf{a}^{[1]}$ is the adjacency matrix of layer one (virtual layer) and $\mathbf{a}^{[2]}$ is the adjacency matrix of layer 2 (physical layer).

The resulting transition trees determining the MMCA are shown in Fig. 13.7.

The dynamical equations for the probabilities $p_i^{AS}, p_i^{AI}, p_i^{US}$ in discrete time are therefore given by

$$p_i^{US}(t+1) = p_i^{AI}(t)\delta\mu + p_i^{US}(t)r_i(t)q_i^U + p_i^{AS}\delta q_i^U(t)$$

$$p_i^{AS}(t+1) = p_i^{AI}(t)(1-\delta)\mu + p_i^{US}\left[1 - r_i(t)\right]q_i^A(t) + p_i^{AS}(t)(1-\delta)q_i^A(t)$$

$$p_i^{AI}(t+1) = p_i^{AS}(t)(1-\mu) + p_i^{US}\left\{ \left[1 - r_i(t)\right]\left[1 - q_i^A(t)\right] + r_i(t)\left[1 - q_i^U(t)\right]\right\}$$

$$p_i^{AS}(t)\left\{ \delta\left[1 - q_i^U(t)\right] + (1-\delta)\left[1 - q_i^A(t)\right]\right\}. \tag{13.58}$$

This system of equations characterizes the onset of the disease epidemics. To this end, it is possible to study the stationary-state solution characterized by having $p_i^{AS}(t+1) = p_i^{AS}(t) = p_i^{AS}$ and equivalently $p_i^{US}(t+1) = p_i^{US}(t) = p_i^{US}$ and $p_i^{AI}(t+1) = p_i^{AI}(t) = p_i^{AI}$ close to the epidemic threshold of the SIS dynamics. In particular, we assume that

$$p_i^{AI} = \epsilon_i \ll 1. \tag{13.59}$$

In this limit we have also

$$q_i^A = 1 - \beta \sum_j a_{ij}^{[2]} \epsilon_j,$$

$$q_i^U = 1 - \beta\gamma \sum_j a_{ij}^{[2]} \epsilon_j. \tag{13.60}$$

By inserting these expressions and neglecting higher-order terms in ϵ, the linearized Eqs (13.58) read

$$p_i^{US} = p_i^{US} r_i + p_i^{AS}\delta$$

$$p_i^{AS} = p_i^{US}(1 - r_i) + p_i^{AS}(1 - \delta)$$

$$\mu\epsilon_i = \beta\left(p_i^{AS}\gamma + p_i^{US}\right) \sum_j a_{ij}^{[2]} \epsilon_j. \tag{13.61}$$

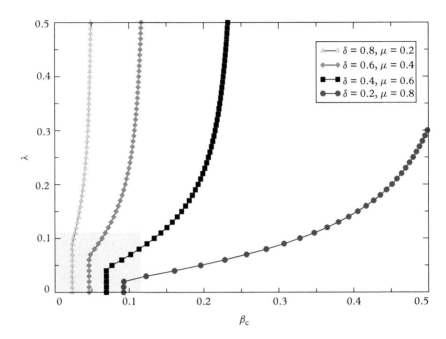

Fig. 13.8 *Dependence of the onset of the epidemics β_c as a function of λ for different values of the recovery rates δ and μ for a duplex network composed of a scale-free network of $N = 10^3$ nodes and exponent 2.5 in the physical contact layer where SIS dynamics takes place, and the same scale-free network of the physical layer plus 400 extra random links (non-overlapping with previous) in the virtual layer, where Aware–Unaware dynamics works. The shaded rectangle shows the free phase (in which all individuals are Unaware and Healthy) when the layers are uncoupled. The boundaries are defined by the bound of the structural characteristics of each layer: $1/\Lambda(\mathbf{a}^{[1]})$ and $1/\Lambda(\mathbf{a}^{[2]})$. Reprinted figure with permission from [140] ©2013 by the American Physical Society.*

Since $p_i^{AI} \ll 1$ we have $p_i^{AS} + p_i^{US} \simeq 1$, and the epidemic spreading in the physical layer close to the epidemic threshold is described by the following equations for $p_i^{AI} = \epsilon_i$

$$\frac{\mu}{\beta}\epsilon_i = \sum_j H_{ij}\epsilon_j, \tag{13.62}$$

where

$$H_{ij} = \left[1 - (1-\gamma)p_i^A\right]a_{ij}^{[2]}. \tag{13.63}$$

Therefore, the value of the infection rate $\beta = \beta_c$ determining the onset of the epidemic threshold is given by

$$\beta_c = \frac{\mu}{\Lambda(H)}, \tag{13.64}$$

where $\Lambda(H)$ is the maximum eigenvalue of the matrix $\mathbf{H}$ of elements H_{ij} given by Eq. (13.63).

The phase diagram of this coupled UAU and SIS dynamics on the duplex network formed by the virtual layer and the physical layer is shown in Fig. 13.8. Here it is shown that the epidemic threshold β_c in the physical layer is independent of the rate of spreading of the awareness λ for small values of λ. Nevertheless, there is a certain point in the phase diagram (β_c, λ), called by Granell et al. a meta-critical point, where the dynamics in the awareness of the disease in the virtual layer can contain and delay the spreading of the disease in the physical layer. This point, depending on the values of δ, μ can be found in a rectangular region of the phase space (λ, β) bounded by the points $(0,0), (0, 1/\Lambda(\mathbf{a}^{[1]}))(1/\Lambda(\mathbf{a}^{[2]}), 0)$ indicated in the Fig. 13.8 as a shaded area. Here $\Lambda(\mathbf{a}^{[\alpha]})$ indicates the largest eigenvalue of the matrix $\mathbf{a}^{[1]}$.

13.5 Competing epidemic spreading on multiplex networks

Characterizing multiple viral spreadings within a single population is a theoretical problem relevant not only for biological pathogens but also for the diffusion of memes and adoption of behaviour. In the context of single layers, modelling the diffusion of two competing viruses has been shown to give rise to a very rich phenomenology [167]. Nevertheless, it is often the case that two competing viruses do not propagate on the same network, but instead take distinct transmission routes, as for example the case of airborne diseases and blood-borne diseases propagating in the two layers of a duplex network without interlinks. In Ref. [268] the Authors have addressed this model by formulating and solving the Markov Chain approach combined with a mean-field approximation, the $SI_1 SI_2 S$ model. This model is a generalization of the SIS model in the presence of two distinct diseases. An agent can be in fact either susceptible (S) or infected by virus 1 (I_1) or virus 2 (I_2). If a node is susceptible it can either be infected by virus 1 (with infection rate β_1) or virus 2 (with infection rate β_2). A node infected by virus 1 (virus 2) can be cured and become susceptible with rate μ_1 (μ_2). If a node is already infected by virus 1 (virus 2) it cannot be also infected by virus 2 (virus 1). The Authors characterize the phase diagram of the model comprising four regions of the parameter space (τ_1, τ_2) where $\tau_1 = \beta_1/\mu_1$ and $\tau_2 = \beta_2/\mu_2$. The phase diagram shown in Fig. 13.9 comprises four regions:

  - in region N none of the viruses are endemic;
  - in region I only virus 1 is endemic;
  - in region II only virus 2 is endemic;
  - in region III both viruses are endemic.

In Ref. [268] these theoretical predictions are compared with the numerical results, finding good agreement.

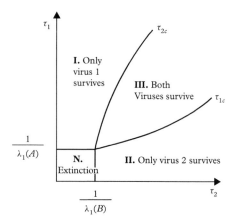

Fig. 13.9 *Phase diagram in the $SI_1 SI_2 S$ dynamics in a duplex network. Four possible scenarios appear: extinction region (N) where both viruses die-out, mutual extinction region I, where virus 1 survives and virus 2 dies out, mutual extinction region II, where only virus 2 survives and virus 1 dies out, and finally the coexistence region III, where both viruses survive and persist in the population. Reprinted figure with permission from [268]©2014 by the American Physical Society.*

13.6   Epidemic spreading on multi-slice temporal networks

Recently, a novel approach based on the individual mean-field approximation has provided an analytical computation of the epidemic threshold on temporal networks [296].

This work addresses the challenging question characterizing the epidemics evolving over timescales that are comparable with the network temporal dynamics.

In fact, the more traditional approaches to studying the epidemic threshold on complex networks usually consider only the two limiting cases of annealed networks and quenched networks. An annealed network is a network that evolves on a timescale that is fast compared with the evolution of the dynamical process. A quenched network is a network that instead evolves over much longer timescales than the ones of the dynamical process.

The approach proposed in Ref. [296] allows us instead to characterize the epidemics in realistic cases where the temporal networks are bursty or have characteristic temporal trends like the interactions of school children. Specifically, the Authors focus on the properties of the Susceptible–Infected–Susceptible (SIS) dynamics.

The temporal network is described by a sequence of $N \times N$ adjacency matrices $\mathbf{a}^{[t]}$ describing the interactions between the N nodes of the network $i = 1, 2, \ldots, N$ taking place at time t. A susceptible individual can become infected if a neighbour node is infected and the infection takes place in that case with probability λ (infection rate). An infected individual can become susceptible with probability μ at any given time.

The individual mean-field approximation describes the epidemic-spreading dynamics in terms of the probability p_{it} that node i is infected at time t. Taking a mean-field approximation, i.e. assuming the absence of dynamical correlations, the equations determining the spread of the disease are

$$p_{it} = 1 - \sum_{j=1}^{N} \left(1 - [\mathbf{M}(t-1)]_{ij} p_{j(t-1)}\right) \tag{13.65}$$

where the matrices $\mathbf{M}(t)$ have elements

$$[\mathbf{M}(t)]_{ij} = (1 - \mu)\delta_{ij} + \lambda a_{ij}^{[t]}. \tag{13.66}$$

Eq. (13.65) always has the trivial solution $p_{i,t} = 0$ for every node i and time t. Nevertheless, for sufficiently high infection rates this solution becomes unstable. The stability of the trivial solution can be studied by linearizing Eq. (13.65) close to the solution $p_{it} = 0$. Therefore, by taking $p_{it} \ll 1$ we obtain

$$p_{it} = \sum_{j=1}^{N} [\mathbf{M}(t-1)]_{ij} p_{j(t-1)}. \tag{13.67}$$

Considering a time-window $[0, T]$ and solving Eqs (13.67) recursively, we get

$$p_{iT} = \sum_{j=1}^{N} P_{ij} p_{j0} \tag{13.68}$$

where the matrix $\mathbf{P}$ is given by the product of the matrices $\mathbf{M}(t)$, i.e.

$$\mathbf{P} = \prod_{t' < T} \mathbf{M}(t') = \prod_{t' < T} \left[(1 - \mu)\mathbf{I} + \lambda \mathbf{a}^{[t']}\right]. \tag{13.69}$$

By imposing periodic boundary conditions $p_{iT} = p_{i0}$ it is possible to close the equations obtaining

$$p_{iT} = \sum_{j=1}^{N} P_{ij} p_{jT}, \tag{13.70}$$

finding that the epidemic threshold, when the trivial solution becomes unstable, is determined by the maximal eigenvalue Λ_P of the matrix $\mathbf{P}$. Specifically, it is determined by the equation

$$\Lambda_P = 1. \tag{13.71}$$

Interestingly, from this expression it is possible to derive the expression of the epidemic spreading both in the quenched and in the annealed network approximations. In the quenched network approximation, the adjacency matrices do not change with time, i.e.

$$\mathbf{a}^{[t]} = \mathbf{a}. \tag{13.72}$$

Therefore, we have that

$$\mathbf{P} = [(1 - \mu) + \lambda \mathbf{a}]^T \tag{13.73}$$

and this matrix has maximal eigenvalue $\Lambda_P = 1$ when the matrix $\mathbf{M} = (1 - \mu) + \lambda \mathbf{a}$ has maximal eigenvalue 1. This gives the known quenched results of single-layer networks

$$\frac{\lambda}{\mu} = \frac{1}{\Lambda(\mathbf{a})} \tag{13.74}$$

where $\Lambda(\mathbf{a})$ is the maximum eigenvalue of the adjacency matrix $\mathbf{a}$.

In the annealed approximation it is possible to assume that $\lambda \ll 1$ and $\mu \ll 1$ which indicates that the dynamics of the epidemic spreading is much slower than the dynamics of the network. In this approximation it is possible to express $\mathbf{P}$ linearly in $\lambda/(1 - \mu)$, obtaining

$$\mathbf{P}_{slow} = \left(1 - \mu T\right) \left[1 + \frac{\lambda}{1 - \mu} \hat{\mathbf{a}}\right] \tag{13.75}$$

where $\hat{\mathbf{a}} = \sum_{t' < t} \mathbf{a}^{[t']}$ is an aggregated, weighted and static representation of the network. Therefore, by imposing in the annealed approximation that the maximum eigenvalue of $\mathbf{P}_{slow}$ is one, we get

$$\frac{\lambda}{\mu} = \frac{T}{\Lambda(\hat{\mathbf{a}})} \tag{13.76}$$

where $\Lambda(\hat{\mathbf{a}})$ is the maximum eigenvalue of the aggregated matrix $\hat{\mathbf{a}}$.

In Ref. [296] the analytical predictions discussed so far have been validated against simulations on multi-slice network models on real-world temporal datasets. The analytical results and the simulation have been found in good agreement in all the studied cases. The specific value of the epidemic threshold has been shown to change with the number of slices T of the temporal network, as expected by the theoretical predictions. However, often a saturation effect is observed for large T when the slices of the temporal network are aggregated over a longer time-window.

14

Diffusion

14.1 The relevance of diffusion on multilayer networks

Diffusion processes are central in network theory as they guarantee the communication between nodes of the networks. In the context of multilayer networks a very crucial question is whether multilayer topology allows for a more efficient network structure, promoting faster diffusion. For instance, in the context of transportation networks it is important to extablish whether by increasing the number of layers the diffusion process speeds up. In this context, it has been shown [133] that not only can increasing the number of layer be beneficial for the diffusion, but also in some conditions the diffusion on a multilayer network can be faster than on each of its single layers, a pheonomenon called super-diffusion. Although this is clearly good news for engineers planning to design a new transportation infrastructure to speed up the commuting time of the inhabitants of a city, this result needs to be compared with studies investigating the onset of congestion states in multilayer networks. Somewhat less intuitively, in this context it is found that adding new layers in a multilayer network can favour congestion events. These two results need therefore to be analysed together when, for instance, designing new transportation infrastructures.

More in general, diffusion on multilayer networks can be characterized by studying random walks on these structures, which can be designed to optimize navigability on multilayer networks. This can be achieved by tuning the probability to hop across links of different layers (with biased random walks) or allowing long jumps in the framework of the Lévy flight random walk.

While it is generally found that in static multilayer networks the addition of new layers favours diffusion, in the context of temporal multi-slice networks it is found that the diffusion is slower than in the aggregated network where the temporal dimension of the network is disregarded. This phenomenon is due to the fact that the number of time-respecting paths is typically smaller than the number of paths in the aggregated networks, and that on temporal networks the random walker can be trapped on nodes that are only seldom connected with the other nodes of the network.

Multilayer Networks. Ginestra Bianconi, Oxford University Press (2018).
© Ginestra Bianconi. DOI: 10.1093/oso/9780198753919.001.0001

14.2 Diffusion on multiplex networks

14.2.1 The general formalism

In order to describe the diffusion within each layer and across different layers of a multiplex network, it is opportune, as first proposed in Ref. [133], to associate a dynamical diffusion state $x_i^{[\alpha]}(t)$ with each replica node (i,α) of the multiplex network with $i = 1, 2, \ldots, N$ and $\alpha = 1, 2, \ldots M$ and to assume as in Ref. [133] that both intralinks and interlinks can sustain the diffusion dynamics.

The overall timescale of the diffusion within each layer α (interlayer diffusion) is determined by the diffusion constant $D^{[\alpha]}$. Similarly, the diffusion across different layers (interlayer diffusion) is dictated by the diffusion constants $D^{[\alpha,\beta]}$ with $D^{[\alpha,\beta]} = D^{[\beta,\alpha]}$. In order to modulate the diffusion across different links of the same layer, a weight $w_{ij}^{[\alpha]}$ is associated with each undirected link from the replica node (i,α) to the replica node (j,α). With this notation, the general diffusion equation in a multiplex network is given by [133]

$$\frac{dx_i^{[\alpha]}}{dt} = D^{[\alpha]} \sum_{j=1}^{N} w_{ij}^{[\alpha]}(x_j^{[\alpha]} - x_i^{[\alpha]}) + \sum_{\beta=1}^{M} D^{[\alpha,\beta]}(x_i^{[\beta]} - x_i^{[\alpha]}), \tag{14.1}$$

where the first term and the second term on the right-hand side account respectively for the intralayer and the interlayer diffusion.

The Eq. (14.1) can be written as a general diffusion equation in an $(N \cdot M)$ dimensional space, i.e.

$$\frac{d\mathbf{X}}{dt} = -\mathcal{L}\mathbf{X}, \tag{14.2}$$

where $\mathcal{L}$ is an $(N \cdot M) \times (N \cdot M)$ matrix called the *supra-Laplacian matrix* and $\mathbf{X}$ is an $N \cdot M$ column vector encoding the dynamical state of each replica node of the multiplex network. In a multiplex network of M layers the supra-Laplacian matrix $\mathcal{L}$ is given by

$$\mathcal{L} = \begin{pmatrix} D^{[1]}\mathbf{L}^{[1]} & 0 & \cdots & 0 \\ 0 & D^{[2]}\mathbf{L}^{[2]} & \cdots & 0 \\ \vdots & \vdots & \ddots & \vdots \\ 0 & 0 & \cdots & D^{[M]}\mathbf{L}^{[M]} \end{pmatrix}$$

$$+ \begin{pmatrix} \sum_{\beta\neq 1} D^{[1,\beta]}\mathbf{I} & -D^{[1,2]}\mathbf{I} & \cdots & -D^{[1,M]}\mathbf{I} \\ -D^{[2,1]}\mathbf{I} & \sum_{\beta\neq 2} D^{[2,\beta]}\mathbf{I} & \cdots & -D^{[2,M]}\mathbf{I} \\ \vdots & \vdots & \ddots & \vdots \\ -D^{[M,1]}\mathbf{I} & -D^{[M,2]}\mathbf{I} & \cdots & \sum_{\beta\neq M} D^{[M,\beta]}\mathbf{I} \end{pmatrix}, \tag{14.3}$$

where $\mathbf{I}$ indicates the $N \times N$ identity matrix and $\mathbf{L}^{[\alpha]}$ indicates the Laplacian matrix in each layer α (whose elements are $L_{ij}^{[\alpha]} = s_i^{[\alpha]} \delta_{ij} - w_{ij}^{\alpha}$, with $s_i^{[\alpha]}$ being the strength of the replica node (i, α), $s_i^{\alpha} = \sum_j w_{ij}^{[\alpha]}$. Similarly, the dynamical vector $\mathbf{X}$ appearing in Eq. (14.3) can be written as

$$\mathbf{X} = \begin{pmatrix} \mathbf{x}^{[1]} \\ \hline \mathbf{x}^{[2]} \\ \hline \vdots \\ \hline \mathbf{x}^{[M]} \end{pmatrix} \tag{14.4}$$

where $\mathbf{x}^{[\alpha]}$ indicates the N column vector of elements $x_i^{[\alpha]}$ with $i = 1, 2 \ldots N$.

In an undirected multiplex network, where the diffusion constant $D^{[\alpha,\beta]} = D^{[\beta,\alpha]}$, the supra-Laplacian matrix $\mathcal{L}$ is symmetric and semi-positive definite. Therefore, its eigenvalues λ_n are real and non-negative, i.e. $\lambda_n \geq 0$ for $n = 1, 2, \ldots, N \cdots M$. Here we assume that the multiplex network, including its interlinks, is connected, i.e. we assume that from a replica node it is possible to reach any other node by following a combination of intralinks and interlinks.

Considering the diffusion Eq. (14.2), it is immediate to realize that, like in single networks, the typical timescale τ of the diffusion in a multiplex network is determined by the smallest non-zero eigenvalue λ_2 also called the *algebraic connectivity* of the supra-Laplacian $\mathcal{L}$, as long as there is a significant spectral gap in the supra-Laplacian spectrum. In this case we have

$$\tau = \frac{1}{\lambda_2}. \tag{14.5}$$

Therefore, we can get physical insights into the diffusion properties of multilayer networks by studying the dependence of the eigenvalue λ_2 as a function of the diffusion constant for intralayer and interlayer diffusion.

14.2.2 The diffusion regimes

In order to characterize the different diffusion regimes that we can expect in a multiplex network, let us follow Refs [133, 279] and consider the simple case of a duplex network (multiplex of $M = 2$ layers). Additionally, we will assume that the interlayer diffusion constants are all the same, i.e. $D^{[1,2]} = D^{[2,1]} = D_x$ and that the interlayer diffusion constants are absorbed by the weights of the single layers. As a consequence of the last assumption, without loss of generality we can set the interlayer diffusion constant to one, i.e. $D^{[1]} = D^{[2]} = 1$.

In these assumptions, we can characterize the diffusion on the duplex network using Eq. (14.2) with the supra-Laplacian matrix $\mathcal{L}$ and the dynamical state vector $\mathbf{X}$ given respectively by

$$\mathcal{L} = \left(\begin{array}{c|c} \mathbf{L}^{[1]} + D_x\mathbf{I} & -D_x\mathbf{I} \\ \hline -D_x\mathbf{I} & \mathbf{L}^{[2]} + D_x\mathbf{I} \end{array} \right) \tag{14.6}$$

and by

$$\mathbf{X} = \left(\begin{array}{c} \mathbf{x}^{[1]} \\ \mathbf{x}^{[2]} \end{array} \right). \tag{14.7}$$

Let us additionally assume that the duplex network is formed by two connected layers. In this set-up, diffusion on a multiplex network displays two regimes depending on the value of the diffusion constant D_x [133] (see Appendix G for a derivation of these results).

For weak interlayer diffusion constant ($D_x \ll D_x^\star$), λ_2 increases linearly with the interlayer coupling D_x and the typical timescale τ for diffusion scales like

$$\tau = \frac{1}{2D_x}. \tag{14.8}$$

This result implies that in this regime the interlayer diffusion is the limiting value for the diffusion spreading.

For the strong interlayer diffusion constant ($D_x \gg D_x^\star$), λ_2 saturates to a finite value $\lambda_2 = \frac{\lambda_s}{2}$ where λ_s is the smallest non-zero eigenvalue of the Laplacian of the aggregated network given by $\mathbf{L}^{[1]} + \mathbf{L}^{[2]}$. The eigenvalue λ_s of the aggregated Laplacian satisfies

$$\frac{\lambda_s}{2} \geq \frac{\lambda_2^{[1]} + \lambda_2^{[2]}}{2} \geq \min(\lambda_2^{[1]}, \lambda_2^{[2]}), \tag{14.9}$$

where $\lambda_2^{[\alpha]}$ is the smallest non-zero eigenvalue of the single layer Laplacian $\mathbf{L}^{[\alpha]}$. Therefore, the diffusion on the multiplex network will always be faster than the diffusion on the slowest layer of the multiplex network. Super diffusion, i.e. the fact that the timescale of the multiplex network is actually faster than the timescale of diffusion in every single layer of the multiplex network, is possible but not guaranteed in general. An upper bound [255] for $D_x^\star$ can be expressed as

$$D_x^\star \leq \frac{1}{4}\lambda_s, \tag{14.10}$$

where λ_s is the algebraic connectivity λ_2 of the Laplacian $\mathbf{L}^{[1]} + \mathbf{L}^{[2]}$ of the aggregated network. In Fig. 14.1 the dependence of the eigenvalue λ_2 is plotted as a function for D_x and the two diffusion regimes are clearly visible.

These results can be obtained by performing perturbative expansions in the two limits $D_x \ll 1$ and $D_x \gg 1$ (see Appendix G) or by finding the algebraic connectivity λ_2 using the Courant and Fisher theorem by directly solving the minimization problem [255]

$$\lambda_2(\mathcal{L}) = \min_{v \in \mathcal{V}} \mathbf{v}^{\mathsf{T}} \mathcal{L}\mathbf{v}. \tag{14.11}$$

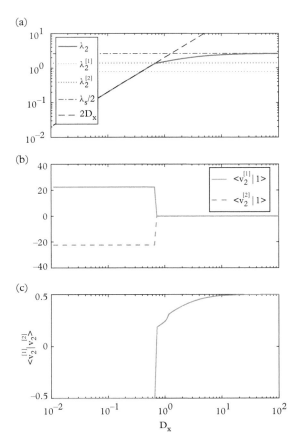

Fig. 14.1 *Evolution of the Fiedler eigenvalue λ_2 of the supra-Laplacian and its corresponding eigenvector $\mathbf{v}_2 = (\mathbf{v}_2^{[1]}|\mathbf{v}_2^{[2]})$ as a function of the diffusion constant D_x for a duplex network whose first layer is a Poisson network with power-law exponent $\gamma = 3$ and whose second layer is a regular network with degree $c = 3$. Panel (a) shows λ_2 as a function of D_x. The discontinuity of the first derivative of λ_2 clearly occurs at the transition between the two different regimes $\lambda_2 \simeq 2\,D_x$ and $\lambda_2 \simeq \lambda_s/2$, where λ_s is the Fiedler eigenvalue of the aggregated network. In this case we observe superdiffusion for large values of D_x, i.e. $\lambda_2 > \max(\lambda_2^{[1]}, \lambda_2^{[2]})$, where $\lambda_2^{[1]}$ and $\lambda_2^{[2]}$ indicate respectively the Fiedler eigenvalues of the Laplacian of the first and the second layer. Panel (b) shows the projection of the unit vector on $\mathbf{v}_2^{[1]}$ and $\mathbf{v}_2^{[2]}$ as functions of D_x. These two projections indicate the sum of all components of $\mathbf{v}_2^{[1]}$ and $\mathbf{v}_2^{[2]}$ respectively. Panel (c) shows the projection of $\mathbf{v}_2^{[1]}$ on $\mathbf{v}_2^{[2]}$ as a function of D_x.*

Here **v** is an 2N column vector that can be decomposed to two blocks, i.e.

$$\mathbf{v} = \begin{pmatrix} \mathbf{v}^{[1]} \\ \mathbf{v}^{[2]} \end{pmatrix} \tag{14.12}$$

with $\mathbf{v}^{[\alpha]}$ associated with the replica nodes in layer α. Further insights on the physical phenomena occurring when the change of transition is observed have been obtained in Ref. [255] by characterizing the structure of the *Fiedler eigenvector* $\mathbf{v}_2$ of the supra-Laplacian matrix, i.e. the eigenvector associated with the eigenvalue λ_2.

For weak coupling $D_x \leq D_x^\star$, the eigenvector $\mathbf{v}_2$ reveals that the duplex network can be partitioned into two layers. Indeed, its block components $\mathbf{v}_2^{[1]}$ and $\mathbf{v}_2^{[2]}$ differ only by an overall change of sign, i.e.

$$\mathbf{v}_2^{[1]} = -\mathbf{v}_2^{[2]} = -\frac{1}{\sqrt{2N}} \mathbf{1}, \tag{14.13}$$

where $\mathbf{1}$ is the N-dimensional column vector of elements $1_i = 1$ for all nodes $i = 1, 2, \ldots, N$.

On the contrary, for strong coupling $D_x \gg D_x^\star$, the block components $\mathbf{v}_2^{[1]}$ and $\mathbf{v}_2^{[2]}$ have the same sign. Thus, in this limit, the strong interlayer coupling does not allow the partition of the multiplex into two layers (see Fig. 14.1). The transition between the two regimes at $D_x = D_x^\star$ is discontinuous.

In Ref. [279], Solé-Ribalta et al. extended these results to the full spectrum of the supra-Laplacian matrix $\mathcal{L}$, $\{\lambda_1 = 0, \lambda_2, \ldots, \lambda_{M \cdot N}\}$, and for an arbitrary number M of layers.

14.3   Random walks on multiplex networks

14.3.1   Walking across the layers

As we navigate a multiplex network, it can be possible to move in different networks. This is, for example, the situation of multiplex transportation networks across large cities that include different layers such as metro, bus and regional train networks. In this context, it is important to characterize the behaviour of random walkers that can jump from one layer to another one, where the hops of the random walker from one node to another node in the same layer (the intralayer transitions) can have a different probability from the transitions between replica nodes in different layers (the interlayer transition). The path of a random walker exploring a multiplex network composed of $N = 7$ nodes and $M = 3$ layers is illustrated in Fig. 14.2, in which intralayer and interlayer transitions are shown. This type of random walk was first proposed in Ref. [91].

In this case, the probability $p_{i\alpha}$ that the random walker is on the replica node (i, α) at time t follows

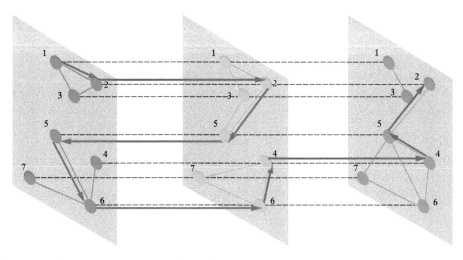

Fig. 14.2 *Path (sequence of arrows) of a random walker navigating a multiplex network composed of N = 7 nodes and M = 3 layers. In this example, the walker is neither allowed to switch between layer 1 and layer 3 in one time step nor to change node and layer simultaneously. Reprinted figure from Ref. [93].*

$$p_{i\alpha}(t+1) = \sum_{j,\beta} \mathcal{P}_{j\beta,i\alpha} p_{j\beta}(t) \tag{14.14}$$

where $p_{i\alpha}(t)$ is the probability of finding the walker at time t at the replica node (i,α) and where $\mathcal{P}_{j\beta,i\alpha}$ is the transition probability from the replica node (j,β) to the replica node (i,α). The transition matrix $\mathcal{P}$ is stochastic, i.e. it satisfies

$$\sum_{i,\alpha} \mathcal{P}_{j\beta,i\alpha} = 1, \tag{14.15}$$

for every replica node (j,β). We notice that Eq. (14.14) can also be expressed in matrix form as

$$\mathbf{p}(t) = \mathbf{p}(t-1)\mathcal{P}, \tag{14.16}$$

where $\mathbf{p}(t)$ is the $N \cdot M$ row vector of elements $p_{i\alpha}(t)$. Given the initial condition $\mathbf{p}(0)$ specifying the probability $p_{i\alpha}(0)$ that the random walker at time $t = 0$ is at the generic replica node (i,α), the state of the random walk at time t follows

$$\mathbf{p}(t) = \mathbf{p}(0)\mathcal{P}^t. \tag{14.17}$$

Until now, we have described the diffusion in a general multilayer network considering it as a larger network formed by $N \cdot M$ distinct replica nodes. Nevertheless, the interest in characterizing the behaviour of this random walk relies on the possibility of inferring the effect of multiplexity. Therefore, we need to take into account two important points:

(i) First, the transition probabilities $\mathcal{P}_{j\beta,i\alpha}$ can and should take into account whether the transition occurs between replica nodes of the same layer or between replica nodes of different layers. This will allow us to modulate the interlayer transition probability with a suitable diffusion parameter. In transportation networks these parameters indicate the cost of changing the transportation network.

(ii) Secondly, it is important to evaluate the probability π_i that the random walk at time t is on a given node i of the multiplex network. This is given by the probability that the random walker is on any of the replica nodes (i,α), i.e.

$$\pi_i(t) = \sum_{\alpha=1}^{M} p_{i\alpha}(t). \tag{14.18}$$

As a consequence of (i), care should be given in assigning the values to the transition probabilities $\mathcal{P}_{j\beta,i\alpha}$, since the different transition probabilities have different meanings for the multiplex network. In particular, for a random walker on the replica node (i,α)

(a) $\mathcal{P}_{i\alpha,i\alpha}$ is the probability that the walker remains on the replica node (i,α);
(b) $\mathcal{P}_{i\alpha,j\alpha}$ is the probability that the walker hops to the neighbour replica node in layer α, and specifically the replica node (j,α) (interlayer transition);
(c) $\mathcal{P}_{i\alpha,i\beta}$ is the probability that the walker remains at the same node i but hops from layer α to layer $\beta \neq \alpha$ (intralayer transition);
(d) $\mathcal{P}_{i\alpha,j\beta}$ is the probability that the walker jumps to a replica node (j,β) with $j \neq i$ and also $\beta \neq \alpha$ (this last movement is not actually represented in Fig. 14.2).

To be concrete, in the case of a multiplex network, the elements of the supra-transition matrix can be modulated by the strength of the coupling $D^{[\alpha,\beta]}$ between distinct layers $\alpha \neq \beta$, and the probability q that the random walker does not remain at the same replica node. Additionally, assuming that the intralayer transition probabilities depend on the weight $\{w_{ij}^{[\alpha]}\}$ of the link between replica nodes (i,α) and (j,α), the generalization of the classical random walk to a multiplex is obtained by setting

$$\mathcal{P}_{i\alpha,j\beta} = \begin{cases} (1-q) & \text{if } \alpha = \beta, i = j, \\ qw_{ij}^{[\alpha]}/\sigma_{i,\alpha} & \text{if } \alpha = \beta, i \neq j, \\ qD^{[\alpha,\beta]}/\sigma_{i,\alpha} & \text{if } \alpha \neq \beta, i = j, \\ 0 & \text{otherwise}, \end{cases} \tag{14.19}$$

where $\sigma_{i,\alpha} = \sum_j w_{ij}^{[\alpha]} + \sum_{\beta \neq \alpha} D^{[\alpha,\beta]}$ ensure that $\mathcal{P}$ is a stochastic matrix, i.e. it satisfies Eq. (14.15). This expression can be simplified further by making the typical assumption $q = 1$ so that the walker never remains in the same position. Additionally, following the simplification done for linear diffusion, one can further assume that $D^{[\alpha,\beta]} = D_x \ \forall \alpha, \beta$ with $\alpha \neq \beta$.

An important question that was addressed in Ref. [93] is the coverage of the multiplex network up to time t. In order to characterize this quantity it is necessary to evaluate the probability $\chi_{ij}(t)$ that the random walker initially at node j is not found at node i up to time t. This probability is given by

$$\chi_{ij}(t) = \prod_{\tau=1}^{t} \big[1 - \pi_i(\tau)\big]\big[1 - \pi_i(0)\big]. \tag{14.20}$$

We note here that $\pi_i(t)$ defined in Eq. (14.18) is given by

$$\pi_i(t) = \mathbf{p}(t)\mathbf{E}_i^T, \tag{14.21}$$

where $\mathbf{E}_i^T$ indicates the transpose of the $N \cdot M$ row supra-vector $\mathbf{E}_i$ which has all the elements (i, α) for $\alpha = 1, 2, \ldots, M$ equal to one and all the other elements equal to zero, i.e.

$$\mathbf{E}_i = (\mathbf{e}_i | \mathbf{e}_i | \ldots | \mathbf{e}_i), \tag{14.22}$$

with $\mathbf{e}_i$ indicating the i-th canonical N-dimensional vector. By inserting Eq. (14.21) and the expression for $\mathbf{p}(t)$ given by Eq. (14.17) we get

$$\chi_{ij}(t) = [1 - \pi_i(0)] \prod_{\tau=1}^{t} \Big[1 - \mathbf{p}(0)\mathcal{P}^\tau \mathbf{E}_i^T\Big], \tag{14.23}$$

where $\mathbf{p}(0)$ characterizes the initial condition of the random walk. Without loss of generality, the initial condition $\mathbf{p}(0)$ can be chosen such that the random walker at time $t = 0$ is at the replica node $(j, 1)$, i.e.

$$p_{i,\alpha}(0) = \delta(i,j)\delta(\alpha, 1), \tag{14.24}$$

where $\delta(x, y)$ indicates the Kronecker delta. Finally, from Eq. (14.23) using the approximation $1 - x \simeq e^{-x}$ valid for $x \ll 1$, we get

$$\chi_{ij}(t) = [1 - \pi_i(0)] \exp\left[-\sum_{\tau=1}^{t} \mathbf{p}(0)\mathcal{P}^\tau \mathbf{E}^T\right]. \tag{14.25}$$

From this expression we can evaluate the coverage of the random walker up to time t. In fact, since $\chi_{ij}(t)$ indicates the probability that node i has not been visited up to time t, by averaging $1 - \chi_{ij}(t)$ it is possible to obtain the coverage $\rho(t)$ of the random walk up to time t, i.e.

$$\rho(t) = 1 - \frac{1}{N} \sum_i \chi_{ij}(t). \tag{14.26}$$

By studying the coverage of the network in the presence of damage (disruption) of nodes, in Ref. [93] De Domenico et al. have investigated the robustness of the public transport of London, focusing on three different layers, i.e tube, overground and Docklands Light Railway. The resulting multiplex network is shown to be more resilient than its single layers taken in isolation. In fact, the interconnected multiplex network provides sufficient redundancy to find paths from apparently isolated parts of single layers.

14.3.2   Lévy flight and other searching strategies

The Lévy flights are a special class of random walks in which the random displacements have a length l following a power-law degree distribution $P(l) \propto l^{-\theta}$. Lévy flights clearly generalize random walks that have steps restricted to the nearest neighbours. In fact, for $\theta \to \infty$ the Lévy flights reduce to the standard random walk while for $\theta = 0$ they describe motion following random jumps of any length. Under very general circumstances, Lévy flights have been proven to be the most efficient strategy for exploration and navigation, outperforming the standard random walk (Brownian motion) [302, 174]. Lévy flights are rather common in the foraging of animals [302, 34], in human mobility [61, 138, 283] and behaviour and in strategies of human mental search [22, 256]. It is therefore interesting to characterize the properties of Lévy flights in complex networks [262] and in multilayer networks [145]. Let us first consider the Lévy flight on a single network. First, we associate with each pair of nodes of the (connected) network i and j the hopping distance d_{ij}. The Lévy flight on a single network has transition probabilities π_{ij} from a node i to a node j determined by the distance d_{ij}. Specifically, they will be given by

$$\pi_{ij} = \frac{d_{ij}^{-\theta}}{\sum_{m \neq i} d_{im}^{-\theta}}. \tag{14.27}$$

The Lévy flight has been studied in Ref.[262] where it has been shown that the average time to join any given node of the network from a random initial condition has a minimum for $\theta = 0$ where the Lévy transition probabilities are independent of the distance between the nodes and the Lévy flight essentially performs random jumps between the nodes of the network.

The situation becomes more interesting when the Lévy flight is placed on a multilayer network [145]. In this case, the Lévy jumps are allowed only between the nodes of the same layer, while the walker can also remain in its current position or switch to another layer (see Fig. 14.3). In this case, the transition probabilities $\mathcal{P}_{i\alpha;j\beta}$ are taken to be

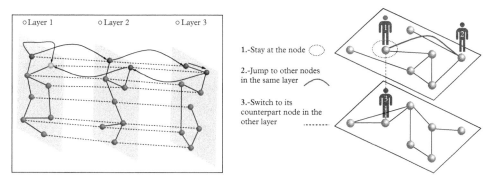

Fig. 14.3 *Illustration of the Lévy flight navigation strategy on a multiplex network. Reprinted figure from Ref. [145].*

$$
\mathcal{P}_{i\alpha;j\beta} = \begin{cases} \left(d_{ij}^{[\alpha]}\right)^{-\theta}/\sigma_{i\alpha} & \text{if } \alpha = \beta, i \neq j \\ D_{ii}^{[\alpha,\alpha]}/\sigma_{i\alpha} & \text{if } \alpha = \beta, i = j \\ D_{ii}^{[\alpha,\beta]}/\sigma_{i\alpha} & \text{if } \alpha \neq \beta, i = j \end{cases} \tag{14.28}
$$

where $d_{ij}^{[\alpha]}$ is the distance between node i and node j in layer α, and

$$
\sigma_{i,\alpha} = \sum_{j \neq i} \left(d_{ij}^{[\alpha]}\right)^{-\theta} + \sum_{\beta=1}^{M} D_{ii}^{[\alpha,\beta]}.
$$

When the parameters $D_{ii}^{[\alpha,\beta]}$ with $\alpha \neq \beta$ are independent of the node i, i.e. $D_{ii}^{[\alpha,\beta]} = D_x$, the Authors of Ref. [145] have found that for very small interlayer cost D_x the best strategy to reach 50% of the nodes in the shortest time is not $\theta = 0$, as for single layers, but a larger non-zero value of θ. This implies that if the cost of following the interlinks is vanishing, the random walk can remain localized among the replica nodes if the other Lévy jumps have a sufficiently small transition probability (such as for the case when $\theta = 0$). Only when D_x is significantly high are the results on multilayer networks consistent with the single-layer scenario and the optimal coverage obtained for $\theta = 0$.

14.3.3 Biased random walks

In multiplex networks it is also possible to explore the case in which the interlinks are absent and the random walk does not move between replica nodes but just hops from a node to any of its neighbour nodes by following links of different types belonging to the different layers of the multiplex network. The main research questions of the works that address this type of random walk is the optimization of the search strategies and the coverage of the random walk when the probabilities of hopping across links of different layers are different. In Ref. [97] a random walker initially on node i has a probability $z^{[\alpha]}$ of hopping to a neighbour node j connected to node i by a link of layer α. Therefore, transition probabilities π_{ij} from node i to node j are given by

$$\pi_{ij} = \sum_{\alpha=1}^{M} z^{[\alpha]} \frac{a_{ij}^{[\alpha]}}{\kappa_i^{[\alpha]}} \tag{14.29}$$

with $\kappa_i^{[\alpha]} = \max(1, \sum_{r=1}^{N} a_{ir}^{[\alpha]})$ and $z^{[\alpha]}$ normalized, i.e.

$$\sum_{\alpha=1}^{M} z^{[\alpha]} = 1. \tag{14.30}$$

In this type of multiplex random walk the Authors of Ref. [97] have shown that there is an optimal choice of the probabilities $z^{[\alpha]}$ that minimizes the search time of the random walk. The specific position of this optimal solution for the probabilities $z^{[\alpha]}$ depends on the topologies of the layers of the multiplex and on their correlations. In conclusion, the different coupling strengths between the layers can be optimized to achieve the best efficiency of the random walk, and this suggests a mechanism to explain the evolutionary advantage of many natural multilayer networks.

Importantly, very efficient centrality measures for multiplex networks including the Functional Multiplex PageRank and the MultiRank (discussed in Sec. 9.5) are based on this biased random walk. In the MultiRank algorithm first the random walk is modified in the spirit of the PageRank algorithm by introducing a teleportation parameter that allows random jumps to occur between any pair of nodes. Secondly, the parameters $z^{[\alpha]}$ of the random walk are interpreted as the *influences* (centralities) of the layers and the steady state of the random walk is interpreted as the centrality of the nodes.

In Ref. [31] a biased random walk is studied on multiplex networks where nodes are connected by different types of links. The biased random walk navigates the multiplex network by performing a walk that is biased according to the properties of the nodes indicated by the vector f_i with $i = 1, 2, \dots N$. The biased walk considered in Ref. [31] is defined on the weighted, aggregated network in which each link has weight v_{ij} given by the multiplicity of the overlap of the link (i, j). Therefore, it has transition probability π_{ij} given by

$$\pi_{ij} = \frac{v_{ij} f_j}{\sum_m v_{im} f_m}. \tag{14.31}$$

Here it is found that the structural properties of the multiplex networks such as the number of layers, the link overlap and interlayer degree correlations have significant impact on the diffusion properties of the random walk.

14.4   Random walks on multi-slice temporal networks

For multiplex networks it is usually the case that multiplexity favours diffusion, as for instance it occurs in transportation networks where the addition of new layers allows a better and faster coverage of the network. When one considers instead multi-slice networks, the temporal nature of these structures slows down the exploration of a given random walk as compared to the aggregated network where the temporal nature of the interaction is disregarded. This effect is due to the fact that actually the random walker on temporal networks has a reduced number of available possible time-respecting paths. Time-respecting paths are defined as sequences of nodes $\mathcal{P}_t = (i_1, i_2, \ldots, i_n)$ where each consecutive node can be reached from the previous one in a temporal sequence by following links of the temporal network, i.e. if (i_r, i_{r+1}) with $r = 1, 2, \ldots, n-2$ is a link of time-slice α corresponding to time $t = t_\alpha$, the link (i_{r+1}, i_{r+2}) should be a link of a time-slice β corresponding to a subsequent time $t = t_\beta > t_\alpha$. Time-respecting paths in a multi-slice network are a subset of the paths of the aggregated network. Assume, for instance, that at time $t = 1$ the network is formed by the link $(2, 3)$ and at time $t = 2$ it is formed by the link $(1, 2)$. While the path $\mathcal{P}_{aggr} = (1, 2, 3)$ exists in the aggregated network, this path is not time-respecting because the link $(1, 2)$ is established only after the link $(2, 3)$. Therefore, it is not possible to send signals in the network between node 1 and node 3. Additionally, in multi-slice networks the random walker can remain trapped at nodes that are only seldom connected to other nodes of the temporal network.

In Ref. [287] an analysis of the properties of a random walker on real temporal datasets was performed, analysing the effect of temporal correlations between consecutive contacts present in the data. Interestingly, in this study it is found that the shortest time-respecting paths have statistical characteristics which are significantly different from the fastest time-respecting paths. Indeed, shortest time-respecting paths are not typically the fastest, as the random walker can remain trapped on some of the nodes for quite some time.

The model for a temporal network with heterogeneous temporal activity of the nodes is a very well-defined modelling framework that captures the consequences of having nodes which are connected in the network more or less frequently. In fact, as discussed in Sec. 10.5.2 each node i is assigned a temporal activity $a_i \in [0, 1]$ drawn from the distribution $F(a)$ and at each time step t it establishes m links with probability a_i. In the framework of this model it is shown [263] that the mean first-passage time indicating the average time a random walker takes to come back to its origin for the first time is inversely proportional to the activity of the node. This property of the random walker on temporal activity-driven networks captures the slowing down of the random walker observed on real temporal networks.

A very successful way to describe random walks on arbitrary temporal networks relies on a non-Markovian description of this diffusion process [272, 181]. According to this non-Markovian random walk the probability of a transition between node i and node j depends on the past history of the random walk. For instance the most simple example of non Markovian dynamics assigns a given probability $\pi(\ell i \rightarrow ij)$ to the hop of the random walk from node i to node j given the the random walk at the previous step was hopping between node ℓ and node i.

By extending the treatment of Markovian random walks on temporal networks to non-Markovian random walks it is possible to observe both a slowing down or a speed up of the diffusion with respect to its Markovian counterpart [272].

14.5 Traffic and congestion

In transportation networks as well as in technological networks it is essential to monitor the traffic and to avoid congestion. Traffic and congestion on single networks have been studied extensively in the literature [299, 142, 113, 95]. Since most transportation and technological networks are multilayer, it is essential to study the effects of multiplexity on traffic and congestion transition.

Traffic problems differ from the previously characterized diffusion processes because the walker does not perform a random exploration of the network but is routed from a source node to a destination node following a given routing algorithm. The most natural routing algorithm for traffic problems is the shortest path between the source and the destination node. On a single network the characterization of the flow passing through each node (or each link), assuming that each node has equal probability to be a source or a sink, has prompted network scientists to introduce the *betweenness centrality* discussed in Sec. 2.7.6.

In multiplex networks the shortest paths can be distinguished between the paths that belong to a single network and paths that include links from several network interlinks connecting replica nodes from different layers (although one might choose to assign a zero or a non-zero length to interlinks).

In order to characterize how the flow is distributed across the different layers of a multiplex network in Ref. [217] the Authors introduced the *weighted interdependence* $\lambda \in [0, 1]$ which is a modification of the interdependence already introduced in Sec. 6.4.1. The weighted interdependence depends on the traffic and is given by

$$\lambda = \sum_{i \neq j} T_{ij} \frac{\psi_{ij}}{\sigma_{ij}} \tag{14.32}$$

where ψ_{ij} are the number of shortest paths between node i and node j including links from different layers, σ_{ij} is the number of shortest paths between node i and node j and T_{ij} is the normalized origin–destination matrix determining the fraction of walkers with source i and destination j. Therefore, the interdependence λ is the fraction of walkers that use more than one layer.

Other general measures indicating how well the system is operating in normal traffic conditions (i.e. in the absence of congestion) are the *average distance travelled* $\bar{d}$ given by

$$\bar{d} = \sum_{i \neq j} T_{ij} d_{ij} \tag{14.33}$$

and the *Gini coefficient G*. The Gini coefficient $G \in [0, 1]$ is a measure used usually in economics for characterizing the distribution of wealth within a nation. In Ref. [217] it has been proposed as a good measure to describe the distribution of traffic flow in a network. When $G = 1$ all the flow concentrates on a single link, while if all the flow were spread evenly across all the links G would be zero.

Simulation results on planar multilayer networks [217] show that as the weighted interdependence increases the average distance travelled decreases, indicating that when the shortest paths take advantage of the multiplexity of the network the average distance travelled can be significantly reduced. Nevertheless, for given network topologies it is possible that as the weighted interdependence increases the Gini coefficient increases, indicating a higher risk for congestion.

When each node of the network has a finite capacity, i.e. can route only a finite number of elements at each time, the network can undergo a congestion phase transition. Congestion occurs when the injection of elements for unit time is above a critical value, and the nodes start to work at their maximum capacity. In these conditions the elements are injected into the network at a faster rate than the rate at which they reach the destination, resulting in a number of elements in transit in the network growing linearly in time.

Therefore, the order parameter of the congestion is given by [8, 143]

$$\rho = \lim_{t \to \infty} \frac{D(t + T) - D(t)}{RT}, \tag{14.34}$$

where $D(t)$ is the number of elements in transit at time t and R is the number of elements injected at each time. In a non-congested regime with $R < R_c$ the number of elements in transit is stationary, therefore $\rho = 0$. On the contrary, in the congested regime with $R \geq R_c$, the elements accumulate in the network and $\rho \neq 0$ measures the normalized $(0 \geq \rho \geq 1)$ increase of elements in transit per unit of time.

In a multiplex network it is possible to observe a cooperative counter-intuitive phenomenon [280]. Despite the fact that the multiplexity increases the number of paths in the system, the multiplex network can be more prone to congestion than the single networks that form its layers. This phenomenon is reminiscent of Braess's paradox [259], in which adding an extra link to a single network can reduce the overall network performance.

The counter-intuitive effect of multiplexity in determining the congestion risk is due to the non-local nature of the routing algorithm that follows the shortest paths of the network. In the case in which one layer is more efficient than the other, i.e. has typically smaller shortest distance between the nodes, the traffic tends to concentrate on a single layer and, depleting the other layers, increases the risk for congestion [280]. In line with this work, other papers have characterized the congestion transition on duplex networks with different degree distributions [323].

In Ref. [292] the effect of degree correlation in determining the congestion threshold has been investigated. This study shows that in multilayer networks with correlated degrees using capacity allocation strategies such as attributing to each node a capacity proportional to its betweenness allows for a higher traffic capacity. We note here that other routing strategies tailored to multiplex networks have been proposed in Ref. [321].

15

Synchronization, Non-linear Dynamics and Control

15.1 Dynamical systems in multilayer networks

Multilayer network topologies can have an important effect on dynamical systems including synchronization dynamics, pattern formation and network control. For example, in a multiplex network synchronization can occur globally among all the nodes of the multilayer structures, it can occur within single layers (intra-synchronization) or else it might occur between replica nodes (inter-synchronization). Additionally, multiplex networks can allow us to couple different dynamical processes occurring in different layers. For example, in the brain the neuronal activity is coupled with the flux (diffusion) of blood in the neuronal cells.

The role of multiplexity on pattern formation can be significant and multiplex networks can display non-trivial patterns under conditions that are different from the ones predicted by the analysis of the single layers taken in isolation.

Finally, the control of multilayer networks can be studied in the framework of structural controllability where there are specific constraints on where the driver nodes can be placed. For instance, it is possible to assume that in a multiplex network driver nodes should be corresponding replica nodes belonging to different layers, or instead that driver nodes can be only replica nodes of a given layer. In most of the cases it is found that controlling multiplex networks is affected by the built-in correlations between the layers.

Interestingly, it has also been found that purely dynamical systems as multivariate time series can be investigated by mapping them to multiplex networks, capturing both the internal correlated dynamics of each single time series and the relations between different time series.

15.2 Synchronization

15.2.1 Master Stability Function

Synchronization on interacting multilayer networks is of relevance for a number of applications, ranging from brain research to applications in power-grid engineering,

Multilayer Networks. Ginestra Bianconi, Oxford University Press (2018).
© Ginestra Bianconi. DOI: 10.1093/oso/9780198753919.001.0001

international business and communication networks. The Master Stability Function (MSF) is a major tool for investigating the stability of global synchronization in dynamical systems. As discussed in Sec. 3.6.1, in the context of single networks the Master Stability Function has been first proposed in Refs [246, 17]. In Ref. [286] Sorrentino extended this framework to the so-called hypernetworks, characterizing a set of nodes connected by different types of interaction. Therefore, hypernetworks are effectively multiplex networks where interlinks are not taken into account. The results obtained in Ref. [286] concern specifically duplex networks satisfying one of the following three cases:

(i)   the Laplacian matrices associated with the two networks commute;

(ii) one of the two networks is unweighted and fully connected;

(iii) one of the two networks is such that the coupling strength from node i to node j is a function of j but not of i.

Recently a more general approach using the Master Stability Function to study synchronization in multiplex networks formed by nodes with multiple interactions has been formulated in Ref. [96]. This approach has since been extended to multiplex networks where interlinks are explicitly taken into account [293] and to stylized networks of networks describing ecological foodwebs [58].

   In the following three paragraphs we will present in some detail the main results achieved so far using the Master Stability Function.

15.2.2   Master Stability Function on multiplex networks without interlinks

In Ref. [96] the Authors study the stability of the fully synchronized state on a multiplex network in which a given set of nodes interacts through different types of connections. A typical multiplex network considered in Ref. [96] is shown in Fig. 15.1 where it is apparent that in this theoretical framework the interlinks are not playing any role. Every node i of the multiplex network is characterized by a dynamical variable $\mathbf{x}_i \in \mathbb{R}^d$ whose evolution is coupled to the dynamical variables $\mathbf{x}_j \in \mathbb{R}^d$ of the nodes j that are connected to node i in any given layer α. However, the nature of the coupling strongly depends on the layer in which two nodes are connected.

   Given a multiplex network with M layers, the dynamics of the variable $\mathbf{x}_i$ associated with node i is assumed to follow

$$\frac{d\mathbf{x}_i}{dt} = \mathbf{f}(\mathbf{x}_i) - \sum_{\alpha=1}^{M} \sigma_\alpha \sum_{j=1}^{N} \mathrm{L}_{ij}^{[\alpha]} \mathbf{H}^{[\alpha]}(\mathbf{x_j}). \qquad (15.1)$$

Here $\mathbf{f}(\mathbf{x}) \in \mathbb{R}^d$ is a continuous and differentiable function that determines the dynamics of each node variable in the absence of any coupling with other nodes of the multiplex network, whereas $\mathbf{H}^{[\alpha]}(\mathbf{x})$ are layer-specific continuous and differentiable functions

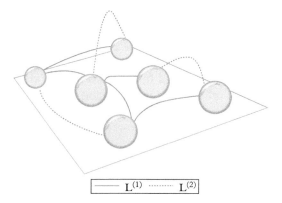

Fig. 15.1 *Schematic representation of a multiplex network formed by two layers* M = 2 *studied in Ref. [96]. Every node of the multiplex network can interact with other nodes through multiple types of interactions here indicated respectively as solid and dashed lines.*

determining the functional nature of the interaction in layer α. The parameters σ_α modulate the strength of the coupling between connected nodes in any given layer α and $\mathbf{L}^{[\alpha]}$ indicates the Laplacian matrix in layer α with elements

$$L_{ij}^{[\alpha]} = \left(\sum_{r=1}^{N} w_{ir}^{[\alpha]} \right) \delta_{ij} - w_{ij}^{[\alpha]} \tag{15.2}$$

where $w_{ij}^{[\alpha]}$ indicates the weight of the interaction between node i and node j in layer α.

We indicate with $\lambda_i^{[\alpha]}$ the eigenvalues of the Laplacian $\mathbf{L}^{[\alpha]}$ ranked in non-decreasing order ($\lambda_1^{[\alpha]} \leq \lambda_2^{[\alpha]} \leq \lambda_3^{[\alpha]} \leq \ldots \lambda_N^{[\alpha]}$) and with $\mathbf{V}^{[\alpha]}$ the corresponding matrix of eigenvalues. Since the Laplacians are all zero-sum-row matrices, the system of equations given by Eq. (15.1) always admits a fully synchronized solution in which all the nodes are in the same dynamical state, i.e. $\mathbf{x}_i = \mathbf{s} \ \forall i = 1, 2, \ldots, N$. In order to study the stability of this solution we linearize the Eqs (15.1) for $\mathbf{x}_i = \mathbf{s} + \delta \mathbf{x}_i$ where $|\delta \mathbf{x}_i| \ll 1$. The resulting equations for $\delta \mathbf{X} = (\delta \mathbf{x}_1, \delta \mathbf{x}_2, \ldots, \delta \mathbf{x}_N)^T$ are given by

$$\frac{d (\delta \mathbf{X})}{dt} = \left(\mathbf{I} \otimes \mathcal{J}\mathbf{f}(\mathbf{s}) - \sum_{\alpha=1}^{M} \sigma_\alpha \mathbf{L}^{[\alpha]} \otimes \mathcal{J}\mathbf{H}^{[\alpha]}(\mathbf{s}) \right) \delta \mathbf{X} \tag{15.3}$$

where $\mathbf{I}$ indicates the identity matrix, $\otimes$ denotes the Kronecker product and $\mathcal{J}$ the Jacobian operator. This equation can be studied by projecting $\delta \mathbf{X}$ into the basis of eigenvectors of an arbitrary Laplacian of one of the layers. Let us assume that we choose layer 1 and let us indicate with $\boldsymbol{\eta}_j$ for $j = 1, 2, \ldots, N$ the vector coefficient of the eigen-decomposition of $\delta \mathbf{X}$. The system of Eqs (15.3) reads then

$$\frac{d\boldsymbol{\eta}_j}{dt} = \left(\mathcal{J}\mathbf{f}(\mathbf{s}) - \sigma_1 \lambda_j^{[1]} \mathcal{J}\mathbf{H}_1(\mathbf{s}) \right) \boldsymbol{\eta}_j - \sum_{\alpha=2}^{M} \sigma_\alpha \sum_{n=2}^{N} \sum_{r=2}^{N} \lambda_r^{[\alpha]} \Gamma_{rn}^{[\alpha]} \Gamma_{rj}^{[\alpha]} \mathcal{J}\mathbf{H}^{[\alpha]}(\mathbf{s})\boldsymbol{\eta}_n, \qquad (15.4)$$

where the matrices $\Gamma^{[\alpha]}$ indicate

$$\Gamma^{[\alpha]} = \mathbf{V}^{[\alpha]\,T}\mathbf{V}^{[1]}, \qquad (15.5)$$

and where we have used the fact that the smallest eigenvalue of each Laplacian is zero. Additionally, we have assumed that all the networks have a single connected component so the degeneracy of this eigenvalue is one. This system of equations for the variable $\boldsymbol{\Omega} = (\boldsymbol{\eta}_2, \boldsymbol{\eta}_3, \ldots, \boldsymbol{\eta}_N)$ has a dynamics dominated by its largest Lyapunov exponent Λ, i.e.

$$\boldsymbol{\Omega} \simeq \exp(\Lambda t). \qquad (15.6)$$

Therefore, the condition determining the stability of the fully synchronized state

$$\mathbf{x}_i = \mathbf{s} \qquad (15.7)$$

against small perturbation is

$$\Lambda < 0. \qquad (15.8)$$

Using this theoretical framework for studying a duplex network of Rössler oscillators of dimension $d = 3$, the Authors of Ref. [96] have revealed that the multiplex network can sustain a stable synchronization dynamics not only when both layers in isolation can sustain a synchronized state but also when only one or when neither of them can sustain a synchronized state. This finding is a surprising and novel dynamical effect that multiplexity has on the synchronization properties of coupled networks.

15.2.3 Master Stability Function on multiplex networks with interlinks

When studying non-linear equations on multiplex networks two different scenarios might occur: either all corresponding replica nodes have the same dynamical state or every replica node has a distinct dynamical state. In the first scenario it is not necessary to associate a particular dynamical role with interlinks, while in the latter interlinks are used to couple the dynamics across the different layers. In the previous paragraph the Master Stability Function was applied to a dynamical system on a multiplex network where each node has the same dynamical state on every layer. In this chapter we will follow Ref. [293] and treat the case in which every replica node has a distinct dynamics.

Interestingly, in this scenario we might observe three different types of synchronization: the complete synchronized state where each replica node is synchronized with

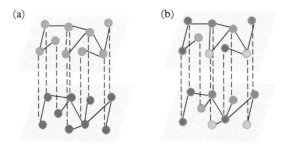

Fig. 15.2 *Schematic representation of intralayer synchronization (panel (a)) and interlayer synchronization in a multiplex network of two layers.*

each other replica node of the multiplex network; the intralayer synchronization in which every replica node is synchronized with every other replica node of the same layer; and interlayer synchronization where every replica node is synchronized with its corresponding replica nodes in the other layers (see Fig. 15.2). The stability of each of the synchronized states mentioned above can be studied using the Master Stability Function. For instance, in Ref. [293] the Authors have considered a multilayer network of M layers having replica nodes and a supernetwork with adjacency matrix $\mathbf{A}$. To each replica node of this multilayer network a d-dimensional dynamical variable $\mathbf{x}_i^{[\alpha]}$ has been assigned. In the absence of interactions the internal dynamics of each replica node is identical and determined by the set of equations

$$\frac{d\mathbf{x}_i^{[\alpha]}}{dt} = \mathbf{f}(\mathbf{x}_i^{[\alpha]}). \tag{15.9}$$

Nevertheless, the internal dynamics at each replica node is coupled to its interlayer and intralayer neighbours according to

$$\frac{d\mathbf{x}_i^{[\alpha]}}{dt} = \mathbf{f}\left(\mathbf{x}_i^{[\alpha]}\right) - \gamma \sum_{j=1}^{N} L_{ij}^{[\alpha]} H \mathbf{x}_j^{[\alpha]} - \sigma \sum_{\beta=1}^{M} \hat{L}_{\alpha\beta} \Gamma \mathbf{x}_i^{\beta}, \tag{15.10}$$

where γ and σ indicate respectively the coupling strength for nodes within and across the layers and H and Γ are $d \times d$ matrices describing respectively the inner coupling functions within and across the layers. Additionally, we have indicated with $\mathbf{L}^{[\alpha]}$ the $N \times N$ Laplacian matrix of each layer α with elements

$$L_{ij}^{[\alpha]} = k_i^{[\alpha]} \delta_{ij} - a_{ij}^{[\alpha]}$$

and with $\hat{\mathbf{L}}$ the $M \times M$ super-Laplacian matrix describing the diffusion across layers. The super-Laplacian of the supernetwork has elements

$$\hat{L}_{\alpha\beta} = \left(\sum_{\beta'} A_{\alpha\beta'}\right)\delta_{\alpha\beta} - A_{\alpha\beta}.$$

The system of equations (15.10) can be equivalently written as a differential equation for the $(N \cdot M \cdot d)$-dimensional vector $\mathbf{X} = \left(\mathbf{X}^{[1]}, \mathbf{X}^{[2]}, \dots \mathbf{X}^{[M]}\right)^T$ with $\mathbf{X}^{[\alpha]}$ indicating the $(N \cdot d)$-dimensional vector $\mathbf{X}^{[\alpha]} = (\mathbf{x}_1^{[\alpha]}, \mathbf{x}_2^{[\alpha]}, \dots \mathbf{x}_N^{[\alpha]})^T$. This equation reads

$$\frac{d\mathbf{X}}{dt} = \mathbf{F}(\mathbf{X}) - \gamma \mathcal{L}^L \otimes H\mathbf{X} - \sigma \mathcal{L}^I \otimes \Gamma\mathbf{X}, \qquad (15.11)$$

where

$$\mathbf{F}(\mathbf{X}) = \begin{pmatrix} \hat{\mathbf{f}}(\mathbf{X}^{[1]}) \\ \hat{\mathbf{f}}(\mathbf{X}^{[2]}) \\ \vdots \\ \hat{\mathbf{f}}(\mathbf{X}^{[M]}) \end{pmatrix}, \quad \text{with} \quad \hat{\mathbf{f}}(\mathbf{X}^{[\alpha]}) = \begin{pmatrix} \mathbf{f}\left(\mathbf{x}_1^{[\alpha]}\right) \\ \mathbf{f}\left(\mathbf{x}_2^{[\alpha]}\right) \\ \vdots \\ \mathbf{f}\left(\mathbf{x}_N^{[\alpha]}\right) \end{pmatrix}, \qquad (15.12)$$

and $\mathcal{L}^L$ stands for the supra-Laplacian of the intralayer connections while $\mathcal{L}^I$ stands for the supra-Laplacian of the interlayer connections. The two $(N \cdot M) \times (N \cdot M)$ supra-Laplacians $\mathcal{L}^L$ and $\mathcal{L}^I$ are given by

$$\mathcal{L}^L = \bigoplus_{\alpha=1}^M \mathbf{L}^{[\alpha]},$$
$$\mathcal{L}^I = \hat{\mathbf{L}} \otimes \mathbf{I}_N, \qquad (15.13)$$

where $\mathbf{I}_N$ indicates the $N \times N$ identity matrix. The stability of synchronous state $\mathbf{X} = \mathbf{1}_M \otimes \mathbf{1}_N \otimes \mathbf{s}$, where $\mathbf{s}$ satisfies $\dot{\mathbf{s}} = \mathbf{f}(\mathbf{s})$ and $\mathbf{1}_N$ indicates the N-dimensional vector with all entries equal to one, is obtained by considering the linearized equations

$$\frac{d\boldsymbol{\xi}}{dt} = \left[\mathbf{I}_{M \times N} \otimes \mathcal{J}\mathbf{f}(\mathbf{s}) - \gamma \mathcal{L}^L \otimes H - \sigma \mathcal{L}^I \otimes \Gamma\right]\boldsymbol{\xi}. \qquad (15.14)$$

Here $\mathcal{J}$ indicates the Jacobian operator and $\boldsymbol{\xi} = \mathbf{X} - \mathbf{1}_M \otimes \mathbf{1}_N \otimes \mathbf{s}$. In the specific case in which $\mathcal{L}^L$ and $\mathcal{L}^I$ commute, this system of equations can be projected into the common diagonalization basis. Therefore, the stability of the synchronous solution $\mathbf{1}_M \otimes \mathbf{1}_N \otimes \mathbf{s}$ to small perturbations can be studied by considering the linear system of equations

$$\frac{d\mathbf{y}}{dt} = [\mathcal{J}\mathbf{f}(\mathbf{s}) - aH - b\Gamma]\,\mathbf{y}, \qquad (15.15)$$

with $a = \gamma\lambda$, $b = \sigma\mu$ and λ, μ indicating respectively the eigenvalues of $\mathcal{L}^L$ and $\mathcal{L}^I$ as long as $\lambda^2 + \mu^2 \neq 0$. By imposing that the maximum Lyapunov exponent of this system of equations is negative we find the stability region (SR) of the fully synchronized state $1_M \otimes 1_N \otimes \mathbf{s}$, i.e.

$$\text{SR} = \{(a, b)|LLE(a, b) < 0, a \geq 0, b \geq 0\}, \tag{15.16}$$

where LLE indicates the largest Lyapunov exponent of the system of Eqs (15.15).

For $\lambda \neq 0$ and $\mu = 0$ the system of Eqs (15.15) describes a system without interlayer couplings ($b = 0$). The largest Lyapunov exponent of this system determines the stability region SR^{Intra} for intralayer synchronization, i.e.

$$\text{SR}^{Intra} = \{(a, \sigma)|LLE(a, 0) < 0, \sigma \geq 0\}. \tag{15.17}$$

Similarly, for $\mu \neq 0$ and $\lambda = 0$ the system of Eqs (15.15) describes a system without intralayer couplings ($a = 0$). The largest Lyapunov exponent of this system determines the stability region SR^{Inter} for interlayer synchronization, i.e.

$$\text{SR}^{Inter} = \{(\gamma, b)|LLE(a, b) < 0, \gamma \geq 0\}. \tag{15.18}$$

15.2.4   Master Stability Function on simple networks of networks

The Master Stability Function approach has also been used on a simple model of networks of networks describing interactions between different ecological patches [58]. The network of networks is formed by M supernodes representing habitat patches of different spatial locations, each one supporting an ecological system described by the very same foodweb network including N species. The different patches are connected by a supernetwork with adjacency matrix $\mathbf{A}$ describing their geographical proximity (see Fig. 15.3). This so-called meta-foodweb is described by a dynamics which includes both the intralayer foodweb population dynamics and the effects due to the migration of species from one patch to the other. Let us indicate with $X_i^{[\alpha]}$ the population of species i on patch α. This population is assumed to change in time according to the differential equations

$$\frac{dX_i^{[\alpha]}}{dt} = f_i\left(\mathbf{X}^{[\alpha]}\right) + \sum_\beta A_{\alpha\beta} c_i\left(\mathbf{X}^{[\alpha]}, \mathbf{X}^{[\beta]}\right), \tag{15.19}$$

with $i = 1, 2, \ldots, N$ and $\alpha = 1, 2, \ldots, M$. Here, $f_i\left(\mathbf{X}^{[\alpha]}\right)$ describes the internal dynamics of a foodweb in a given patch and $c_i\left(\mathbf{X}^{[\alpha]}, \mathbf{X}^\beta\right)$ describes the migration flow of species i from the generic patch β to its connected patch α. In Ref. [58] it is assumed that this dynamics admits a stationary state in which each species has the same population in each

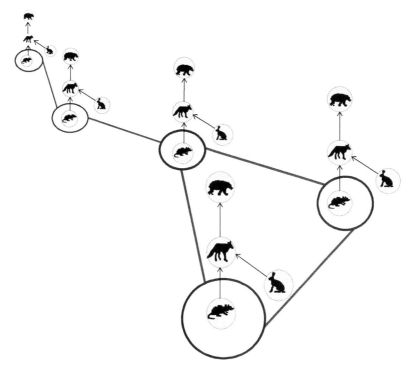

Fig. 15.3 *Example of meta-foodweb with four species and five patches studied in Ref. [58].*

patch $\left(\text{i.e.} X_i^{[\alpha]} = X_i^\star\right)$ and the stability of such a solution is investigated using the Master Stability Function approach.

This study shows that for a given foodweb and migration dynamics the spatial proximity of patches can modify the stability of the ecosystem. Therefore in this framework it is found that some foodwebs could be stable in some geographical networks of patches and unstable in others. This result is even more relevant when considering realistic foodwebs that induce a complex set of contraints on the spectral properties of the supernetwork between the ecological patches.

15.2.5   Discontinuous synchronization transitions

The synchronization transition in a system of Kuramoto oscillators had been considered to be continuous until the recent discovery of a new mechanism that has been shown to trigger a discontinuous, 'explosive' synchronization transition by assigning to the nodes of the network an internal frequency dependent on the node degree [136].

In Ref. [313] an alternative mechanism for triggering a discontinuous 'explosive' synchronization has been proposed and applied both to single and multilayer networks. This approach relies exclusively on an adaptive mechanism in which a fraction f of the nodes have a coupling constant proportional to their local order parameter. On a single layer this novel type of synchronization process is described by the following modified Kuramoto equations

$$\frac{d\theta_i}{dt} = \omega_i + \lambda\sigma_i \sum_{j=1}^{N} a_{ij} \sin\left(\theta_j - \theta_i\right), \tag{15.20}$$

where a_{ij} indicates the adjacency matrix of the network and σ_i are given with probability $1 - f$ by $\sigma_i = 1$ and with probability f by $\sigma_i = \hat{r}_i$ where $\hat{r}_i$ is the normalized local order parameter given by

$$\hat{r}_i e^{\hat{\psi}_i} = \frac{1}{k_i} \sum_{j=1}^{N} a_{ij} e^{i\theta_j}. \tag{15.21}$$

This adaptive mechanism yields a discontinuous synchronization transition already in a single layer for large enough values of the parameter f. In Ref. [313] this framework has been extended to duplex networks in which the dynamics on each replica node follows a generalized Kuramoto dynamics in which the coupling constant depends on the local order parameter of the corresponding replica node in the other layer. The dynamical system of equations reads

$$\frac{d\theta_i^{[\alpha]}}{dt} = \omega_i^{[\alpha]} + \lambda\sigma_i^{[\alpha]} \sum_{j=1}^{N} a_{ij}^{[\alpha]} \sin(\theta_j^{[\alpha]} - \theta_i^{[\alpha]}), \tag{15.22}$$

with $\alpha = 1, 2$. In this case the local order parameters for the synchronization read

$$\hat{r}_i^{[\alpha]} e^{\hat{\psi}_i^{[\alpha]}} = \frac{1}{k_i^{[\alpha]}} \sum_{j=1}^{N} a_{ij}^{[\alpha]} e^{i\theta_j^{[\alpha]}}, \tag{15.23}$$

and most crucially the coupling parameters $\sigma_i^{[1]}$ and $\sigma_i^{[2]}$ of any node i that implement the adaptive dynamics are chosen to be given by

$$\sigma_i^{[1]} = \sigma_i^{[2]} = 1$$

with probability $1 - f$ and by

$$\sigma_i^{[1]} = \hat{r}_i^{[2]}, \sigma_i^{[2]} = \hat{r}_i^{[1]}$$

with probability f. The synchronized state emerging in this dynamical system can be studied by considering the two global order parameters R_1, R_2 for the synchronization in each layer $\alpha = 1, 2$. These are defined by

$$R_1 e^{i\Psi_1} = \sum_{j=1}^{N} e^{i\theta_j^{[1]}},$$

$$R_2 e^{i\Psi_2} = \sum_{j=1}^{N} e^{i\theta_j^{[2]}}. \tag{15.24}$$

In this way it is found that this dynamical system, for sufficiently high values of f, yields a sudden and simultaneous emergence of the synchronized state in both layers of the duplex network.

A different mechanism producing discontinuous synchronization transition is proposed in Ref. [231], in which a duplex network whose first layer sustains the dynamics of coupled oscillators while the second layer supports the dynamics of a biased random walk is considered. This model is formulated to describe a possible mechanism for onset of synchronization in the brain where the dynamics at the level of brain regions is also influenced by the flux of blood. The dynamics on the first layer is a Kuramoto model whose dynamics reads

$$\frac{d\theta_i}{dt} = \omega_i + \lambda \sum_{j=1}^{N} a_{ij}^{[1]} \sin(\theta_j - \theta_i) \tag{15.25}$$

where θ_i is the dynamical variable associated with each node $i = 1, 2, \ldots, N$ in layer 1. The dynamics on the second layer is a diffusion dynamics dictated by the equation

$$\frac{dy_i}{dt} = \frac{1}{\tau_y} \sum_{j=1}^{N} \tilde{L}_{ij} y_j \tag{15.26}$$

where y_i is the dynamical variable in layer 2 and the generalized Laplacian describes the diffusion taking place according to a biased random walk, i.e.

$$\tilde{L}_{ij} = \frac{a_{ji}^{[2]} \chi_i^{\eta}}{\sum_{r=1}^{N} a_{jr}^{[2]} \chi_r^{\eta}} - \delta_{ij}. \tag{15.27}$$

Here the bias of the random walk is indicated with χ_i and η is a parameter of the model. The coupling between the two layers is modelled by assuming that the internal frequencies of the Kuramoto oscillators ω_i depend on the dynamical state of the diffusion process in layer 2 and that the bias χ_i of the diffusion dynamics depends on the local synchronization of the node i in layer 1. Therefore, ω_i and χ_i obey two dynamical equations that couple the dynamics in the two layers. Specifically, we have

$$\frac{d\omega_i}{dt} = \frac{1}{\tau_\omega} (N y_i - \omega_i), \tag{15.28}$$

where τ_ω is the typical timescale of this relaxation process, and

$$\frac{d\chi_i}{dt} = \frac{1}{\tau_\chi}\left(s_i^{dyn} - \chi_i\right),\qquad(15.29)$$

where τ_χ is the typical timescale of this relaxation process. Here the synchronization strength s_i^{dyn} measures how much node i is in synchrony with its neighbours and is given by

$$s_i^{dyn} = r_i \cos(\psi_i - \theta_i)\qquad(15.30)$$

where $r_i e^{i\psi_i} = \sum_{j=1}^N a_{ij}^{[1]} e^{i\theta_j}$. This model, as a function of the parameter λ that measures the strength of the coupling between the oscillators and the parameter η modulating the bias of the random walk, displays a rich phase diagram including a bistable region. The analysis of this model indicates that a synchronized state emerges discontinuously as a function of λ. Therefore, the model presented in Ref. [231] reveals that coupling the synchronization dynamics occurring on one layer to the diffusion dynamics occurring in the other layer of a duplex network can be an important mechanism for the emergence of a discontinuous synchronization transition.

15.2.6 Outer synchronization, time delays

The literature on synchronization on multilayer networks is vast and to a large extent preceding the formalization of multilayer network structures. Here it is not our intention to give a full account of all the dynamical systems studied so far, but we believe that the following papers are worth a particular mention. Early results [191, 192] investigate two coupled networks of dynamical oscillators formed by a master layer and a coupled slave layer and study their global 'outer synchronization' both numerically and analytically. The Authors find numerical evidence that outer synchronization can be achieved by the interacting multilayer networks and in the case in which the two coupled networks have the same adjacency matrix derive the sufficient analytical conditions for synchronization. Other works introduce time delays in the interlayer couplings of a duplex network [198]. In this paper the Authors show evidence of a breathing regime where two groups of nodes belonging to the two networks synchronize at different frequencies. By tuning the coupling strength the phase diagram of the dynamical model is significantly modified.

15.3 Pattern formation

Two works have investigated the properties of pattern formation in multiplex (specifically duplex) networks. In Ref. [10] each layer sustains a reaction–diffusion dynamics and the activator (inhibitor) is allowed to diffuse across layers with a tunable diffusion constant. On the contrary, in Ref. [176] the situation in which the activator diffuses in one layer and the inhibitor diffuses in another layer is considered as a function of their different diffusion constants.

15.3.1   Reaction–diffusion dynamics in each layer

Let us follow Ref. [10] and consider a duplex network in which every layer sustains a reaction–diffusion dynamics but the two layers are coupled by allowing diffusion of activator and inhibitor across the layers with a tunable diffusion constant. In this case, calling $u_i^{[\alpha]}/v_i^{[\alpha]}$ the concentrations of activator/inhibitor at node i in layer α and indicating with $D_u^{[\alpha]}/D_v^{[\alpha]}$ the diffusion constant of the activator/inhibitor within layer α and with $D_u^{[12]}/D_v^{[12]}$ the diffusion constant of the activator/inhibitor across the layers, the dynamical equations of the coupled reaction–diffusion dynamics read

$$\frac{du_i^{[\alpha]}}{dt} = f\left(u_i^{[\alpha]}, v_i^{[\alpha]}\right) + D_u^{[\alpha]} \sum_{j=1}^{N} L_{ij}^{[\alpha]} u_j^{[\alpha]} + D_u^{[12]} \left(u_i^{[\beta]} - u_i^{[\alpha]}\right),$$

$$\frac{dv_i^{[\alpha]}}{dt} = g\left(u_i^{[\alpha]}, v_i^{[\alpha]}\right) + D_v^{[\alpha]} \sum_{j=1}^{N} L_{ij}^{[\alpha]} v_j^{[\alpha]} + D_v^{[12]} \left(v_i^{[\beta]} - v_i^{[\alpha]}\right). \tag{15.31}$$

Here $f(u, v)$ and $g(u, v)$ are non-linear functions that specify respectively the dynamics of the activator that auto-catalytically enhances its own production and of the inhibitor that contrasts the activator growth. The matrices $\mathbf{L}^{[\alpha]}$ are the Laplacian matrices of layers $\alpha = 1, 2$ and we have indicated with β the layer different from α, i.e. $\beta \neq \alpha$.

In the case $D_u^{[12]} = D_v^{[12]} = 0$ the layers are decoupled, while for $D_u^{[12]} \neq 0, D_v^{[12]} \neq 0$ they are coupled. In Ref. [10] it has been found, using analytical approaches and direct simulations, that from the multiplex nature of this problem interesting and surprising results can be found by sufficiently increasing the diffusion constants $D_u^{[12]}, D_v^{[12]}$. In fact, it is possible to show that the system can develop self-organized patterns which result from the positive interference between the layers of the duplex network. By starting from a dynamic configuration that prevents the onset of self-organized patterns in each layer when the layers are decoupled, i.e. when $D_u^{[12]} = D_v^{[12]} = 0$, by sufficiently raising the values of the diffusion constants across the layers it is possible to observe that the stable homogeneous solution of the Eq. (15.31) becomes unstable and that Turing patterns set in. Conversely, the opposite phenomenon can also result from coupling the diffusion across the layers of the duplex networks, and self-organized patterns present in the absence of the coupling can fade away when the diffusion constant between the layers is switched on.

15.3.2   Activator and inhibitor dynamics layers

In a reaction–diffusion dynamics it can also occur that the activator and the inhibitor diffuse over different networks. For studying this theoretical scenario, in Ref. [176] pattern formation is characterized in a duplex network. The first layer of the duplex network indicates the network on which the activator diffuses, the second layer indicates the network in which the inhibitor diffuses. The resulting pattern formation is studied as a function of the network topologies and the diffusion constants of the reactants in the two layers. By indicating with u_i and v_i respectively the fraction of activators and inhibitors at node i, and with D_u and D_v their diffusion constant, the dynamical equations determining the reaction–diffusion dynamics are given by

$$\frac{du_i}{dt} = f(u_i, v_i) + D_u \sum_{j=1}^{N} L_{ij}^{[1]} u_j,$$

$$\frac{dv_i}{dt} = g(u_i, v_i) + D_v \sum_{j=1}^{N} L_{ij}^{[2]} v_j, \qquad (15.32)$$

where $\mathbf{L}^{[\alpha]}$ indicates the Laplacian matrices in layer α with $\alpha = 1, 2$. When $\mathbf{L}^{[1]} = \mathbf{L}^{[2]}$, this dynamical problem reduces to the well-known Turing instability which might occur when the inhibitor diffuses much faster than the activator. In Ref. [176], by studying the stability of the homogeneous solutions on scale-free duplex networks the Authors have shown that a different instability mechanism can be at work when the activator and the inhibitor diffuse on different layers, i.e. $\mathbf{L}^{[1]} \neq \mathbf{L}^{[2]}$, potentially yielding instabilities also when $D_u = D_v$. Specifically, in Ref. [176] it is shown that the instability can be triggered by nodes with degree $(k^{[1]}, k^{[2]}) = (k^{[u]}, k^{[v]})$, satisfying

$$k^{[u]} = \frac{f_u g_v - f_v g_u - f_u D_v k^{[v]}}{g_v D_u - D_u D_v k^{[v]}}, \qquad (15.33)$$

where f_u, f_v, g_u, g_v indicates the partial derivatives of the functions $f = f(u, v)$ and $g = g(u, v)$.

15.4   Multiplex visibility graphs

The multilayer network framework has been shown to be useful for characterizing multivariate time sequences [180]. Time-series analysis is of relevance in a variety of disciplines including biomedical fields, finance and climate science [126]. In these contexts it is the norm rather than the exception that the time series are multivariate. For instance, in climate the weather can be characterized by different physical quantities and in finance a financial market is characterized by the time series of different stock prices. Multivariate time series can be studied using a weighted network derived by the correlations among the time series. A more recent approach to studying multivariate time series combines the so-called visibility graphs defined to characterize univariate time series with a multiplex network analysis [180]. Given a univariate time series of elements $x(i)$ with $i = 1, 2, \ldots, T$ the visibility graph [179] (specifically the horizontal visibility graph) is a directed network of $N = T$ nodes having the directed link (i, j) whereas for each $\ell \in (i, j)$

$$x(\ell) < \min(x(i), x(j)). \tag{15.34}$$

In the context of univariate time series the analysis of the visibility graphs has been shown to provide relevant insights that cannot be deducted with more traditional methods for time series analysis. Given M timeseries forming a multivariate time series of elements $x^{[\alpha]}(i)$ with $i = 1, 2, \ldots, T$ and $\alpha = 1, 2, \ldots, M$, the multivariate time series analysis can be recast into the characterization of the structure of a multiplex network having each layer α given by the visibility graph of the time series $\mathbf{x}^{[\alpha]}$. By performing a multilayer analysis of the resulting multiplex network, the Authors of Ref. [180] are able to characterize both the temporal structure of each time series and the correlation existing among different time series. This approach that does not assume stationarity has shown to quantify non-trivial properties of chaotic maps and the onset of various types of synchronization. Finally, this method has been shown to be useful for discriminating crises from periods of financial stability starting from financial multivariate time series.

15.5   Control of multilayer networks

The controllability of a network is a theoretical problem of relevance in a variety of contexts ranging from financial markets to the brain. Until now, network controllability has been characterized mostly on isolated networks (see Sec. 3.7), while few works have started to tackle the richer problem of controllability of multilayer networks.

Correlated driver nodes

The controllability of a multiplex network can be studied assuming that each layer sustains a distinct dynamical process. However, typically we expect that in real multiplex networks it will not in general be possible to choose the driver nodes of the network in an arbitrary way. Rather, in a number of cases there will be some constraints on which nodes can be the driver nodes of the network and receive directly the external signals.

This problem was first addressed in Ref. [208] where network control was considered over a duplex network where *each node is either a driver node in each layer or it is not a driver node in any layer* (see Fig. 15.4).

In this framework it has been assumed that each replica node (i, α) of a duplex network formed by layers A and B is characterized by a different dynamical variable $x_i^{[\alpha]} \in \mathbb{R}$ with $\alpha = A, B$. The dynamical state of the network at time t is assumed to follow

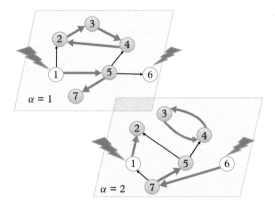

Fig. 15.4 *Control of a duplex network by correlating the position of driver nodes in different layers. The dynamics in the two layers is assumed to be independent but it is imposed that driver nodes are corresponding replica nodes. The controllability of this duplex network can be mapped to a Maximum Matching Problem in which the unmatched nodes (indicated with a white circle) are the driver nodes of the duplex network. Here red thick links indicate matched links and black thin links indicate unmatched links. Reprinted figure from Ref. [208].*

$$\frac{d\mathbf{X}(t)}{dt} = \mathcal{G}\mathbf{X} + \mathcal{B}\mathbf{u}, \tag{15.35}$$

where $\mathbf{X}(t)$ is the $2N$-dimensional vector describing the dynamical state of each replica node,

$$\mathbf{X} = \begin{pmatrix} \mathbf{x}^{[A]} \\ \mathbf{x}^{[B]} \end{pmatrix}, \tag{15.36}$$

where $\mathbf{x}^{[A]}/\mathbf{x}^{[B]}$ is the N-dimensional vector (with elements $x_i^{[A]}/x_i^{[B]}$) describing the dynamical state of nodes in layer A/B.

The matrix $\mathcal{G}$ is a $2N \times 2N$ (asymmetric) matrix and $\mathcal{K}$ is a $2N \times P$ matrix. They have the block structure

$$\mathcal{G} = \begin{pmatrix} \mathbf{a}^{[A]} & 0 \\ 0 & \mathbf{a}^{[B]} \end{pmatrix}, \quad \mathcal{B} = \begin{pmatrix} \mathbf{b}^{[A]} & 0 \\ 0 & \mathbf{b}^{[B]} \end{pmatrix}, \tag{15.37}$$

in which $\mathbf{a}^{[\alpha]}$ are the $N \times N$ directed and weighted matrices describing the directed weighted interactions within the layers and $\mathbf{b}^{[\alpha]}$ are the $N \times P^{[\alpha]}$ matrices describing the coupling between the nodes of each layer α and $P^{[\alpha]} \leq N$ external signals. The latter are represented by a vector $\mathbf{u}(t)$ of elements u_γ and $\gamma = 1, 2 \ldots P = P^{[A]} + P^{[B]}$.

If the driver nodes in the two layers can be chosen arbitrarily, the problem of controlling the duplex network can be recast into the one of controlling the two layers considered in isolation. However, if we cannot choose the position of the driver nodes independently in the two different layers the controllability of this system presents some relevant effect due to multiplexity.

Specifically, in Ref. [208] it has been assumed that the driver nodes must be corresponding replica nodes of the duplex network.

Using the approach of structural controllability, it is possible to show along the lines of Ref. [197], valid for the controllability of single layers (see Sec. 3.7), that finding the driver nodes of the duplex network can be mapped into a generalization of the Maximum Matching Problem. In this algorithm every link is either matched or unmatched and every node has at most one matched incoming link. The imposed condition on the possible choice of the driver nodes is reflected in an additional condition that the Maximum Matching Problem needs to satisfy. Namely, any two replica nodes either both have one matched incoming link or none of them has any matched incoming link (see Fig. 15.4). The driver nodes are the replica nodes that have no matched incoming links. Among the matching configurations that satisfy the above constraints the Maximum Matching Problem chooses the configuration that minimizes the number of driver nodes.

This optimization algorithm can be studied using statistical mechanics techniques such as the cavity method and the Belief Propagation (BP) algorithm. In Ref. [208] it is shown that the multiplexity of the problem has the following consequences:

(a) Controlling the dynamics of multiplex networks is more costly than controlling the single layers taken in isolation. In fact, the number of driver replica nodes in the multiplex network is in general higher than the sum of the number of driver nodes in the single layers taken in isolation (see Fig. 15.5).

(b) By tuning the degree correlations among different layers the number of driver nodes can be modulated. In particular, in the duplex network, by coupling the low-degree nodes in layer A with the low-degree nodes in layer B the number of driver nodes is minimized, whereas if the low-degree nodes in layer A have replica nodes with high in- and out-degree in the other layer B the number of driver nodes is maximized. This result is in line with the results obtained in single layers [207] indicating that the low-degree nodes (nodes with in- and out-degree smaller than 3) determine the number of driver nodes in the network.

(c) Multiplexity can contribute to stabilizing the fully controllable configuration. In fact, on duplex networks the fully controllable configuration can be stable even if it is not stable in the isolated networks that form the multilayer structure.

Moreover, the controllability of multiplex networks displays unexpected new phenomena. In fact, these multiplex networks can become extremely sensitive to damage in conjunction with a discontinuous phase transition characterized by a jump in the number of driver nodes (see Fig. 15.5). This phase transition is, for instance, observed on duplex networks in which each layer has Poisson degree distribution with the same average degree c. A careful investigation of the critical behaviour reveals that this is a hybrid phase transition with a square-root singularity, and therefore in the

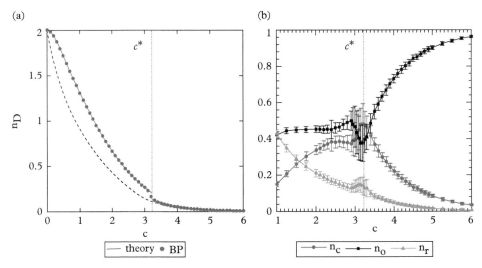

Fig. 15.5 *Controllability of a directed multiplex network formed by two Poisson layers (layer A and layer B) with in and out average degrees equal to c. In panel (a) the fraction n_D of driver nodes in a Poisson duplex network is plotted as a function of the average degree c. The points indicate the average Belief Propagation (BP) results obtained over five single realizations of the Poisson duplex networks with average degree c and $N = 10^4$, the solid line is the theoretical expectation. The dashed line represents twice the density of driver nodes for a single Poisson network with the same average degree. In panel (b) the densities n_c, n_r and n_o respectively of critical redundant and ordinary nodes are shown as functions of c for the same type of duplex networks with $N = 10^3$, where each point is the average over 100 different instances. In both panels the dot-dashed vertical line indicates the phase transition average degree $c^* = 3.22\ldots$. Reprinted figure from Ref. [208].*

same universality class as the emergence of the mutually connected component in multiplex networks. In correspondence to this phase transition the network responds non-trivially to perturbations. This is observed by performing a numerical calculation of the robustness of the networks [197]. The nodes are classified into three categories: *critical nodes, redundant nodes* and *ordinary nodes*. When a critical node is removed from the (multiplex) network, controllability is sustained at the cost of increasing the number of driver nodes. If the number of driver nodes decreases or is unchanged, the removed nodes are classified as redundant and ordinary respectively. Fig. 15.5B shows that in Poisson duplex networks displaying the phase transition in the number of driver nodes [208] the fraction of critical nodes reaches a maximum at the transition, revealing an increased fragility of the duplex network to random damage with respect to single layers. While an abrupt change in the number of driver nodes can result from a small change in the network topology, it is important to stress that the non-monotonic behaviour of these quantities around the critical average degree value could be interpreted as a precursor to the discontinuity.

15.5.1 Role of peripheral nodes

Sometimes it is reasonable to assume that the driver nodes of a multilayer network can only belong to a given layer. This is the case when it is possible to apply an external signal only to the nodes of a given 'accessible' layer, whereas the node of other layers cannot receive input signals directly. This scenario has been considered in Ref. [314]. In this paper the Authors have studied a multiplex network with two layers (layer A and layer B) with interlinks coupling the dynamics of the two layers where only nodes in layer A can be driver nodes and receive external signals. The linear dynamical process is determined by the system of equations

$$\frac{d\mathbf{X}(t)}{dt} = \mathcal{A}\mathbf{X} + \mathcal{B}\mathbf{u}, \tag{15.38}$$

in which the $2N$-dimensional vector $\mathbf{X}(t)$ describes the dynamical state of each node,

$$\mathbf{X} = \begin{pmatrix} \mathbf{x}^{[A]} \\ \mathbf{x}^{[B]} \end{pmatrix}, \tag{15.39}$$

where $\mathbf{x}^{[A]}/\mathbf{x}^{[B]}$ is the N-dimensional vector describing the dynamical state of nodes in layer A/B. Here $\mathcal{A}$ is the directed and weighted supra-adjacency matrix of the multilayer network, indicating the full set of interactions within each layer and across layers and $\mathcal{B}$ enforces the constraints that only the nodes in layer A can receive external input signals, i.e. $\mathcal{B}$ has the following block structure:

$$\mathcal{B} = \begin{pmatrix} \mathbf{b} \\ 0 \end{pmatrix}, \tag{15.40}$$

where $\mathbf{b}$ is a $N \times P$ matrix and P is the number of external signals.

Given a fixed number of driver nodes in layer A in Ref. [314] it has been found that the fraction n_D of controlled nodes can be tuned by placing the interlinks between nodes in layer A and nodes in layer B according to different coupling strategies.

In particular, in Ref. [314] the Authors rank each node i in layer α according to a measure of centrality given by $r_{i\alpha}$ that is lower for nodes with both low in-degree and low out-degree. Specifically, $r_{i\alpha}$ is given by

$$r_{i\alpha} = \left(k_{i,in}^{[\alpha]} \right)^{\lambda} \left(k_{i,out}^{[\alpha]} \right)^{1-\lambda} \tag{15.41}$$

where λ is a parameter between zero and one. As λ increases, more importance is attributed to the in-degree in determining the centrality of the nodes.

The different coupling strategies include the PP strategy coupling peripheral nodes in layer A to peripheral nodes in layer B, the CP strategies coupling central nodes in layer A to peripheral nodes in layer B and the CC strategy coupling the central nodes in layer A to central nodes in layer B.

It is found that the PP strategy optimizes the fraction of controlled nodes n_b for any given number of driver nodes. This result obtained numerically using linear programming techniques is totally in line with the analytical result obtained for single networks in Ref. [207] where it is shown that the low in-degree and low out-degree nodes determine the structural controllability of networks. Therefore, in the context of multiplex network controllability, low in-degree and low out-degree nodes should be treated with particular care for optimizing the controllability of networks.

15.5.2   Control of multi-slice temporal networks

Controlling multiplex networks can also be affected by the separation of the typical timescales at which different layers operate. In order to tackle this relevant question in Ref. [254] the Authors have considered the controllability of a multi-timescale duplex network. It is assumed also in this case that the driver nodes can be chosen only from the replica nodes of a given layer, called the master layer. It is found that if the typical timescales of the two layers are equal, the number of driver nodes is greater or equal to the number of nodes in each isolated layer. The number of driver nodes changes if the two layers have two different timescales for their dynamics. It is found that if the master layer is also the one supporting a faster dynamics, the number of driver nodes decreases as the difference of timescales between the two layers increases. For very large differences of the timescales between the two layers the number of driver nodes reaches a plateau whose value is determined by the number of driver nodes of the faster layer. Therefore, in this case controllability is enhanced. On the contrary, if the master layer is the slowest layer, the number of driver nodes increases as the difference between the layers' timescales increases. Therefore the controllability of the multilayer network is reduced.

16

Opinion Dynamics and Game Theory

16.1 Modelling social network dynamics

In the last decade, there has been a surge in interest in modelling social network dynamics [72, 248], including opinion dynamics, election models and game theory approaches. As multiplexity is a prominent property of social networks, it is of fundamental importance to test how the results on social network dynamics obtained in the single network framework are modified on multilayer networks. In this chapter firstly we will cover the voter model which is a very stylized model for adoption of behaviour and for opinion dynamics, with a very interesting phenomenology and of special theoretical interest. Secondly we will study the dynamics of competing layers in the framework of (a) an election model in which different layers correspond to competing political parties and (b) a model in which layers compete for the centrality of their nodes. Finally, we will briefly discuss the rich field of game theory on multilayer networks, showing evidence that multiplexity can capture several important mechanisms for explaining the emergence of the cooperative behaviour in social systems.

16.2 Voter model

16.2.1 Definition and results on single networks

The voter model describes opinion dynamics in a network where nodes do not have individual convictions and are exclusively influenced by their neighbours. From any given initial condition the dynamics evolves until a final frozen state is reached, in which any two neighbour nodes share the same opinion. In a connected network this is a state of global consensus. The model can be formulated for a number of opinions equal or greater than two. However, since the model always reaches a final consensus, the two-state voter model is the most frequently analysed case as it is a simplification that preserves most of the relevant phenomenology. Let us here discuss briefly the voter model on single layers [177]. In a two-state voter model each node i can have two opinions indicated by the spin values $s_i = 1$ or $s_i = -1$. Starting from a given initial configuration, at each time step we update the spins and the time clock as follows:

Multilayer Networks. Ginestra Bianconi, Oxford University Press (2018).
© Ginestra Bianconi. DOI: 10.1093/oso/9780198753919.001.0001

- a random node i is picked with uniform probability;
- the selected node adopts the opinion of a randomly selected neighbour node j, i.e.

$$s_i \rightarrow s_j, \tag{16.1}$$

- the time t is incremented by an interval of time $\Delta t = 1/N$, i.e.

$$t \rightarrow t + \Delta t. \tag{16.2}$$

The algorithm stops when no neighbour nodes have a different opinion and the consensus is reached.

The voter model displays a phenomenology that depends on the underlying networks where it has been defined. The main dynamical properties that are analysed as a function of the network topology are the time T_N to reach consensus in a network of N nodes and the temporal behaviour of the interfacial density $\rho(t)$ given by the fraction of links between nodes of different opinions, i.e.

$$\rho(t) = \frac{\sum_{i<j} a_{ij}|s_i - s_j|}{\sum_{i<j} a_{ij}}. \tag{16.3}$$

On d-dimensional lattices of N nodes we observe a significant dependence of these quantities on the dimension d, (see Ref. [177] for the derivation of these results)

$$T_N \simeq \begin{cases} N^2 & \text{for } d = 1, \\ N \ln N & \text{for } d = 2, \\ N & \text{for } d > 2, \end{cases} \qquad \rho(t) \simeq \begin{cases} t^{-1/2} & \text{for } d = 1, \\ (\ln t)^{-1} & \text{for } d = 2, \\ \mathcal{O}(1) & \text{for } d > 2. \end{cases} \tag{16.4}$$

The two-dimensional voter model has also been analysed in Ref. [104] as an important non-equilibrium transition solely driven by interface noise. The active interface displays a temporal evolution called coarsening that is common to a large class of non-equilibrium models where the phase ordering takes place without surface tension. Therefore, the coarsening dynamics of the voter model in two dimensions is distinct from the domain growth in Glauber dynamics for simulating the Ising model (see Fig. 16.1).

The voter model has been extensively studied in complex networks including small-world networks [301, 73] and scale-free networks [285, 284].

On power-law networks the average time T_N to reach global consensus in a network of N nodes scales like

$$T_N \simeq N \frac{\langle k \rangle}{\langle k^2 \rangle}. \tag{16.5}$$

Therefore, on scale-free networks with power-law exponent $\gamma \leq 3$ the voter model converges faster than on homogeneous networks because $\langle k^2 \rangle \rightarrow \infty$ as $N \rightarrow \infty$.

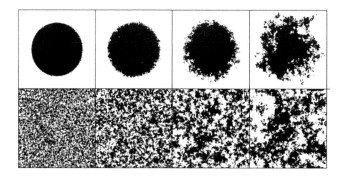

Fig. 16.1 *Domain growth of the two-dimensional voter mode of system size* N = *256* × *256*. *The figure displays snapshot of the voter dynamics at times* t = *4, 16, 64, 256 starting from an initial bubble of radius* r_0 = *180 (top panels) or starting with a random initial condition (bottom panels). Reprinted figure with permission from [104] ©2012 by the American Physical Society.*

16.2.2   Voter model on static multilayer networks

The voter model has been extensively used to model opinion dynamics in social networks. Given that the vast majority of social networks have a multilayer nature, it is then natural to ask how multiplexity will change the dynamic properties of this dynamical process.

One of the first works [203] on the subject describes a minimal model allowing for a multilayer network interpretation. Specifically, it characterizes the two-state voter model on a multilayer network formed by two layers, each one formed by a clique of N nodes and connected by a number m of interlinks with $0 \leq m \leq N^2$. Initially, each clique is formed by $N/2$ nodes in state $s = 1$ and $N/2$ nodes in state $s = -1$. At each given time a random link is chosen with uniform probability and a random node at the end of the link is selected with probability $1/2$. Then the selected node copies the state of the node at the other end of the link. Subsequently, the global time is increased by $\Delta t = \frac{1}{2N}$. The analytical solution of this model predicts that for large density of interlinks, the mean consensus time grows linearly with the network size, i.e. $T_N \simeq N$, while for low density of interlinks when $m = O(1)$ the dynamics slows down and $T_N \simeq N^2$. Since the cross-over between the two scaling regimes occurs for a number of interlinks m close to one, this work concludes that the slowdown of the ordering dynamics is a phenomenon that is marginal for the two-cliques model since it is observed only for very weakly connected layers.

If the two-cliques voter model does present only marginal dependence on the multilayer structure of the network, a recently proposed variation of the voter model [98] is able to capture the exclusive properties that opinion dynamics can present in a multiplex network. The social network is taken to be a duplex network where only a fraction q of replica nodes are connected. The replica nodes in different layers represent the same individual if the replica nodes are connected while they correspond to different individuals if they are unconnected. The multiplex two-state voter model proposed in Ref. [98] assumes that, if connected, two replica nodes always take the same opinion.

Therefore, if a replica node changes state its connected replica node in the other layer (if it exists) also changes its state accordingly and adopts the new opinion.

Let us indicate by $\alpha = 1, 2$ the two layers of the duplex network and by $s_i^{[\alpha]} = 1, -1$ the opinion of node i in layer α. The multiplex two-state voter model is defined as follows. Given an initial configuration of the spins, at each time step

- a random layer α is selected;
- one node i of layer α is selected at random and changes its state according to the single-layer voter dynamics, i.e. it adopts the state of a random neighbour j in layer α,

$$s_i^{[\alpha]} \to s_j^{[\alpha]}; \qquad (16.6)$$

- if the selected node is present in both layers, also its replica node in the other layer $\beta \neq \alpha$ is updated,

$$s_i^{[\beta]} \to s_i^{[\alpha]}; \qquad (16.7)$$

- the time is incremented by $\Delta t = 1/N$,

$$t \to t + \Delta t. \qquad (16.8)$$

The algorithm stops when global consensus is achieved and there is no pair of connected nodes with different opinions (spin states).

A numerical investigation of the multiplex voter model [98] reveals that this dynamics has an exclusive multilayer nature and cannot be reduced to a voter dynamics on a single aggregated network.

16.2.3   Coevolving voter model on multiplex networks

Adaptive multiplex networks have been studied in Ref. [99] from the perspective of the coevolving voter model, in which each layer evolves according to a voter model while rewiring its links in order to maximize the number of neighbour nodes with the same opinion. In single networks the coevolving voter model has been shown [300] to give rise to a fragmentation transition in which the system is frozen into two disconnected components, each one formed by nodes sharing the same opinion. This transition occurs when the probability p of rewiring a link connecting two nodes with opposite opinions is greater than a critical value $p > p_c$.

In a duplex, the two layers can be allowed to have very different timescales, i.e. very different probabilities $p^{[\alpha]}$ of rewiring links among nodes of opposite opinion. In this scenario it is possible to study the interplay between the dynamical state of the nodes and the topology of the resulting networks as a function of the fraction q of nodes that are present in both layers. Specifically, the model proposed in Ref. [99] implements the following adaptive multilayer network dynamics. At each time step:

(1) a random layer is picked;

(2) the chosen layer is updated using the coevolving voter model node update;

(3) the state of nodes present in both layers is updated in both layers.

Here we assume that the coevolving voter model node update of a given layer α consists in the following dynamics:

(a) a random node i of the chosen layer α is selected;

(b) its state is compared with the state of a random neighbour j on the same layer α;

(c) if node i and node j are in the same dynamical state nothing happens; otherwise with probability $1 - p^{[\alpha]}$ node i copies the state of node j, or else with probability $p^{[\alpha]}$ it severs the connection with node j.

The extremely asymmetric scenario occurs when one layer (network 1) has $p^{[1]} = 1$ and it is therefore a very dynamical network, while the other layer (network 2) has rewiring probability $p^{[2]} = 0$, resulting in a layer supporting the voter model dynamics without rewiring of the links. In this case the first network is also called the *dynamic layer* while the second network is called the *voter layer*. For $q = 1$ the two layers are fully coupled: in both layers consensus is reached, therefore the dynamic layer is formed just by one component. For $q = 0$ the two layers are fully decoupled; therefore the dynamic layer is fragmented into two components formed by nodes having opposite opinions. For intermediate values of q the system undergoes an anomalous transition called *shattered fragmentation* in which the dynamic layer is formed by many disconnected components that can be interpreted as the precursor of the fragmentation transition observed for $q = 0$.

For less asymmetric scenarios where $\left(p^{[1]}, p^{[2]}\right) \neq (1, 0)$ a rich phase diagram can be observed, and the system can include both frozen layers, both active layers or one layer active, one layer frozen.

16.3 Competing networks

Cooperative effects correlating nodes among different layers such as interlink interdependencies have attracted large interest in the context of multilayer network theory. However, competing interactions can also play a major role in determining the multilayer network dynamics. In the preceding chapters we have already described models of competing dynamics as, for instance, antagonistic percolation (Sec. 12.4) or competing epidemic spreading (Sec. 13.5). Here we discuss a model of competing networks in the context of election models [149] and a model for competing centralities [2].

16.3.1 Election model

In election models, we can assume that each party represents a layer of social interaction through which a contagion dynamics can take place. Every individual can have an opinion or propensity to vote for a party that changes during the election campaign. An individual is represented by a node and whereas a node has a propensity to vote for a party it is considered active on the corresponding social network. The election campaign can therefore be described [149] as a competition between layers trying to maximize the number of active nodes at the end of the dynamical evolution of the model (election day).

Specifically, in Ref. [149] the Authors have proposed a model for describing the election campaign in which there are just two parties (party A and party B) and in which each individual node has three possible dynamical states at the end of the campaign: it might be active either on the first or the second layer, representing his vote for the corresponding party, or might be inactive in both layers, representing his decision not to vote. Crucially, the nodes' opinions are affected by the opinions of their neighbours through an opinion-spreading dynamics. As there is evidence that the chance to change opinion decreases as the decision moment approaches, the spreading dynamics during the election campaign is modelled by a simulated annealing algorithm that implies an uncertainty reduction mechanism as the election day comes closer. Therefore, at the initial stage of the election campaign the nodes are likely to change their opinions while as the election campaign proceeds their dynamics is slowed down until it freezes at the election day.

A different rule applies, however, for a small fraction f_A, f_B of nodes that represent committed communities of individuals that vote either for party A or party B and never change their opinion.

This election model has been studied on duplex networks having layers formed by Poisson networks with average degree given by z_A and z_B respectively. The main result of this model is that in the large-network limit the most-connected network is more likely to win the election independently of the initial condition of the system (see Fig. 16.2). However, when the two parties have a comparable density of links small

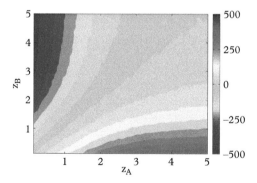

Fig. 16.2 *Difference between the number of people* m_A *that at the end of the election campaign vote for party A and the number of people* m_B *that vote for party B as a function of the average degrees* z_A *and* z_B *of the social networks A and B when there are no committed agents* $f_A = f_B = 0$. *The data is simulated for two networks for* N = 500 *nodes and averaged 60 times. The more connected the social network connected to party A, the larger the difference of votes* $m_A - m_B$. *Reprinted figure from Ref. [149].*

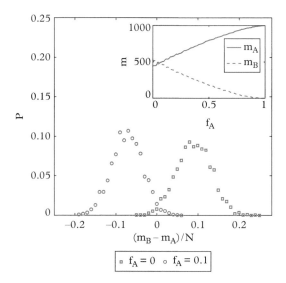

Fig. 16.3 *A fraction* f *of committed agents is shown to be able to reverse the outcome of the election. The histogram of the difference between the fraction of agents* m_B/N *voting for party B and the fraction of agents* m_A/N *voting for party A is plotted for a fraction* f_A *of committed agents to party A, (with* $f_A = 0$ *and* $f_A = 0.1$*) and average connectivities of the networks* $z_A = 2.5, z_B = 4$ *of the two layers. The histogram is performed for 1000 realizations of two networks of size* N = *1000. In the inset the average number of agents in network A* (m_A) *and agents in network B* (m_B) *is shown as a function of the fraction of committed agents* f_A*. A small fraction of agents* ($f_A \simeq 0.1$) *is sufficient to reverse the outcome of the elections. The data in the inset is simulated for two networks for* N = *1000 nodes and averaged ten times. Reprinted figure from Ref. [149].*

perturbations in the system could alter the results. In this context, it has been observed that a small minority of committed agents can reverse the outcome of the election result (see Fig. 16.3).

16.3.2   Competing for centrality

Another scenario of a competing dynamics has been considered in Ref. [2] where it is assumed that two layers of a multilayer network structure with $M = 2$ compete for resources. Specifically, the layers compete for having nodes of higher eigenvector centrality. Here the eigenvector centrality is perceived as a measure of importance of the nodes in the network, but also as a measure directly related to the properties of linear dynamical processes occurring on it.

Considering a multilayer network formed by two layers $\alpha = 1, 2$, each with N_α nodes, the eigenvector centrality $\mathbf{u}$ satisfies

$$\lambda_1 \mathbf{u} = \mathcal{A}\mathbf{u} \tag{16.9}$$

where $\mathcal{A}$ is the supra-adjacency matrix of the multilayer network, λ_1 is its maximum eigenvalue and the eigenvector centrality $\mathbf{u}$ has the following block-diagonal structure

$$\mathbf{u} = \begin{pmatrix} \mathbf{u}^{[1]} \\ \mathbf{u}^{[2]} \end{pmatrix}, \tag{16.10}$$

where $\mathbf{u}^{[\alpha]}$ is the eigenvector centrality of the nodes in layer α.

In Ref. [2] different coupling strategies between the two layers have been considered with the goal of optimizing the centrality of a network with respect to the other one. The centrality of a network α is assumed to be measured by the normalized sum of the eigenvector centralities, i.e.

$$C^{[\alpha]} = \frac{\sum_{i=1}^{N_\alpha} u_i^{[\alpha]}}{\sum_{\alpha=1}^{2} \sum_{i=1}^{N_\alpha} u_i^{[\alpha]}}. \tag{16.11}$$

The result of this competing dynamics is studied from multilayer networks with identical layers as a function of the strategy adopted for placing the interlinks. Assuming that the coupling between the layers includes exclusively undirected interactions, in Ref. [2] it is shown that the CC strategy coupling central nodes of a layer with central nodes of the other layer strongly favours the weak network. On the contrary, the PP strategy that couples peripheral nodes of one layer with peripheral nodes of the other layer enhances the centrality of the stronger layer.

16.4   Game theory on multilayer networks

16.4.1   Two-player games on single networks

Game theory is a mathematical subject that was first formulated to model social and economical choices [131] and was then applied to evolutionary dynamics [232] and more recently to Network Science [248].

One of the major questions in game theory is the emergence of cooperative behaviour which is found both in human and animal societies. In order to study how cooperating behaviour can result from game theory, it is possible to consider two player games where each individual can either adopt strategy C (Cooperate) or D (Defect). Each player adopts a given strategy to optimize his payoff without knowing the other player's strategy. In this framework, game theory assumes that each player will play rationally. The payoff of the agent is represented by the following matrix Π

$$\Pi = \begin{array}{c} \\ C \\ D \end{array} \begin{array}{c} C \quad D \\ \begin{pmatrix} R & S \\ T & P \end{pmatrix} \end{array} \tag{16.12}$$

where

- *T temptation to defect* is the payoff of a defector when the other player cooperates;
- *R reward for mutual cooperation* is the payoff of a cooperator when the other player cooperates;
- *P punishment for mutual defection* is the payoff of a defector when the other player also is a defector;
- *S sucker's payoff* is the payoff of a cooperator when the other player is a defector.

As a function of the relative values of these parameters we distinguish between the following four games:

(I) **Harmony game with** $S > P, R > T$.
In this case cooperating is always the best strategy, independently of the strategy of the other player.

(II) **Snow drift game with** $T > R > S > P$.
In this case defecting is advantageous only if the opponent is cooperating, and therefore constitutes a risky strategy. The best strategy is to do the opposite of the other player.

(III) **Stag hung game with** $R > T > P > S$.
In this case cooperating is only advantageous if the other player is also cooperating and therefore constitutes a risky strategy. The best strategy is to adopt the same strategy as the other player.

(IV) **Prisoner's Dilemma with** $T > R > P > S$.
In this case the rational choice for each player is to defect despite the fact that both players would gain more if they both cooperated.

In between the games mentioned above special attention has been given to the Prisoner's Dilemma which describes the conflict between rational behaviour and the cooperative behaviour that maximizes the overall gain in the population. Specifically, large attention has been devoted to the study of the Prisoner's Dilemma described generally by the payoff matrix

$$\Pi = \begin{matrix} & C & D \\ C & \\ D & \end{matrix} \begin{pmatrix} 1 & 0 \\ b > 1 & 1 > \epsilon \geq 0 \end{pmatrix}, \tag{16.13}$$

where b indicates the temptation to defect.

The outcome of the game when both players are rational is described by the Nash equilibrium in which each player plays the best strategy given the other player's strategies. In the case of the Prisoner's Dilemma the Nash equilibrium is reached when both players defect.

A different approach for studying the two-player games is provided by evolutionary dynamics [232] in which it is assumed that the system is formed by a population of players, each one playing a given strategy. By playing repeatedly with the other players each player accumulates his own individual payoff. Individuals (or their strategies) that are more successful replicate faster, as their payoff is treated as their fitness.

There are different mechanisms used to study how the population evolves. This includes processes in which each player imitates the strategies of other successful players. In the case of the Prisoner's Dilemma, when the evolutionary dynamics takes place on a well-mixed population in which every player can play with every other player the population asymptotically in time is formed exclusively by defectors [155, 273]. However, the evolution of the population can also be studied in structured populations in which players are playing exclusively with players in their proximity. For instance players can be placed in a two-dimensional space where their proximity is measured through their Euclidean distance or they can be placed on a network where typically it is assumed that players interact exclusively with their neighbouring players at distance one on the network. When the Prisoner's Dilemma takes place on structured populations, the cooperator strategy can be maintained also in the presence of a temptation to defect $b > 1$. This mechanism promoting cooperation was coined as *network reciprocity* [233]. Interestingly, on scale-free networks the cooperator's strategies can be supported and enhanced [270, 134].

16.4.2   Two-player games on multiplex networks

Two-player games can be considered over multilayer networks [307]. As multilayer networks describe locally structured interactions the general expectation is that multiplexity will enhance cooperation. However, two-player games can be defined on multilayer networks in a large variety of ways and the results obtained are often non-trivial. Here we will provide a discussion of two interesting theoretical frameworks in which the two-player game takes place on multiplex networks in which interlinks are not explicitly taken into account. In this setting the evolutionary dynamics describes a population of players interacting pairwise via multiple layers of interactions while imitating strategies of successful neighbour nodes.

In Ref. [135] a population of players of the Prisoner's Dilemma is considered on a multiplex network and it is additionally assumed that the strategies of the same player in different layers can be different. Therefore, for each layer α the player i can either cooperate $s_{i\alpha} = 1$ or defect $s_{i\alpha} = 0$. Each player collects in each layer α a payoff $U_i^{[\alpha]}$ given by the sum of the payoffs obtained by playing the Prisoner's Dilemma with all his neighbours in layer α. Finally, the overall payoff $U_i = \sum_{\alpha=1}^{M} U_i^{[\alpha]}$ is calculated for each player i.

The strategies of each player evolve by a synchronous dynamics. For each player a layer α is chosen at random and in that layer a neighbour node j is chosen randomly. If the payoff of node i is bigger than the payoff of node j, i.e. $U_i > U_j$, nothing happens. Instead, if $U_i \leq U_j$ node i in layer α will adopt the strategy of node j in layer α with probability

$$\Pi_{i \to j} = \frac{U_j(t) - U_i(t)}{b \max{(S_i, S_j)}}, \qquad (16.14)$$

where $S_i = \sum_{\alpha=1}^{M} k_i^{[\alpha]}$.

This model shows that multiplexity can enhance the cooperative behaviour in particular for high values of the temptation to defect b. Indeed, if one considers the fraction
of cooperators over entire multiplex networks (see Fig. 16.4) one observes that for high
temptation to defect b the number of cooperators increases with the number of layers of
the multiplex network. However, one can also notice that the opposite trend is observed
for low temptation to defect b.

In this model it is of crucial importance for sustaining cooperative behaviour that
players can play different strategies in different layers, as it is shown that if the players
adopt homogeneous strategies among all the layers the population of cooperators
dramatically decreases.

A different theoretical framework is presented in Ref. [306] where players of the
Prisoner's Dilemma are placed on a duplex network formed by two square lattices which
are only partially coupled, i.e. only a fraction ρ of the nodes of a layer has a corresponding
node in the other layer. This framework allows the Authors to investigate the role of
interlinks in the game theory settings. Every player on a node (i, α) is either a cooperator
$s_{i\alpha} = 1$ or a defector $s_{i\alpha} = 0$. At each time each player acquires a payoff $\pi_{i\alpha}$ which is
equal to the sum of the payoffs obtained by playing with every neighbour node in the
same layer. The total payoff of the player (i, α), nevertheless, is affected by the presence

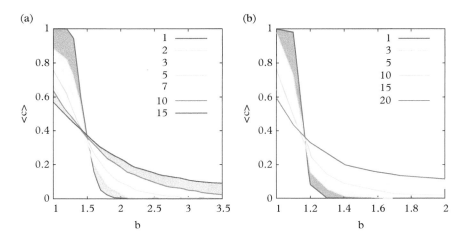

Fig. 16.4 *Cooperation diagrams of multiplex networks. Average level of cooperation ⟨c⟩ as a function of
the temptation b to defect for several multiplex networks with different number of layers M (the number
of layers is indicated in the legend). In panel (**a**) the network layers are ER graphs with ⟨k⟩ = 3 (sparse
graphs) while in panel (**b**) we have ⟨k⟩ = 20. In both cases N = 250 nodes. As can be observed, the
resilience of cooperation increases remarkably as the number of layers M grows. Reprinted by permission
from Macmillan Publishers Ltd: [135] ©2012.*

of the interlinks. If node (i, α) is coupled with the corresponding node (i, β) in the other layer β, its payoff is given by

$$U_{i\alpha} = \pi_{i\alpha} + a\pi_{i\beta}, \tag{16.15}$$

where a indicates the strength of the coupling between the two layers. If, on the contrary, node (i, α) is not coupled with the corresponding node in the other layer we have

$$U_{i\alpha} = \pi_{i\alpha}. \tag{16.16}$$

The strategies of the players evolve by picking a random node (i, α) of the duplex network and a random neighbour (j, α) of the same layer and making the change $s_{i\alpha} = s_{j\alpha}$ with probability

$$W_{i \to j} = \frac{1}{1 + \exp\left[(U_{i\alpha} - U_{j\alpha})/D\right]}, \tag{16.17}$$

where D is taken to be $D = 0.1$.

In this context it is shown that for sufficiently high values of the coupling strength a there is always an optimal density of interlinks ρ for sustaining the cooperative behaviour. This surprising result shows that a moderate coupling between the layers can be actually more beneficial than a complete coupling.

Finally, we have seen two scenarios for the implementation of the two-player game of multiplex networks. While in both cases it is shown that multiplexity can be beneficial for sustaining the cooperation behaviour, it is also evident that the results obtained are far from trivial and that the study of game theory on multilayer networks can be highly surprising.

Appendix A
The Barabási–Albert model:
the Master Equation

A.1   The master equation

In this appendix we show the derivation of the exact asymptotic expression of the degree distribution of the Barabási–Albert model defined in Sec. 2.8.4 using the master equation approach [108, 178]. The master equation is the equation describing the evolution of the average number $N_k(t)$ of nodes that at time t have degree k. Specifically, it indicates that the average number $N_k(t+1)$ of nodes of degree k at time $t+1$ is equal to the average number $N_k(t)$ of nodes of degree k at time t plus the average number of nodes that at time t have degree $k-1$ and acquire a link, minus the average number of nodes that at time t have degree k and acquire a link. Therefore, this equation reads for the Barabási–Albert model,

$$
\begin{aligned}
N_k(t+1) &= N_k(t) + m\Pi(k-1)N_{k-1}(t) - m\Pi(k)N_k(t) \quad \text{for } k > m, \\
N_k(t+1) &= N_k(t) - m\Pi(k)N_k(t) + 1 \qquad\qquad\qquad\quad \text{for } k = m
\end{aligned}
\tag{A.1}
$$

where $\Pi(k)$ is given by the probability that a node of degree k acquires a new link, i.e. enforces the linear preferential attachment

$$
\Pi(k) = \frac{k}{\sum_k kN_k(t)}.
\tag{A.2}
$$

A.2   Derivation of the exact degree distribution

Considering the Eq. (A.2) enforcing the preferential-attachment mechanism, we observe that for $t \gg 1$ we can approximate

$$
\Pi(k) = \frac{k}{\sum_k kN_k(t)} \simeq \frac{k}{2mt}.
\tag{A.3}
$$

In fact, the sum in the denominator is equal to twice the total number of links, therefore since at each time we add m links for large times $t \gg 1$, we can put

$$
\sum_k kN_k(t) \simeq 2mt.
\tag{A.4}
$$

Using this approximation valid for $t \gg 1$ we can write the Eq. (A.1) as

$$
\begin{aligned}
N_k(t+1) &= N_k(t) + \frac{k-1}{2t}N_{k-1}(t) - \frac{k}{2t}N_k(t) \quad \text{for } k > m, \\
N_k(t+1) &= N_k(t) - \frac{k}{2t}N_k(t) + 1 \qquad\qquad\qquad \text{for } k = m.
\end{aligned}
\tag{A.5}
$$

Now we observe that for sufficiently large values of $t \gg 1$ we have that the fraction of nodes of degree k converges to the degree distribution $P(k)$, i.e.

$$N_k(t) \simeq tP(k) \tag{A.6}$$

where the total number of nodes in the network is given by $N \simeq t$. Substituting Eq. (A.6) into the master equation (A.5), we obtain

$$
\begin{aligned}
(t+1)P(k) &= tP(k) + \tfrac{(k-1)}{2}P(k-1) - \tfrac{k}{2}P(k) \quad \text{for } k > m \\
(t+1)P(k) &= tP(k) - \tfrac{k}{2}P(k) + 1 \qquad\qquad\quad \text{for } k = m.
\end{aligned} \tag{A.7}
$$

Simplifying the equation valid for $k > m$ we obtain

$$P(k) = \frac{k-1}{2+k}P(k-1) \tag{A.8}$$

for $k > m$. This recursive equation for $k > m$ can be solved in terms of $P(m)$ giving

$$P(k) = \prod_{j=1+m}^{k} \left[\frac{j-1}{2+j} \right] P(m) = \frac{m(m+1)(m+2)}{k(k+1)(k+2)} P(m). \tag{A.9}$$

Taking Eq. (A.7) as valid for $k = m$ we obtain

$$P(m) = \frac{2}{2+m}. \tag{A.10}$$

Therefore, it follows that the degree distribution of the Barabási–Albert model in the limit $t \gg 1$ is given by

$$P(k) = \frac{2m(m+1)}{k(k+1)(k+2)}. \tag{A.11}$$

Appendix B
Entropy and Null Models of Single Networks

B.1 Maximum-entropy network ensembles

B.1.1 General remaks

In this appendix we provide the details for fully characterizing maximum-entropy network ensembles by following Refs [39, 40, 6, 46, 7]. As discussed in Sec. 2.8.5 these network ensembles constitute the least biased network models satisfying a given set of constraints. Therefore, this framework is widely used to construct null models of networks.

B.1.2 Definitions

Given the set of all the networks $G = (V, E)$ with $|V| = N$ nodes, a network ensemble is specified when a probability $P(G)$ is given to each network of this set.

The entropy S of an ensemble is the logarithm of the number of typical networks in the ensemble and is given by

$$S = - \sum_G P(G) \ln P(G). \tag{B.1}$$

A maximum-entropy network ensemble satisfying a given set of constraints is the most unbiased ensemble displaying the desired properties.

B.1.3 Constraints

We distinguish between *soft constraints* and *hard constraints*.
The soft constraints are the constraints satisfied on average over the ensemble of networks. The hard constraints are the constraints satisfied by each network in the ensemble.

Examples of hard constraints are constraints of the type

$$F_\mu(G) = C_\mu, \tag{B.2}$$

where $F_\mu(G)$ is a network measure whose value is fixed to be C_μ and each independent constraint imposed on the network is indicated by $\mu = 1, 2, \ldots, N_C$. A constraint of this type can fix the total number of links L or the degree k_i of a node i. Indicating with a_{ij} the adjacency matrix of the network we have for these two types of constraint

$$F(G) = \sum_{i<j} a_{ij} = L,$$

$$F_i(G) = \sum_{j=1}^{N} a_{ij} = k_i. \tag{B.3}$$

Soft constraints are constraints that instead fix the average value of a network measure $F_\mu(G)$ in the network ensemble and are therefore of the type

$$\sum_G P(G)F_\mu(G) = C_\mu. \tag{B.4}$$

A constraint of this type can fix the average number of links L of network in the ensemble of the average degree k_i of node i. Therefore, corresponding to these two types of constraint we will have

$$\sum_G P(G)\left[\sum_{i<j} a_{ij}\right] = L,$$

$$\sum_G P(G)\left[\sum_{j=1}^{N} a_{ij}\right] = k_i. \tag{B.5}$$

B.2 Canonical and microcanonical network ensembles

B.2.1 Canonical network ensemble

The probability $P_C(G)$ in a canonical network ensemble satisfying N_C constraints

$$\sum_G P_C(G)F_\mu(G) = C_\mu \tag{B.6}$$

with $\mu = 1, 2 \ldots, N_C$ has the exponential form given by

$$P_C(G) = \frac{1}{Z_C} e^{-\sum_{\mu=1}^{N_C} \lambda_\mu F_\mu(G)} \tag{B.7}$$

where Z_C is the normalization sum and where the Lagrangian multipliers λ_μ are fixed by the constraints in Eq. (B.6).

In order to find $P_C(G)$ we maximize the entropy of the ensemble S under the constraints given by Eq. (B.6) and the condition that $P_C(G)$ is normalized. We define a functional $\mathcal{F}$ in which we have introduced the Lagrangian multipliers λ_μ,

$$\mathcal{F} = -\sum_G P_C(G)\ln P_C(G) - \sum_\mu \lambda_\mu \left(\sum_G P_C(G)F_\mu(G) - C_\mu\right)$$

$$- \nu \left(\sum_G P_C(G) - 1\right). \tag{B.8}$$

Differentiating with respect to $P_C(G)$ and setting the partial derivative to zero,

$$\frac{\partial \mathcal{F}}{\partial P_C(G)} = -\ln P_C(G) - 1 - \sum_\mu \lambda_\mu F_\mu(G) - \nu = 0,$$ (B.9)

we get

$$P_C(G) = e^{-\sum_\mu \lambda_\mu F_\mu(G) - \nu - 1}.$$ (B.10)

Differentiating $\mathcal{F}$ with respect to the Lagrangian multipliers λ_μ and the Lagrangian multiplier ν and setting the partial derivatives to zero, we get the constraint defined by Eq. (B.6) and the normalization condition for $P_C(G)$. Imposing the normalization condition we fix the chemical potential ν and have

$$P_C(G) = \frac{1}{Z_C} e^{-\sum_\mu \lambda_\mu F_\mu(G)},$$ (B.11)

where

$$Z_C = \sum_G e^{-\sum_\mu \lambda_\mu F_\mu(G)}.$$ (B.12)

Here the Lagrangian multipliers λ_μ are determined imposing the constraints

$$\sum_G P_C(G) F_\mu(G) = \sum_G \frac{1}{Z} e^{-\sum_{\mu=1}^{N_C} \lambda_\mu F_\mu(G)} F_\mu(G) = C_\mu.$$ (B.13)

B.2.2 Examples of canonical network ensembles

First let us consider the canonical network ensemble in which the average number of links are fixed. Therefore, the ensemble satisfies only the constraint

$$\sum_G P_C(G) \left[\sum_{i<j} a_{ij} \right] = L.$$ (B.14)

The probability $P_C(G)$ in this ensemble is given by

$$P_C(G) = \frac{1}{Z_C} e^{-\lambda \sum_{i<j} a_{ij}}.$$ (B.15)

Since the sum over all graphs G can be expressed as a sum over all elements of the adjacency matrix $\mathbf{a}$, the normalization sum Z_C is given by

$$Z_C = \sum_{\mathbf{a}} e^{-\lambda \sum_{i<j} a_{ij}} = \left(\sum_{a_{ij}=0,1} e^{-\lambda a_{ij}} \right)^{N(N-1)/2} = \left(1 + e^{-\lambda}\right)^{N(N-1)/2}.$$ (B.16)

The probability of each link is given by

$$p = \sum_{\mathbf{a}} a_{ij} P_C(G) = \frac{e^{-\lambda}}{1 + e^{-\lambda}}.$$ (B.17)

The quantity p, or equivalently the quantity λ, is fixed by the condition

$$pN(N-1)/2 = L. \tag{B.18}$$

Therefore we have that this is the $G(N, p)$ ensemble, in which the probability $P_C(G)$ is given by

$$P_C(G) = p^L(1-p)^{N(N-1)/2-L}. \tag{B.19}$$

Let us now consider the case in which we fix the average degree sequence $\{k_i\}_{i=1,2,\dots,N}$. The network ensemble satisfies the N soft constraints

$$\sum_G P_C(G)\left[\sum_{j=1}^{N} a_{ij}\right] = k_i \tag{B.20}$$

with $i = 1, 2\dots, N$. The probability of a network in the ensemble is given by

$$P_C(G) = \frac{1}{Z_C} e^{-\sum_{i=1}^{N} \lambda_i \sum_{j=1}^{N} a_{ij}} \tag{B.21}$$

where the normalization constant Z_C is given by

$$Z_C = \sum_{\mathbf{a}} e^{-\sum_{i=1}^{N} \lambda_i \sum_{j=1}^{N} a_{ij}} = \sum_{\mathbf{a}} e^{-\sum_{i<j}(\lambda_i+\lambda_j)a_{ij}}$$

$$= \prod_{i<j}\left(1 + e^{-\lambda_i-\lambda_j}\right). \tag{B.22}$$

The probability p_{ij} of a link between node i and node j is given by

$$p_{ij} = \sum_{\mathbf{a}} a_{ij}\left[\frac{1}{Z_C} e^{-\sum_{r<s}(\lambda_r+\lambda_s)a_{rs}}\right] = \frac{e^{-\lambda_i-\lambda_j}}{1 + e^{-\lambda_i-\lambda_j}}. \tag{B.23}$$

The Lagrangian multipliers are fixed by the conditions

$$\sum_{j=1}^{N} p_{ij} = \sum_{j=1}^{N} \frac{e^{-\lambda_i-\lambda_j}}{1 + e^{-\lambda_i-\lambda_j}} = k_i. \tag{B.24}$$

The probability $P_C(G)$ takes the form

$$P_C(G) = \prod_{i<j} p_{ij}^{a_{ij}}\left(1 - p_{ij}\right)^{1-a_{ij}}. \tag{B.25}$$

The entropy takes the form

$$S = -\sum_{i<j} p_{ij} \ln p_{ij} - \sum_{i<j}\left(1 - p_{ij}\right)\ln\left(1 - p_{ij}\right). \tag{B.26}$$

If the maximum degree K of the network is below the structural cutoff, i.e. if

$$K \ll \sqrt{\langle k\rangle N}, \tag{B.27}$$

we have that in the first approximation

$$e^{-\lambda_i} = \frac{k_i}{\sqrt{\langle k \rangle N}} \ll 1, \tag{B.28}$$

and therefore

$$p_{ij} = \frac{k_i k_j}{\langle k \rangle N}. \tag{B.29}$$

Therefore, in this limit the network ensemble is a random uncorrelated network with degree sequence $\{k_i\}$.

B.2.3 Microcanonical network ensemble

The maximum-entropy network ensemble in which every network with non-zero probability satisfies the hard constraints

$$F_\mu(G) = C_\mu, \tag{B.30}$$

with $\mu = 1, 2, \ldots, N_C$ assigns to each network G of N nodes the probability $P_M(G)$ given by

$$P_M(G) = \frac{1}{Z_M} \prod_{\mu=1}^{N_C} \delta \left(F_\mu(G), C_\mu \right), \tag{B.31}$$

where $\delta(x, y)$ is the Kronecker delta and Z_M indicates the total number of networks satisfying the hard constraints defined in Eq. (B.30). In fact, the most unbiased ensemble satisfying the hard constraints defined above is the ensemble that attributes the same non-zero probability to each network satisfying the constraints and zero probability to all the other networks.

The entropy Σ of the microcanonical ensemble is given by

$$\Sigma = \ln Z_M. \tag{B.32}$$

The relation between Σ and the entropy S of the conjugated canonical ensemble given by

$$S = -\sum_G P_C(G) \ln P_C(G) \tag{B.33}$$

is simply given by

$$\Sigma = S - \Omega \tag{B.34}$$

where Ω is given by the absolute value of the logarithm of the probability that networks of the canonical ensembles satisfy the hard constraints, i.e.

$$\Omega = -\ln \left[\sum_G P_C(G) \delta \left(F_\mu(G), C_\mu \right) \right]. \tag{B.35}$$

In order to derive this result let us first show that since the probability $P_C(G)$ is given by Eq. (B.7) the entropy S of the canonical ensemble is given by [7]

$$S = \sum_G P_C(G) \left[\sum_\mu \lambda_\mu F_\mu(G) + \ln Z_C \right]$$
$$= \sum_\mu \lambda_\mu C_\mu + \ln Z_C. \tag{B.36}$$

If we start from the definition of Ω given by Eq. (B.35) we have

$$e^{-\Omega} = \sum_G P_C(G) \delta \left(F_\mu(G), C_\mu \right)$$
$$= \sum_G \frac{1}{Z_C} e^{-\sum_\mu \lambda_\mu F_\mu(G)} \delta \left(F_\mu(G), C_\mu \right)$$
$$= \frac{1}{Z_C} e^{-\sum_\mu \lambda_\mu C_\mu} \sum_G \delta \left(F_\mu(G), C_\mu \right)$$
$$= e^{-S+\Sigma}. \tag{B.37}$$

Therefore we get

$$\Sigma = S - \Omega. \tag{B.38}$$

When the number of imposed constraints is growing linearly with the number of nodes, it is found that Ω is extensive and cannot be neglected in the large network limit, proving the non-equivalence of the canonical and microcanonical network ensemble. In the case in which the imposed constraints are linear it is possible to calculate Ω [46, 7]. In the case in which the constraints fix the degree sequence of the network Ω is given by Eq. (2.96) as long as the network is uncorrelated.

Appendix C
Growing Multiplex Networks: the Master Equation

C.1   Master equation

In this appendix we follow Ref. [228] and we will discuss the general formalism to derive the exact joint degree distribution of growing multiplex networks with linear generalized preferential attachment defined in Sec. 10.2.2. Furthermore, we will provide detail for the explicit derivation of the joint degree distribution of the growing duplex network model with $c_{1,1} = c_{2,2} = 1$, i.e. Eq. (10.9).

To derive the exact degree distribution of growing multiplex network models we will extend the master equation approach defined in Appendix A to the multiplex network scenario. For simplicity, we will consider a duplex network with $M = 2$ layers. We start from a small connected network and at each time we add a node which brings, at the same time, m new links in layer 1 and m new links in layer 2. We assume that at each time the expected number of new links in layer one attached to a node i of degree $k_i^{[1]}$ in layer 1 and degree $k_i^{[2]}$ in layer 2 is given by $m\Pi^{[1]}(k_i^{[1]}, k_i^{[2]})$. Similarly, the expected number of new links in layer 2 attached to a node i is given by $m\Pi^{[2]}(k_i^{[1]}, k_i^{[2]})$. By assuming that the probability of attaching a new link to a node i either in layer 1 or in layer 2 is small, i.e. $\Pi^{[1]}(k_i^{[1]}, k_i^{[2]}) \ll 1$ and $\Pi^{[2]}\left(k_i^{[1]}, k_i^{[2]}\right) \ll 1$, we can neglect the probability that a single node i acquires a link in both layers in the same time step. In these conditions the master equation of the average number of nodes $N_{k,q}(t)$ that at time t have degrees $k^{[1]} = k$ and $k^{[2]} = q$ is given by

$$
\begin{aligned}
N_{k,q}(t+1) = {} & N_{k,q}(t) + m\Pi^{[1]}(k-1, q)N_{k-1,q}(t)[1 - \delta(k, m)] \\
& + m\Pi^{[2]}(k, q-1)N_{k,q-1}(t)[1 - \delta(q, m)] \\
& - m\left[\Pi^{[1]}(k, q) + \Pi^{[2]}(k, q)\right]N_{k,q}(t) + \delta(k, m)\delta(q, m)
\end{aligned}
\tag{C.1}
$$

for $k \geq m$ and $q \geq m$, where $\delta(x, y)$ indicates the Kronecker delta. By considering the model with linear generalized preferential attachment defined in Eq. (10.1) it is straightforward to derive the expression of $\Pi^{[\alpha]}(k, q)$ given by

$$
\Pi^{[1]}(k, q) = \frac{c_{1,1}k + c_{1,2}q}{\mathcal{N}^{[1]}}
$$

$$
\Pi^{[2]}(k, q) = \frac{c_{2,1}k + c_{2,2}q}{\mathcal{N}^{[2]}},
\tag{C.2}
$$

where $c_{\alpha,\beta} \in [0, 1]$ with $c_{1,1} + c_{1,2} = c_{2,1} + c_{2,2} = 1$ and $\mathcal{N}^{[1]}, \mathcal{N}^{[2]}$ are normalization constants equal to the double of the number of links in each layer.

C.2   Derivation of the exact degree distribution

C.2.1   General case

Since at each time we add exactly m links to each layer, we have that $\mathcal{N}^{[1]}, \mathcal{N}^{[2]}$ can be approximated for large times $t \gg 1$ by

$$\mathcal{N}^{[\alpha]} \simeq 2mt, \text{ for } \alpha = 1, 2. \tag{C.3}$$

Therefore, by putting

$$\Pi^{[1]}(k, q) = \frac{c_{1,1}k + c_{1,2}q}{2mt} = \frac{A_{k,q}}{mt},$$

$$\Pi^{[2]}(k, q) = \frac{c_{2,1}k + c_{2,2}q}{2mt} = \frac{B_{k,q}}{mt}. \tag{C.4}$$

The master equation reads

$$N_{k,q}(t+1) = N_{k,q}(t) + \frac{A_{k-1,q}}{t} N_{k-1,q}(t)[1 - \delta(k, m)] + \frac{B_{k,q-1}}{t} N_{k,q-1}(t)[1 - \delta(q, m)]$$

$$- \left[\frac{A_{k,q}}{t} + \frac{B_{k,q}}{t} \right] N_{k,q}(t) + \delta(k, m)\delta(q, m), \tag{C.5}$$

for $k \geq m$ and $q \geq m$. Assuming that $N_{k,q} = tP(k, q)$ is valid in the large time limit $t \gg 1$, we can solve for the combined degree distribution $P(k, q)$ indicating the probability that a node has at the same time degree k in layer 1 and degree q in layer 2. We get that the joint degree distribution is determined by the recursive equations

$$P(m, q) = \left(\prod_{j=m+1}^{q} \frac{B_{m,j-1}}{1 + A_{m,j} + B_{m,j}} \right) P(m, m),$$

$$P(k, q) = \sum_{r=m}^{q} \left(\prod_{j=r+1}^{q} \frac{B_{k,j-1}}{1 + A_{k,j} + B_{k,j}} \right) \frac{A_{k-1,r}}{1 + A_{k,r} + B_{k,r}} P(k - 1, r). \tag{C.6}$$

C.2.2   Case $c_{1,1} = c_{2,2} = 1$

Let us now consider the linear generalized preferential attachment, in which $c_{1,1} = c_{2,2} = 1$ and $c_{1,2} = c_{2,1} = 0$. In this case we have

$$A_{k,q} = \frac{k}{2},$$

$$B_{k,q} = \frac{q}{2}. \tag{C.7}$$

The recursive Eqs (C.6) read

$$P(m, q) = \frac{\Gamma(q)\Gamma(3 + 2m)}{\Gamma(m)\Gamma(3 + q + m)} P(m, m),$$

$$P(k, q) = \sum_{r=m}^{q} \left(\frac{\Gamma(q)\Gamma(3 + r + k)}{\Gamma(r)\Gamma(3 + q + k)} \right) \frac{k - 1}{2 + k + q} P(k - 1, r) \tag{C.8}$$

where $P(m, m)$ is fixed by the normalization condition $\sum_{k=m}^{\infty} \sum_{q=m}^{\infty} P(k, q) = 1$. Using the relation

$$\sum_{r=m}^{q} \frac{\Gamma(k + r - 2m)}{\Gamma(r - m + 1)\Gamma(k - m)} = \frac{\Gamma(k + q - 2m + 1)}{\Gamma(k - m + 1)\Gamma(q - m + 1)} \tag{C.9}$$

it can be proved recursively that the joint degree distribution $P(k, q)$ is given by

$$P(k, q) = \frac{2\Gamma(2 + 2m)}{\Gamma(m)\Gamma(m)} \frac{\Gamma(k + q - 2m + 1)}{\Gamma(k + q + 3)} \frac{\Gamma(q)}{\Gamma(q - m + 1)} \frac{\Gamma(k)}{\Gamma(k - m + 1)}. \tag{C.10}$$

Summing over the degree in layer two we can find the degree distribution $P(k)$ in layer one, i.e. $P(k) = \sum_{q=m}^{\infty} P(k, q)$ obtaining the known result for a single layer,

$$P(k) = \frac{2m(1 + m)}{k(k + 1)(k + 2)}. \tag{C.11}$$

Finally, the conditional average $\langle k|q \rangle = \langle k^{[1]}|k^{[2]} \rangle$ quantifying the degree correlations among the two different networks is given by

$$\langle k|q \rangle = \frac{\sum_{k=m}^{\infty} kP(k, q)}{\sum_{k=m}^{\infty} P(k, q)} = \frac{m}{1 + m}(q + 2). \tag{C.12}$$

Similar expressions are obtained for $P(q)$ and $\langle q|k \rangle$ by summing Eq. (C.10) over k.

Appendix D
Percolation of interdependent networks

D.1 General remarks

In this Appendix we follow Ref. [76] and we derive the equations determining the fraction of nodes in the MCGC in a generic multiplex network with link overlap when the probability that each node is damaged is given by $f = 1 - p$. Additionally, we predict the behaviour of this percolation transition on multiplex network ensembles with given multidegree distribution $P(\{k^{\vec{m}}\})$. Therefore, this Appendix complements the discussion presented in Sec. 11.4.

D.2 The message-passing algorithm

We consider a multiplex network with M layers and adjacency matrix $\mathbf{a}^{[\alpha]}$ in each layer $\alpha = 1, 2, \ldots, M$ which is locally tree-like. Initially, we assume that we know the set of nodes that are initially damaged. The configuration of the initial damage is indicated by the variables $\{s_i\}$ where $s_i = 0$ ($s_i = 1$) if node i is (is not) damaged. The message-passing algorithm for given initial damage configuration (see Sec. 11.4) determines whether node i belongs ($\sigma_i = 1$) or does not belong ($\sigma_i = 0$) to the Mutually Connected Giant Component (MCGC), as long as the multiplex network is locally tree-like. The algorithm requires the determination of the set of messages

$$\vec{n}_{i \to j} = \left(n_{i \to j}^{[1]}, n_{i \to j}^{[2]}, \ldots, n_{i \to j}^{[\alpha]}, \ldots, n_{i \to j}^{[M]} \right) \tag{D.1}$$

going from node i to node j connected at least in one layer. Each message $n_{i \to j}^{[\alpha]}$ indicates whether $\left(n_{i \to j}^{[\alpha]} = 1 \right)$ or not $\left(n_{i \to j}^{[\alpha]} = 0 \right)$ node i connects node j to the MCGC through links in layer α. The messages $\vec{n}_{i \to j}$ and the indicator function σ_i are determined by the message-passing algorithm defined by the Eqs (11.32), (11.33) and (11.34).

For the further derivation it is useful to consider an alternative formulation of the message-passing algorithm for a given configuration of the initial disorder. This alternative formulation will allow us to perform easily the average over random initial damage configurations. To this end, we introduce the variable $\sigma_{i \to j}^{\vec{m}, \vec{n}}$ which indicates whether $\left(\sigma_{i \to j}^{\vec{m}, \vec{n}} = 1 \right)$ or not $\left(\sigma_{i \to j}^{\vec{m}, \vec{n}} = 0 \right)$ node i sends to node j the messages

$$\vec{n}_{i \to j} = \vec{n}$$

given that node i and node j are linked by a multilink

$$\vec{m} = \vec{m}_{ij} = \left(a_{ij}^{[1]}, a_{ij}^{[2]}, \ldots, a_{ij}^{[\alpha]}, \ldots, a_{ij}^{[M]} \right). \tag{D.2}$$

According to Eqs (11.32) and (11.33) a node i, in order to send a message $\vec{n} \neq \vec{0}$, should be connected to the MCGC by nodes different from node j in all the layers where $n^{[\alpha]} = 1$ and in all the layers where $m^{[\alpha]} = 0$. In fact, the first requirement is necessary for having $n^{[\alpha]} = 1$ and the second requirement is necessary for having $v_{i \to j} = M$ as requested by Eq. (11.32) because $m^{[\alpha]} = a_{ij}^{[\alpha]} = 0$. Additionally, in every layer α in which $m^{[\alpha]} = a_{ij}^{[\alpha]} = 1$ and $n^{[\alpha]} = 0$, node i must not receive any positive messages from neighbouring nodes different from node j. Therefore, we have for $\vec{n} \neq \vec{0}$

$$\sigma_{i \to j}^{\vec{m},\vec{n}} = s_i \prod_{\alpha=1}^{M} \left\{ \left(m^{[\alpha]}\right)^{n^{[\alpha]}} \left[1 - \prod_{\ell \in N(i) \setminus j} \left(1 - n_{\ell \to i}^{[\alpha]}\right)\right]^{n^{[\alpha]} m^{[\alpha]} + (1 - m^{[\alpha]})} \right.$$

$$\left. \times \left[\prod_{\ell \in N(i) \setminus j} \left(1 - n_{\ell \to i}^{[\alpha]}\right)\right]^{(1 - n^{[\alpha]}) m^{[\alpha]}} \right\}, \tag{D.3}$$

while for $\vec{n} = \vec{0}$ we have

$$\sigma_{i \to j}^{\vec{m},\vec{0}} = 1 - \sum_{\vec{n} \neq \vec{0}} \sigma_{i \to j}^{\vec{m},\vec{n}}. \tag{D.4}$$

Note that out of the messages $\sigma_{i \to j}^{\vec{m},\vec{n}}$ with different value of $\vec{n}$ only one (corresponding to $\vec{n} = \vec{n}_{i \to j}$) has value one and all the others are zero, therefore

$$\vec{n}_{i \to j} = \mathrm{argmax}_{\vec{n}} \sigma_{i \to j}^{\vec{m},\vec{n}}. \tag{D.5}$$

Let us now further modify Eq. (D.3) by using the identity valid for $q^{[\alpha]}$ taking values $q^{[\alpha]} = 0, 1$

$$\prod_{\alpha=1}^{M} (1 - z_\alpha)^{q^{[\alpha]}} = \prod_{\alpha \mid q^{[\alpha]} > 0} (1 - z_\alpha) = \sum_{\vec{r} \mid r^{[\alpha]} = 0 \text{ if } q^{[\alpha]} = 0} (-1)^{\sum_{\alpha=1}^{M} r^{[\alpha]}} (z_\alpha)^{r^{[\alpha]}}, \tag{D.6}$$

where the sum in the last term is over all the vectors

$$\vec{r} = \left(r^{[1]}, r^{[2]}, \ldots, r^{[\alpha]}, \ldots, r^{[M]}\right) \tag{D.7}$$

of elements $r^{[\alpha]} = 0, 1$ for $q^{[\alpha]} = 1$ and $r^{[\alpha]} = 0$ for $q^{[\alpha]} = 0$. Using this relation in Eq. (D.3) we obtain

$$\sigma_{i \to j}^{\vec{m},\vec{n}} = s_i \sum_{\vec{r} \mid r^{[\alpha]} = 0 \text{ if } (1 - n^{[\alpha]}) m^{[\alpha]} = 1} \left\{ \left[\prod_{\alpha=1}^{M} \left(m^{[\alpha]}\right)^{n^{[\alpha]}}\right] \right.$$

$$\left. \times (-1)^{\sum_{\alpha=1}^{M} r^{[\alpha]}} \prod_{\ell \in N(i) \setminus j} \prod_{\alpha=1}^{M} \left(1 - n_{\ell \to i}^{[\alpha]}\right)^{r^{[\alpha]} + m^{[\alpha]} (1 - n^{[\alpha]})} \right\}. \tag{D.8}$$

Since between all the messages $\sigma_{i \to j}^{\vec{m},\vec{n}}$ sent between node i to node j only one message is equal to one, we have

$$\sigma_{i\rightarrow j}^{\vec{m},\vec{n}} = s_i \sum_{\vec{r}|r^{[\alpha]}=0 \text{ if } (1-n^{[\alpha]})m^{[\alpha]}=1} \left\{ \left[\prod_{\alpha=1}^{M} \left(m^{[\alpha]} \right)^{n^{[\alpha]}} \right] \right.$$

$$\times (-1)^{\sum_{\alpha=1}^{M} r^{[\alpha]}} \prod_{\ell\in N(i)\backslash j} \left(1 - \sum_{\vec{n}'|\sum_\alpha (n')^{[\alpha]}[r^{[\alpha]}+(1-n^{[\alpha]})m^{[\alpha]}]>0} \sigma_{\ell\rightarrow i}^{\vec{m}_{\ell i},\vec{n}'} \right) \right\}. \tag{D.9}$$

Let us now assume that the initial configuration of the damage $\{s_i\}$ is drawn from probability $P(\{s_i\})$ given by

$$P(\{s_i\}) = \prod_{i=1}^{N} p^{s_i} (1-p)^{1-s_i}. \tag{D.10}$$

By averaging the messages $\sigma_{i\rightarrow j}^{\vec{m},\vec{n}}$ over the distribution $P(\{s_i\})$ we can formulate a different message-passing algorithm able to predict the probability $\hat{\sigma}_i$ that a random node belongs to the MCGC for a random realization of the initial disorder. In this case the generic message $\hat{\sigma}_{i\rightarrow j}^{\vec{m}_{ij},\vec{n}}$ indicates the probability that node i connects node j to the MCGC in the layers α where $n^{[\alpha]} = 1$. These messages $\hat{\sigma}_{i\rightarrow j}^{\vec{m}_{ij},\vec{n}}$ are given by the average of the messages $\sigma_{i\rightarrow j}^{\vec{m},\vec{n}}$ over the random realization of the initial disorder. Therefore, they satisfy the following recursive equations:

$$\hat{\sigma}_{i\rightarrow j}^{\vec{m},\vec{n}} = p \sum_{\vec{r}|r^{[\alpha]}=0 \text{ if } (1-n^{[\alpha]})m^{[\alpha]}=1} \left\{ \left[\prod_{\alpha=1}^{M} \left(m^{[\alpha]} \right)^{n^{[\alpha]}} \right] (-1)^{\sum_\alpha r^{[\alpha]}} \right.$$

$$\times \prod_{\ell\in N(i)\backslash j} \left(1 - \sum_{\vec{n}'|\sum_\alpha (n')^{[\alpha]}[r^{[\alpha]}+(1-n^{[\alpha]})m^{[\alpha]}]>0} \hat{\sigma}_{\ell\rightarrow i}^{\vec{m}_{\ell i},\vec{n}'} \right) \right\}. \tag{D.11}$$

Similarly, the probability $\hat{\sigma}_i$ that node i is in the MCGC is the average of the indicator function σ_i over the distribution $P(\{s_i\})$, and satisfies

$$\hat{\sigma}_i = p \sum_{\vec{r}} (-1)^{\sum_\alpha r^{[\alpha]}} \left[\prod_{\ell\in N(i)} \left(1 - \sum_{\vec{n}'|\sum_\alpha (n')^{[\alpha]} r^{[\alpha]}>0} \hat{\sigma}_{\ell\rightarrow i}^{\vec{m}_{\ell i},\vec{n}'} \right) \right]. \tag{D.12}$$

Further averaging these equations over the ensemble of multiplex networks with multidegree distribution $P(\{k^{\vec{m}}\})$ gets that the probability $S_{\vec{m},\vec{n}}$ indicating the average of the messages $\hat{\sigma}_{i\rightarrow j}^{\vec{m}_{ij},\vec{n}}$ over the network ensemble is given by

$$S_{\vec{m},\vec{n}} = p \sum_{\{k^{\vec{m}}\}} \frac{k^{\vec{m}}}{\langle k^{\vec{m}} \rangle} P\left(\{k^{\vec{m}}\}\right) \sum_{\vec{r}|r^{[\alpha]}=0 \text{ if } m^{[\alpha]}(1-n^{[\alpha]})=1} \left\{ (-1)^{\sum_{\alpha=1}^{M} r^{[\alpha]}} \right.$$

$$\times \prod_{\vec{m}'\neq\vec{m}} \left(1 - \sum_{\vec{n}'|\sum_\alpha n^{[\alpha]'}[r^{[\alpha]}+m^{[\alpha]}(1-n^{[\alpha]})]>0} S_{\vec{m}',\vec{n}'} \right)^{k^{\vec{m}'}}$$

$$\times \left(1 - \sum_{\vec{n}'|\sum_\alpha n^{[\alpha]'}[r^{[\alpha]}+m^{[\alpha]}(1-n^{[\alpha]})]>0} S_{\vec{m}',\vec{n}'} \right)^{k^{\vec{m}}-1} \right\}, \tag{D.13}$$

as long as $\vec{m} \neq \vec{0}$. Similarly, the probability S that a random node is in the MCGC is given by

$$S = p \sum_{\{k^{\vec{m}}\}} P\left(\{k^{\vec{m}}\}\right) \sum_{\vec{r}} \left\{ (-1)^{\sum_{\alpha=1}^{M} r^{[\alpha]}} \prod_{\vec{m}} \left(1 - \sum_{\vec{n}| \sum_{\alpha} n^{[\alpha]} r^{[\alpha]} > 0} S_{\vec{m},\vec{n}}\right)^{k^{\vec{m}}} \right\}. \tag{D.14}$$

For networks with an uncorrelated multidegree distribution $P(\{k^{\vec{m}}\})$ given by Eq. (10.53) we get

$$S_{\vec{m},\vec{n}} = p \sum_{\vec{r}|r_\alpha=0 \text{ if } m^{[\alpha]}(1-n^{[\alpha]})=1} \left\{ (-1)^{\sum_{\alpha=1}^{M} r^{[\alpha]}} G_{\vec{m}}^1 \left(1 - \sum_{\vec{n}'| \sum_{\alpha} n^{[\alpha]'}[r^{[\alpha]}+m^{[\alpha]}(1-n^{[\alpha]})]>0} S_{\vec{m}',\vec{n}'}\right) \right.$$

$$\left. \times \prod_{\vec{m}' \neq \vec{m}} G_{\vec{m}'}^0 \left(1 - \sum_{\vec{n}'| \sum_{\alpha} n^{[\alpha]'}[r^{[\alpha]}+m^{[\alpha]}(1-n^{[\alpha]})]>0} S_{\vec{m}',\vec{n}'}\right) \right\} \tag{D.15}$$

with $\vec{m} \neq \vec{0}$ and

$$S = p \sum_{\{k^{\vec{m}}\}} \sum_{\vec{r}} (-1)^{\sum_{\alpha=1}^{M} r^{[\alpha]}} \prod_{\vec{m}} G_{\vec{m}}^0 \left(1 - \sum_{\vec{n}| \sum_{\alpha} n^{[\alpha]} r^{[\alpha]}>0} S_{\vec{m},\vec{n}}\right). \tag{D.16}$$

Here the generating function $G_0^{\vec{m}}(z)$ and $G_1^{\vec{m}}(z)$ are given by

$$G_0^{\vec{m}}(z) = \sum_k P^{\vec{m}}(k) z^k,$$

$$G_1^{\vec{m}}(z) = \sum_k \frac{k}{\langle k^{\vec{m}}\rangle} P^{\vec{m}}(k) z^{k-1}. \tag{D.17}$$

D.3 Percolation of duplex network

In this section we apply the formalism presented in the previous paragraph to study the percolation of an interdependent duplex network with given multidegree distribution. We consider here a duplex network with Poisson multidegree distribution and $\langle k^{(1,0)}\rangle = \langle k^{(0,1)}\rangle = c_1$ with $\langle k^{(1,1)}\rangle = c_2$. In this case, the dynamical variables that we have to consider are

$$S = S_{\vec{m},\vec{m}} = x/p \tag{D.18}$$

with $\vec{m} \neq \vec{0}$ and

$$S_{\vec{1},(1,0)} = S_{\vec{1},(0,1)} = x_{2,1}/p. \tag{D.19}$$

The equations for x and $x_{2,1}$ read

$$\begin{pmatrix} F_1(\mathbf{x}) \\ F_2(\mathbf{x}) \end{pmatrix} = \mathbf{F}(\mathbf{x}) = 0,$$

where the functions $F_1(\mathbf{x})$ and $F_2(\mathbf{x})$ are given by

$$F_1(\mathbf{x}) = x - \left(1 - 2e^{-\hat{c}_1 x - \hat{c}_2(x+x_{2,1})} + e^{-2\hat{c}_1 x - \hat{c}_2(x+2x_{2,1})}\right),$$

$$F_2(\mathbf{x}) = u - \left(e^{-\hat{c}_1 x - \hat{c}_2(x+x_{2,1})} - e^{-2\hat{c}_1 x - \hat{c}_2(x+2x_{2,1})}\right), \tag{D.20}$$

the vector $\mathbf{x}$ is given by

$$\mathbf{x} = \begin{pmatrix} x \\ x_{2,1} \end{pmatrix}$$

and $\hat{c}_1 = c_1 p$ and $\hat{c}_2 = c_2 p$.

The points of discontinuous and hybrid phase transition can be found by imposing the set of equations

$$\mathbf{F}(\mathbf{x}^{\star}) = \mathbf{0}$$

$$\det \mathbf{J}|_{\mathbf{x}=\mathbf{x}^{\star}} = 0, \tag{D.21}$$

where $\mathbf{J}$ is the Jacobian matrix of $\mathbf{F}(\mathbf{x})$.

The critical point of continuous phase transition can be found by imposing

$$\det \mathbf{J}|_{\mathbf{x}=\mathbf{0}} = 0. \tag{D.22}$$

This equation can be expressed explicitly as

$$1 - 2\hat{c}_2 + \hat{c}_2^2 = 0 \tag{D.23}$$

and has a unique real solution for $\hat{c}_2 = 1$. The resulting phase diagram is shown in Fig. 11.11). In this case the only duplex network displaying a continuous transition is the one with complete overlap, i.e. having $c_1 = 0$, for which we observe the emergence of the MCGC at $pc_2 = 1$.

Appendix E
Directed Percolation of Interdependent Networks

E.1 General remarks

In this Appendix we provide the derivation of the equations determining the Directed Mutually Connected Giant Component (DMCGC) introduced in Sec. 12.3 for a multiplex network with link overlap when the probability that each node is damaged is given by $f = 1 - p$. Additionally, we discuss the directed percolation transition in multiplex network ensembles with given multidegree distribution $P\left(\{k^{\vec{m}}\}\right)$. In this Appendix we follow mainly Ref. [77].

E.2 The message-passing algorithm

Assuming that the multiplex network under consideration is locally tree-like and that the initial configuration of the damage of the nodes $\{s_i\}$ is known, it is possible, using the message-passing algorithm determined by Eqs (12.40) and (12.41) to predict which nodes belong and which nodes do not belong to the DMCGC.

By using the Eq. (D.6) in the message-passing equations Eq. (12.41) we can express the message $\sigma_{i \to j}$ sent from node i to node j as

$$\sigma_{i \to j} = s_i \sum_{\vec{r}} (-1)^{\sum_\alpha r^{[\alpha]}} \prod_{\alpha=1}^{M} \prod_{\ell \in N_\alpha(i) \setminus j} \left(1 - \sigma_{\ell \to i}\right)^{r^{[\alpha]}}, \tag{E.1}$$

where $N_\alpha(i)$ indicates the set of nodes that are neighbours of node i in layer α. Note that Eq. (E.1) the sum is performed over all vectors

$$\vec{r} = \left(r^{[1]}, r^{[2]}, \ldots, r^{[\alpha]}, \ldots, r^{[M]}\right) \tag{E.2}$$

with elements $r^{[\alpha]} = 0, 1$. Since a node ℓ is a neighbour of node i in layer α if and only if $m_{\ell i}^{[\alpha]} = 1$, we have

$$\sigma_{i \to j} = s_i \sum_{\vec{r}} (-1)^{\sum_\alpha r^{[\alpha]}} \prod_{\ell \neq j} \left(1 - \sigma_{\ell \to i}\right)^{\sum_{\alpha=1}^{M} r^{[\alpha]} m_{\ell i}^{[\alpha]}}. \tag{E.3}$$

By following similar steps it can be shown that the indicator function σ_i given by Eq. (12.40) can be also expressed as

$$\sigma_i = s_i \sum_{\vec{r}} (-1)^{\sum_\alpha r^{[\alpha]}} \prod_{\ell=1}^{N} \left(1 - \sigma_{\ell \to i}\right)^{\sum_{\alpha=1}^{M} r^{[\alpha]} m_{\ell i}^{[\alpha]}}. \tag{E.4}$$

Therefore, when the initial configuration of the damage $\{s_i\}$ is drawn for the distribution

$$P(\{s_i\}) = \prod_{i=1}^{N} p^{s_i}(1-p)^{1-s_i}, \tag{E.5}$$

the messages $\hat{\sigma}_{i\to j}$ indicating the average of $\sigma_{i\to j}$ and the probabilities $\hat{\sigma}_i$ that node i belongs to the DMCGC derived by averaging σ_i over the distribution $P(\{s_i\})$ satisfy the following set of recursive equations:

$$\hat{\sigma}_{i\to j} = p \sum_{\vec{r}} (-1)^{\sum_\alpha r^{[\alpha]}} \prod_{\ell \neq j} \left(1 - \hat{\sigma}_{\ell \to i}\right)^{\sum_{\alpha=1}^{M} r^{[\alpha]} m_{\ell i}^{[\alpha]}},$$

$$\hat{\sigma}_i = p \sum_{\vec{r}} (-1)^{\sum_\alpha r^{[\alpha]}} \prod_{\ell=1}^{N} \left(1 - \hat{\sigma}_{\ell \to j}\right)^{\sum_{\alpha=1}^{M} r^{[\alpha]} m_{\ell i}^{[\alpha]}}. \tag{E.6}$$

Finally, we can consider a random multiplex network with given multidegree distribution $P(\{k^{\vec{m}}\})$. In this random multiplex network the probability $S'_{\vec{m}}$ that by following a multilink $\vec{m}$ we reach a node in the DMCGC can be found by averaging the messages $\hat{\sigma}_{i\to j}$ over all multilinks $\vec{m}_{ij} = \vec{m}$ and the probability S that a random node is in the DMCGC can be found by averaging $\hat{\sigma}_i$. Therefore these quantities satisfy

$$S_{\vec{m}} = p \sum_{\{k^{\vec{m}'}\}} \frac{k^{\vec{m}}}{\langle k^{\vec{m}} \rangle} P\left(\{k^{\vec{m}}\}\right) \sum_{\vec{r}} (-1)^{\sum_{\alpha=1}^{M} r^{[\alpha]}} \prod_{\vec{m}' \mid \sum_\alpha m^{[\alpha]'} r^{[\alpha]} > 0} (1 - S_{\vec{m}'})^{k^{\vec{m}'} - \delta(\vec{m}, \vec{m}')},$$

$$S = p \sum_{\{k^{\vec{m}'}\}} P\left(\{k^{\vec{m}'}\}\right) \sum_{\vec{r}} (-1)^{\sum_{\alpha=1}^{M} r^{[\alpha]}} \prod_{\vec{m}' \mid \sum_\alpha m^{[\alpha]'} r^{[\alpha]} > 0} (1 - S_{\vec{m}'})^{k^{\vec{m}'}}. \tag{E.7}$$

For uncorrelated multidegrees of the nodes, when the distribution $P\left(\{k^{\vec{m}}\}\right)$ follows Eq. (10.53), these equations read

$$S_{\vec{m}} = p \sum_{\{k^{\vec{m}'}\}} \sum_{\vec{r}} (-1)^{\sum_{\alpha=1}^{M} r_\alpha} \left[G_1^{\vec{m}} (1 - S_{\vec{m}}) \right]^{f(\vec{m}, \vec{r})} \prod_{\vec{m}' \mid \sum_\alpha m^{[\alpha]'} r^{[\alpha]} > 0 \,\&\, \vec{m}' \neq \vec{m}} G_0^{\vec{m}'} (1 - S_{\vec{m}'}),$$

$$S = p \sum_{\{k^{\vec{m}'}\}} \sum_{\vec{r}} (-1)^{\sum_{\alpha=1}^{M} r^{[\alpha]}} \prod_{\vec{m} \mid \sum_\alpha m^{[\alpha]'} r^{[\alpha]} > 0} G_0^{\vec{m}'} (1 - S_{\vec{m}'}) \tag{E.8}$$

where $f(\vec{m}, \vec{r}) = 1$ if $\sum_\alpha r^{[\alpha]} m^{[\alpha]} > 0$ and otherwise $f(\vec{m}, \vec{r}) = 0$ and the generating function $G_0^{\vec{m}}(z)$ and $G_1^{\vec{m}}(z)$ are given by

$$G_0^{\vec{m}}(z) = \sum_k P^{\vec{m}}(k) z^k,$$

$$G_1^{\vec{m}}(z) = \sum_k \frac{k}{\langle k^{\vec{m}} \rangle} P^{\vec{m}}(k) z^{k-1}. \tag{E.9}$$

Note that this algorithm and therefore Eqs (E.8) reduce to the equations found in Sec. 11.3 in the absence of link overlap, i.e. where the only non-trivial multilinks are the ones with $\sum_{\alpha=1}^{M} m^{[\alpha]} = 1$.

E.3 Directed percolation of duplex network

We consider now the case of a duplex $M = 2$ in which the multidegree distributions are Poisson with $\langle k^{(1,1)} \rangle = c_2$, and $\langle k^{(0,1)} \rangle = \langle k^{(1,0)} \rangle = c_1$. Due to the properties of the Poisson distribution, we have $S = S_{\vec{m}}$, for every $\vec{m} \neq \vec{0}$, where S satisfies the equation

$$S = p\left[1 - 2e^{-(c_1+c_2)S} + e^{-(2c_1+c_2)S}\right]. \tag{E.10}$$

By setting $x = S/p$, and $\hat{c}_1 = c_1 p, \hat{c}_2 = c_2 p$ we can study the solutions of the equivalent equation

$$f(x) = x - \left[1 - 2e^{-(\hat{c}_1+\hat{c}_2)x} + e^{-(2\hat{c}_1+\hat{c}_2)x}\right] = 0 \tag{E.11}$$

in the $(\hat{c}_1, \hat{c}_2)$ parameter plane. The critical line of discontinuous and hybrid phase transition is found by solving the system of equations

$$f(x) = 0,$$
$$\left.\frac{df(x)}{dx}\right|_{x=x_c} = 0. \tag{E.12}$$

The critical line of continuous second-order phase transition is found by solving the equation

$$\left.\frac{df(x)}{dx}\right|_{x=0} = 0. \tag{E.13}$$

We note that there is a tricritical point for $c_2 p = 1, c_2/c_1 = \sqrt{2}$ characterized by satisfying

$$\left.\frac{df(x)}{dx}\right|_{x=0} = \left.\frac{d^2 f(x)}{dx^2}\right|_{x=0} = 0. \tag{E.14}$$

This point separates the line of discontinuous hybrid phase transitions and the line of continuous second-order phase transitions in the plane $(c_1 p, c_2 p)$ (see Fig. 12.4).

Appendix F
Immunization Strategies on Multiplex Networks

F.1 General remarks

In this Appendix we describe how the SIR model on multiplex networks discussed in Sec. 13.3.2 is modified when the effect of immunization strategies is taken into account.

Here we follow Ref. [315] and we characterize the effect of immunization strategies in the framework of the SIR epidemic spreading on multiplex networks with multiplex degree distribution $P(\mathbf{k})$. We therefore assume as in Sec. 13.3.2 that each node has the same dynamical state across layers and that each node has a given probability to be immunized, i.e. we consider a so-called node-based immunization.

F.2 Node-based immunization

In the node-based immunization strategy, each node i of the multiplex network is immunized with probability $\phi(\mathbf{k}_i)$ where $\mathbf{k}_i$ indicates the multiplex degree of node i. The node-based immunization can be random if the immunization probability is independent of the multiplex degree, i.e. $\phi(\mathbf{k}) = \phi_R$; alternatively, the node-based immunization can be targeted when $\phi(\mathbf{k})$ is a non-constant function of the multiplex degree $\mathbf{k}$.

When this immunization strategy is adopted, the SIR equations in a multiplex network without degree correlation can be written by mapping the epidemic spreading to bond percolation, generalizing the framework discussed in Sec. 13.3.2. In this case the equations for the fraction S of removed nodes at the end of the epidemics in a multiplex network without link overlap can be obtained to be

$$S = 1 - \sum_{\mathbf{k}} P(\mathbf{k})(1 - \phi(\mathbf{k})) \prod_{\alpha=1}^{M} \left(1 - T^{[\alpha]} S'_\alpha\right)^{k^{[\alpha]}}, \tag{F.1}$$

where $T^{[\alpha]}$ is the transmissibility of the infection in layer α and where S'_α satisfies

$$S'_\alpha = 1 - \sum_{\mathbf{k}} \frac{k^{[\alpha]}}{\langle k^{[\alpha]} \rangle} P(\mathbf{k})(1 - \phi(\mathbf{k})) \prod_{\beta=1}^{N} \left(1 - T^{[\beta]} S'_\beta\right)^{k^{[\beta]} - \delta(\alpha,\beta)}. \tag{F.2}$$

By indicating with Λ, the maximum eigenvalue, the Jacobian matrix of Eq. (F.2), we can predict that an epidemic will affect a finite fraction of the networks if and only if

$$\Lambda > 1. \tag{F.3}$$

By performing similar analytic steps discussed in the absence of the immunization, for a duplex network the condition for having an epidemic reads

$$\Lambda = \frac{1}{2}\left[T^{[1]}\kappa_1 + T^{[2]}\kappa_2 + \sqrt{\left(T^{[1]}\kappa_1 - T^{[2]}\kappa_2\right)^2 + 4T^{[1]}T^{[2]}\mathcal{K}_1\mathcal{K}_2} \right] > 1, \qquad (\text{F.4})$$

where now the parameters $\kappa_\alpha, \mathcal{K}_\alpha$ take into account the immunization protocols and are given by

$$\kappa_\alpha = \frac{\left\langle k^{[\alpha]}\left(k^{[\alpha]} - 1\right)(1 - \phi\,(\mathbf{k}))\right\rangle}{\langle k^{[\alpha]}\rangle},$$

$$\mathcal{K}_\alpha = \frac{\left\langle k^{[1]}k^{[2]}\,(1 - \phi\,(\mathbf{k}))\right\rangle}{\langle k^{[\alpha]}\rangle}. \qquad (\text{F.5})$$

Appendix G
Spectrum of the Supra-Laplacian

G.1 Diffusion equations

In this Appendix we follow Ref. [133] and we discuss in detail the different regimes of the diffusion process in multiplex networks described in Sec. 14.2. The diffusion equation in a multiplex network of M layer and N nodes is given by

$$\frac{d\mathbf{X}}{dt} = -\mathcal{L}\mathbf{X}, \tag{G.1}$$

where $\mathcal{L}$ is the supra-Laplacian matrix and $N \cdot M$-dimensional vector $\mathbf{X}$ indicates the dynamical state of the replica nodes. For simplicity, here we focus on the diffusion properties of duplex network with $M = 2$ layers. The supra-Laplacian of a duplex networks is a $2N \times 2N$ matrix of block structure given by

$$\mathcal{L} = \begin{pmatrix} \mathbf{L}^{[1]} + D_x\mathbf{I} & -D_x\mathbf{I} \\ -D_x\mathbf{I} & \mathbf{L}^{[2]} + D_x\mathbf{I} \end{pmatrix}, \tag{G.2}$$

where $L^{[1]}$ and $L^{[2]}$ are respectively the Laplacian matrices in layer 1 and in layer 2 and D_x is the interlayer diffusion constant. The dynamical state of the nodes can be written by distinguishing two blocks indicating respectively the dynamical state of the replica node in layer 1 and the dynamical state of the nodes in layer 2, i.e.

$$\mathbf{X} = \begin{pmatrix} \mathbf{x}^{[1]} \\ \mathbf{x}^{[2]} \end{pmatrix}. \tag{G.3}$$

Let us indicate with $0 = \lambda_1 \leq \lambda_2 \leq \lambda_3 \cdots \leq \lambda_N$ the N eigenvalues of the supra-Laplacian.

The typical timescale τ for relaxation of the diffusion process is given by the inverse of the smallest eigenvalue of the supra-Laplacian that is different from zero, i.e.

$$\tau = \frac{1}{\lambda_2}, \tag{G.4}$$

as long as there is a finite spectral gap in the supra-Laplacian spectrum. This result is a direct extension of the result obtained for single layers in Sec. 3.5.1. Let us discuss in the subsequent paragraphs the two distinct diffusion regime in the assumption that both layer 1 and layer 2 are formed by connected networks.

G.2 Small interlayer diffusion constant D_x

Let us first show that for small interlayer diffusion constant ($D_x \ll D_x^*$) the typical timescale for relaxation is given by

$$\tau = \frac{1}{2D_x}. \tag{G.5}$$

In order to explore this regime, let us consider the limiting case $D_x = 0$. In this case the supra-Laplacian has a block diagonal structure formed by the two Laplacians $\mathbf{L}^{[1]}, \mathbf{L}^{[2]}$ of the two layers. The zero eigenvalue of this matrix is twofold degenerate and we have $\lambda_1 = \lambda_2 = 0$. Since $\mathbf{L}^{[1]}\mathbf{1} = 0$ $\mathbf{L}^{[2]}\mathbf{1} = 0$, where $\mathbf{1}$ indicates the N-dimensional column vector of ones, it is possible to choose the two eigenvectors of the kernel of the supra-Laplacian corresponding to the eigenvalues $\lambda_1 = \lambda_2 = 0$ as

$$\mathbf{v}_1 = \begin{pmatrix} 1 \\ 1 \end{pmatrix}, \quad \mathbf{v}_2 = \begin{pmatrix} 1 \\ -1 \end{pmatrix}. \tag{G.6}$$

When a small intralayer diffusion constant $D_x \ll D_x^\star$ is introduced, the two eigenvectors $\mathbf{v}_1$ and $\mathbf{v}_2$ remain eigenvectors of the supra-Laplacian but the degeneracy of the zero eigenvalue is lifted. In fact, it is easy to show that $\mathcal{L}\mathbf{v}_1 = 0$, i.e. the eigenvector $\mathbf{v}_1$ corresponds to the eigenvalue $\lambda_1 = 0$. Instead, $\mathbf{v}_2$ corresponds to the eigenvalue

$$\lambda_2 = 2D_x.$$

In fact, we have

$$\mathcal{L}\mathbf{v}_1 = \begin{pmatrix} \mathbf{L}^{[1]} + D_x\mathbf{I} & -D_x\mathbf{I} \\ -D_x\mathbf{I} & \mathbf{L}^{[2]} + D_x\mathbf{I} \end{pmatrix} \begin{pmatrix} 1 \\ -1 \end{pmatrix} = 2D_x \begin{pmatrix} 1 \\ -1 \end{pmatrix}. \tag{G.7}$$

As long as D_x is small enough this eigenvalue is the smallest non-zero eigenvalue of the supra-Laplacian and determines the typical relaxation time of the diffusion process according to Eq. (G.5). Therefore, for small diffusion constant $D_x \ll D_x^\star$ the typical timescale τ for diffusion is given by

$$\tau = \frac{1}{2D_x}. \tag{G.8}$$

This result implies that in this regime the interlayer diffusion is the limiting value for the diffusion spreading.

G.3   Large interlayer diffusion constant D_x

For large interlayer diffusion constant $(D_x \gg D_x^\star)$, λ_2 saturates to a finite value

$$\lambda_2 = \frac{\lambda_s}{2} \tag{G.9}$$

where λ_s is the smallest non-zero eigenvalue of the Laplacian of the aggregated network given by $\mathbf{L}^{[1]} + \mathbf{L}^{[2]}$. This result can be easily obtained perturbatively using an expansion in $\epsilon = 1/D_x$. In fact, the supra-Laplacian $\mathcal{L}$ can be written as

$$\mathcal{L} = D_x \left[\begin{pmatrix} \mathbf{I} & -\mathbf{I} \\ -\mathbf{I} & \mathbf{I} \end{pmatrix} + \epsilon \begin{pmatrix} \mathbf{L}^{[1]} & 0 \\ 0 & \mathbf{L}^{[2]} \end{pmatrix} \right] = D_x\hat{\mathcal{L}}. \tag{G.10}$$

For $\epsilon = 0$ the eigenvalues $\hat{\mathcal{L}}$ are 0 and 2, both N fold degenerate. The eigenvector corresponding to the eigenvalues 0 and 2 are respectively given by the vectors $(\mathbf{u}|\mathbf{u})$ and $(\mathbf{u}|-\mathbf{u})$, i.e. vectors having identical or opposite values in the i-th element and the $(N+i)$-th element. For a non-zero $\epsilon \ll 1$, it is possible to find the eigenvalues λ_i and the eigenvectors $\mathbf{v}_i$ of the Laplacian $\hat{\mathcal{L}}$ by performing an expansion in ϵ. Let us put

$$\lambda_i = \lambda_i^{(0)} + \epsilon\lambda_i^{(1)} + O\left(\epsilon^2\right)$$
$$\mathbf{v}_i = \mathbf{v}_i^{(0)} + \epsilon\mathbf{v}_i^{(1)} + O\left(\epsilon^2\right) \tag{G.11}$$

where $\lambda_i^{(0)}$ and $\mathbf{v}_i^{(0)}$ are the eigenvalues and the eigenvectors of $\hat{\mathcal{L}}$ for $\epsilon = 0$. Let us consider the eigenvalues $\lambda_i^{(0)} = 0$ and the corresponding eigenvectors $\mathbf{v}_i^{(0)} = (\mathbf{u}|\mathbf{u})$; the expansion in ϵ reads

$$\lambda_i = 0 + \epsilon\tilde{\lambda}$$
$$\mathbf{v} = \begin{pmatrix} \mathbf{u} \\ \mathbf{u} \end{pmatrix} + \epsilon\begin{pmatrix} \tilde{\mathbf{u}}_1 \\ \tilde{\mathbf{u}}_2 \end{pmatrix} \tag{G.12}$$

where the vector $(\tilde{\mathbf{u}}_1|\tilde{\mathbf{u}}_2)$ is assumed to be orthogonal to $(\mathbf{u}|\mathbf{u})$. By plugging these expressions into the eigenvalue equation, it can be obtained that

$$\mathbf{L}^{[1]}\mathbf{u} + \tilde{\mathbf{u}}_1 - \tilde{\mathbf{u}}_2 = \tilde{\lambda}\mathbf{u},$$
$$\mathbf{L}^{[2]}\mathbf{u} + \tilde{\mathbf{u}}_2 - \tilde{\mathbf{u}}_1 = \tilde{\lambda}\mathbf{u}. \tag{G.13}$$

Summing and subtracting these two equations, this system of equation can be rewritten as

$$\left(\mathbf{L}^{[1]} + \mathbf{L}^{[2]}\right)\mathbf{u} = 2\tilde{\lambda}\mathbf{u},$$
$$\left(\mathbf{L}^{[1]} - \mathbf{L}^{[2]}\right)\mathbf{u} = 2(\tilde{\mathbf{u}}_1 - \tilde{\mathbf{u}}_2). \tag{G.14}$$

From the first equation it emerges that $\mathbf{u}$ is an eigenvector of the aggregated Laplacian $\mathbf{L}^{[1]} + \mathbf{L}^{[2]}$ with eigenvalue $\lambda_s = 2\tilde{\lambda}$. From the second equation, considering the fact that $(\tilde{\mathbf{u}}_1|\tilde{\mathbf{u}}_2)$ must be perpendicular to $(\mathbf{u}|\mathbf{u})$ and therefore $\tilde{\mathbf{u}}_2 = -\tilde{\mathbf{u}}_1 = -\tilde{\mathbf{u}}$, it is possible to obtain an expression for $\tilde{\mathbf{u}}$ as a function of $\mathbf{u}$, i.e.

$$\left(\mathbf{L}^{[1]} - \mathbf{L}^{[2]}\right)\mathbf{u} = 4\tilde{\mathbf{u}}. \tag{G.15}$$

Therefore, we have

$$\tilde{\lambda} = \frac{\lambda_s}{2}$$
$$\mathbf{v} = \begin{pmatrix} \mathbf{u} + \epsilon\tilde{\mathbf{u}} \\ \mathbf{u} - \epsilon\tilde{\mathbf{u}} \end{pmatrix} \tag{G.16}$$

where λ_s and $\mathbf{u}$ are respectively the eigenvalue and the corresponding eigenvector of the aggregated Laplacian $\mathbf{L}^{[1]} + \mathbf{L}^{[2]}$ and $\tilde{\mathbf{u}}$ is given by Eq. (G.15). The eigenvalue λ_s of the aggregated Laplacian satisfies

$$\frac{\lambda_s}{2} \geq \frac{\lambda_2^{[1]} + \lambda_2^{[2]}}{2} \geq \min\left(\lambda_2^{[1]}, \lambda_2^{[2]}\right), \tag{G.17}$$

where $\lambda_2^{[\alpha]}$ is the smallest non-zero eigenvalue of the single layer Laplacian $\mathbf{L}^{[\alpha]}$. Therefore the diffusion on the multiplex network will always be faster than the diffusion on the slowest layer of the multiplex network. Super-diffusion, i.e. the fact that the timescale of the multiplex network is actually faster than the timescale of diffusion in every single layer of the multiplex network, is possible but not guaranteed in general.

References

[1] Abdi, H. and Williams, L. J. (2010). Principal component analysis. *Wiley Interdisciplinary Reviews: Computational Statistics*, **2**, 433.

[2] Aguirre, J., Papo, D. and Buldu, J. M. (2013). Successful strategies for competing networks. *Nature Phys.*, **9**, 230.

[3] Ahn, Y.-Y., Bagrow, J. P. and Lehmann, S. (2010). Link communities reveal multiscale complexity in networks. *Nature*, **466**, 761.

[4] Almaas, E., Kovacs, B., Vicsek, T., Oltvai, Z. N. and Barabási, A.-L. (2004). Global organization of metabolic fluxes in the bacterium Escherichia coli. *Nature*, **427**, 839.

[5] Altarelli, F., Braunstein, A., DallAsta, L., Wakeling, J. R. and Zecchina, R. (2014). Containing epidemic outbreaks by message-passing techniques. *Phys. Rev. X*, **4**, 021024.

[6] Anand, K. and Bianconi, G. (2009). Entropy measures for networks: toward an information theory of complex topologies. *Phys. Rev. E*, **80**, 045102.

[7] Anand, K. and Bianconi, G. (2010). Gibbs entropy of network ensembles by cavity methods. *Phys. Rev. E*, **82**, 011116.

[8] Arenas, A., Díaz-Guilera, A. and Guimerà, R. (2001). Communication in networks with hierarchical branching. *Phys. Rev. Lett.*, **86**, 3196.

[9] Arenas, A., Díaz-Guilera, A., Kurths, J., Moreno, Y. and Zhou, C. (2008). Synchronization in complex networks. *Phys. Rep.*, **469**, 93.

[10] Asilani, M., Busiello, D. M., Carletti, T., Fanelli, D. and Planchon, G. (2014). Turing patterns in multiplex networks. *Phys. Rev. E*, **90**, 042814.

[11] Azimi-Tafreshi, N., Gómez-Gardeñes, J. and Dorogovtsev, S. N. (2014). k-Core percolation on multiplex networks. *Phys. Rev. E*, **90**, 032816.

[12] Bak, P., Tang, C. and Wiesenfeld, K. (1988). Self-organized criticality. *Phys. Rev. A*, **38**, 364.

[13] Barabási, A.-L. (2005). The origin of bursts and heavy tails in human dynamics. *Nature*, **435**, 207.

[14] Barabási, A.-L. (2016). *Network Science*. Cambridge University Press, Cambridge.

[15] Barabási, A.-L. and Albert, R. (1999). Emergence of scaling in random networks. *Science*, **286**, 509.

[16] Barabási, A.-L., Gulbahce, N. and Loscalzo, J. (2011). Network medicine: a network-based approach to human disease. *Nature Rev. Gen.*, **12**, 56.

[17] Barahona, M. and Pecora, L. M. (2002). Synchronization in small-world systems. *Phys. Rev. Lett.*, **89**, 054101.

[18] Bargigli, L., Di Iasio, G., Infante, L., Lillo, F. and Pierobon, F. (2015). The multiplex structure of interbank networks. *Quantitative Finance*, **15**, 673.

[19] Bargigli, L., Di Iasio, G., Infante, L., Lillo, F. and Pierobon, F. (2016). Interbank markets and multiplex networks: centrality measures and statistical null models. In *Interconnected Networks*, p. 179. Springer.

[20] Barigozzi, M., Fagiolo, G. and Garlaschelli, D. (2010). Multinetwork of international trade: a commodity-specific analysis. *Phys. Rev. E*, **81**, 046104.

[21] Barigozzi, M., Fagiolo, G. and Mangioni, G. (2011). Identifying the community structure of the international-trade multi-network. *Physica A*, **390**, 2051.

[22] Baronchelli, A. and Radicchi, F. (2013). Lévy flights in human behavior and cognition. *Chaos, Solitons & Fractals*, **56**, 101.

[23] Barrat, A., Barthélemy, M., Pastor-Satorras, R. and Vespignani, A. (2004). The architecture of complex weighted networks. *PNAS*, **101**, 3747.

[24] Barrat, A., Barthélemy, M. and Vespignani, A. (2008). *Dynamical Processes on Complex Networks*. Cambridge University Press, Cambridge.

[25] Barthélemy, M. (2011). Spatial networks. *Phys. Rep.*, **499**, 1.

[26] Bascompte, J., Jordano, P., Melián, C. J. and Olesen, J. M. (2003). The nested assembly of plant–animal mutualistic networks. *PNAS*, **100**, 9383.

[27] Bashan, A., Parshani, R. and Havlin, S. (2011). Percolation in networks composed of connectivity and dependency links. *Phys. Rev. E*, **83**, 051127.

[28] Bassett, D. S., Wymbs, N. F., Porter, M. A., Mucha, P. J., Carlson, J. M. and Grafton, S. T. (2011). Dynamic reconfiguration of human brain networks during learning. *PNAS*, **108**, 7641.

[29] Battiston, F., Iacovacci, J., Nicosia, V., Bianconi, G. and Latora, V. (2016). Emergence of multiplex communities in collaboration networks. *PLoS ONE*, **11**, e0147451. https://doi.org/10.1371/journal.pone.0147451.

[30] Battiston, F., Nicosia, V. and Latora, V. (2014). Structural measures for multiplex networks. *Phys. Rev. E*, **89**, 032804.

[31] Battiston, F., Nicosia, V. and Latora, V. (2016). Efficient exploration of multiplex networks. *New J. Phys.*, **18**, 043035.

[32] Baxter, G. J., Bianconi, G., da Costa, R. A., Dorogovtsev, S. N. and Mendes, J. F. F. (2016). Correlated edge overlaps in multiplex networks. *Phys. Rev. E*, **94**, 012303.

[33] Baxter, G. J., Dorogovtsev, S. N., Goltsev, A. V. and Mendes, J. F. F. (2012). Avalanche collapse of interdependent networks. *Phys. Rev. Lett.*, **109**, 248701.

[34] Benhamou, S. (2007). How many animals really do the Levy walk? *Ecology*, **88**, 1962.

[35] Bennett, J. and Lanning, S. (2007). The Netflix prize. In *Proc. of the KDD Cup Workshop 2007*, New York, p. 3. ACM.

[36] Bennett, L., Kittas, A., Muirhead, G., Papageorgiou, L. G. and Tsoka, S. (2015). Detection of composite communities in multiplex biological networks. *Sci. Rep.*, **5**, 10345.

[37] Bentley, B., Branicky, R., Barnes, C. L., Chew, Y. L., Yemini, E., Bullmore, E. T., Vértes, P. E. and Schafer, W. R. (2016). The multilayer connectome of Caenorhabditis elegans. *PLoS Comp. Bio.*, **12**, e1005283.

[38] Berezin, Y., Bashan, A., Danziger, M. M., Li, D. and Havlin, S. (2015). Localized attacks on spatially embedded networks with dependencies. *Sci. Rep.*, **5**, 8934.

[39] Bianconi, G. (2007). Entropy of randomized network ensembles *EPL (Europhysics Letters)*, **81**, 28005.

[40] Bianconi, G. (2009). Entropy of network ensembles. *Phys. Rev. E*, **79**, 036114.

[41] Bianconi, G. (2013). Statistical mechanics of multiplex networks: entropy and overlap. *Phys. Rev. E*, **87**, 062806.

[42] Bianconi, G. (2015). Supersymmetric multiplex networks described by coupled Bose and Fermi statistics. *Phys. Rev. E*, **91**, 012810.

[43] Bianconi, G. (2017). Epidemic spreading and bond percolation on multilayer networks. *JSTAT*, **2017**, 034001. https://doi.org/10.1088/1742-5468/aa5fd8.

[44] Bianconi, G. and Barabási, A.-L. (2001). Bose-Einstein condensation in complex networks. *Phys. Rev. Lett.*, **86**, 5632.

[45] Bianconi, G. and Barabási, A.-L. (2001). Competition and multiscaling in evolving networks. *EPL*, **54**, 436.

[46] Bianconi, G., Coolen, A. C. C. and Vicente, C. J. P. (2008). Entropies of complex networks with hierarchically constrained topologies. *Phys. Rev. E*, **78**, 016114.

[47] Bianconi, G., Darst, R. K., Iacovacci, J. and Fortunato, S. (2014). Triadic closure as a basic generating mechanism of communities in complex networks. *Phys. Rev. E*, **90**, 042806.

[48] Bianconi, G. and Dorogovtsev, S. N. (2014). Multiple percolation transitions in a configuration model of network of networks. *Phys. Rev. E*, **89**, 062814. https://doi.org/10.1103/PhysRevE.89.062814.

[49] Bianconi, G., Dorogovtsev, S. N. and Mendes, J. F. F. (2015). Mutually connected component of network of networks with replica nodes. *Phys. Rev. E*, **91**, 012804. https://doi.org/10.1103/PhysRevE.91.012804.

[50] Bianconi, G., Pin, P. and Marsili, M. (2009). Assessing the relevance of node features for network structure. *PNAS*, **106**, 11433.

[51] Bianconi, G. and Rahmede, C. (2016). Network geometry with flavor: from complexity to quantum geometry. *Phys. Rev. E*, **93**, 032315.

[52] Bianconi, G. and Rahmede, C. (2017). Emergent hyperbolic network geometry. *Sci. Rep.*, 7, 41974.

[53] Blondel, V. D., Guillaume, J.-L., Lambiotte, R. and Lefebvre, E. (2008). Fast unfolding of communities in large networks. *JSTAT*, P10008.

[54] Boccaletti, S., Bianconi, G., Criado, R., Del Genio, C. I., Gómez-Gardenes, J., Romance, M., Sendina-Nadal, I., Wang, Z. and Zanin, M. (2014). The structure and dynamics of multilayer networks. *Phys. Rep.*, **544**, 1.

[55] Boguñá, M., Castellano, C. and Pastor-Satorras, R. (2013). Nature of the epidemic threshold for the susceptible-infected-susceptible dynamics in networks. *Phys. Rev. Lett.*, **111**, 068701.

[56] Bollobás, B. (1998). Random graphs. In *Modern Graph Theory*, p. 215. Springer.

[57] Bonacich, P. (1987). Power and centrality: a family of measures. *Am. Jour. Soc.*, **92**, 1170.

[58] Brechtel, A., Gramlich, P., Ritterskamp, D., Drossel, B. and Gross, T. (2016). Master stability functions reveal diffusion-driven instabilities in multi-layer networks. *arXiv preprint arXiv:1610.07635*.

[59] Breiger, R. L. and Pattison, P. E. (1986). Cumulated social roles: the duality of persons and their algebras. *Social Networks*, 8, 215.

[60] Brin, S. and Page, L. (1998). The anatomy of a large-scale hypertextual web search engine. *Comput. Netw.*, **30**, 107.

[61] Brockmann, D., Hufnagel, L. and Geisel, T. (2006). The scaling laws of human travel. *Nature*, **439**, 462.

[62] Bródka, P., Musial, K. and Kazienko, P. (2010). A method for group extraction in complex social networks. In *Knowledge Management, Information Systems, E-Learning, and Sustainability Research* (ed. M. Lytras, P. Ordonez De Pablos, A. Ziderman, A. Roulstone, H. Maurer and J. B. Imber), Volume 111, Communications in Computer and Information Science, p. 238. Springer Berlin Heidelberg.

[63] Brummitt, C. D., D'Souza, R. M. and Leicht, E. A. (2012). Suppressing cascades of load in interdependent networks. *PNAS*, **109**, 680.

[64] Brummitt, C. D. and Kobayashi, T. (2015). Cascades in multiplex financial networks with debts of different seniority. *Phys. Rev. E*, **91**, 062813.

[65] Brummitt, C. D., Lee, K. M. and Goh, K. I. (2012). Multiplexity-facilitated cascades in networks. *Phys. Rev. E*, **85**, 045102.

[66] Buldyrev, S. V., Parshani, R., Paul, G., Stanley, H. E. and Havlin, S. (2010). Catastrophic cascade of failures in interdependent networks. *Nature*, **464**, 1025.

[67] Buldyrev, S. V., Shere, N. W. and Cwilich, G. A. (2011). Interdependent networks with identical degrees of mutually dependent nodes. *Phys. Rev. E*, **83**, 016112.

[68] Bullmore, E. and Sporns, O. (2009). Complex brain networks: graph theoretical analysis of structural and functional systems. *Nat. Rev. Neurosci.*, **10**(3), 186–98.

[69] Buono, C., Alvarez-Zuzek, L. G, Macri, P. A. and Braunstein, L. A. (2014). Epidemics in partially overlapped multiplex networks. *PLoS ONE*, **9**, e92200.

[70] Cantini, L., Medico, E., Fortunato, S. and Caselle, M. (2015). Detection of gene communities in multi-networks reveals cancer drivers. *Sci. Rep.*, **5**, 17386.

[71] Cardillo, A., Gómez-Gardeñes, J., Zanin, M., Romance, M., Papo, D., del Pozo, F. and Boccaletti, S. (2013). Emergence of network features from multiplexity. *Sci. Rep.*, **3**, 1344.

[72] Castellano, C., Fortunato, S. and Loreto, V. (2009). Statistical physics of social dynamics. *Rev. Mod. Phys.*, **81**, 591.

[73] Castellano, C., Vilone, D. and Vespignani, A. (2003). Incomplete ordering of the voter model on small-world networks. *Eurphys. Lett.*, **63**, 153.

[74] Cattuto, C., Van den Broeck, W., Barrat, A., Colizza, V., Pinton, J, F. and Vespignani, A. (2010). Dynamics of person-to-person interactions from distributed rfid sensor networks. *PLoS ONE*, **5**, e11596.

[75] Cellai, D. and Bianconi, G. (2016). Multiplex networks with heterogeneous activities of the nodes. *Phys. Rev. E*, **93**, 032302.

[76] Cellai, D., Dorogovtsev, S. N. and Bianconi, G. (2016). Message passing theory for percolation models on multiplex networks with link overlap. *Phys. Rev. E*, **94**, 032301.

[77] Cellai, D., Lopez, E., Zhou, J., Gleeson, J. P. and Bianconi, G. (2013). Percolation in multiplex networks with overlap. *Phys. Rev. E*, **88**, 052811.

[78] Cohen, R., Erez, K., ben Avraham, D. and Havlin, S. (2000). Resilience of the internet to random breakdowns. *Phys. Rev. Lett.*, **85**, 4626.

[79] Cohen, R., Erez, K., ben Avraham, D. and Havlin, S. (2001). Breakdown of the internet under intentional attack. *Phys. Rev. Lett.*, **86**, 3682–5.

[80] Cohen, R. and Havlin, S. (2003). Scale-free networks are ultrasmall. *Phys. Rev. Lett.*, **90**, 058701.

[81] Cover, T. M. and Thomas, J. A. (2012). *Elements of Information Theory*. John Wiley & Sons.

[82] Cozzo, E., Banos, R. A., Meloni, S. and Moreno, Y. (2013). Contact-based social contagion in multiplex networks. *Phys. Rev. E*, **88**, 050801.

[83] Cozzo, E., Kivelä, M., De Domenico, M., Solé-Ribalta, A., Arenas, A., Gómez, S., Porter, M. A. and Moreno, Y. (2015). Structure of triadic relations in multiplex networks. *New J. Phys.*, **17**, 073029.

[84] Criado, R., Flores, J., García del Amo, A., Gómez-Gardeñes, J. and Romance, M. (2012). A mathematical model for networks with structures in the mesoscale. *International Journal of Computer Mathematics*, **89**, 291.

[85] Danon, L., Diaz-Guilera, A., Duch, J. and Arenas, A. (2005). Comparing community structure identification. *JSTAT*, **2005**, P09008.

[86] Danziger, M. M., Shekhtman, L. M., Berezin, Y. and Havlin, S. (2016). The effect of spatiality on multiplex networks. *EPL*, **115**, 36002.

[87] De Arruda, G. Ferraz, Cozzo, E., Peixoto, T. P., Rodrigues, F. A. and Moreno, Y. (2017). Disease localization in multilayer networks. *Phys. Rev. X*, 7, 011014. https://doi.org/10.1103/PhysRevX.7.011014.

[88] De Domenico, M., Granell, C., Porter, M. A. and Arenas, A. (2016). The physics of spreading processes on multilayer networks. *Nature Phys.*, 12, 901.

[89] De Domenico, M., Lancichinetti, A., Arenas, A. and Rosvall, M. (2015). Identifying modular flows on multilayer networks reveals highly overlapping organization in interconnected systems. *Phys. Rev. X*, 5, 011027.

[90] De Domenico, M., Nicosia, V., Arenas, A. and Latora, V. (2015). Structural reducibility of multilayer networks. *Nature Comm.*, 6, 6864. https://doi.org/10.1038/ncomms7864.

[91] De Domenico, M., Sole, A., Gómez, S. and Arenas, A. (2013). Random Walks on Multiplex Networks. *arXiv e-prints*.

[92] De Domenico, M., Solé-Ribalta, A., Cozzo, E., Kivelä, M., Moreno, Y., Porter, M. A., Gómez, S. and Arenas, A. (2013). Mathematical formulation of multilayer networks. *Phys. Rev. X*, 3, 041022.

[93] De Domenico, M., Solé-Ribalta, A., Gómez, S. and Arenas, A. (2014). Navigability of interconnected networks under random failures. *PNAS*, 111, 8351.

[94] De Domenico, M., Solé-Ribalta, A., Omodei, E., Gómez, S. and Arenas, A. (2015). Ranking in interconnected multilayer networks reveals versatile nodes. *Nature Comm.*, 6, 6868.

[95] De Martino, D., Dall'Asta, L., Bianconi, G. and Marsili, M. (2009). Congestion phenomena on complex networks. *Phys. Rev. E*, 79, 015101.

[96] Del Genio, C. I., Gómez-Gardeñes, J., Bonamassa, I. and Boccaletti, S. (2016). Synchronization in networks with multiple interaction layers. *Sci. Adv.*, 2, e1601679.

[97] Di Patti, F., Fanelli, D. and Piazza, F. (2015). Optimal search strategies on complex multilinked networks. *Sci. Rep.*, 5, 9869.

[98] Diakonova, M., Nicosia, V., Latora, V. and San Miguel, M. (2016). Irreducibility of multilayer network dynamics: the case of the voter model. *New J. Phys.*, 18, 023010.

[99] Diakonova, M., San Miguel, M. and Eguíluz, V. M. (2014). Absorbing and shattered fragmentation transitions in multilayer coevolution. *Phys. Rev. E*, 89, 062818.

[100] Diez, I., Bonifazi, P., Escudero, I., Mateos, B., Muñoz, M. A., Stramaglia, S. and Cortes, J. M. (2015). A novel brain partition highlights the modular skeleton shared by structure and function. *Sci. Rep.*, 5, 10532.

[101] Donges, J. F., Schultz, H. C. H., Marwan, N., Zou, Y. and Kurths, J. (2011). Investigating the topology of interacting networks. *Eur. Phys. J. B*, 84, 635.

[102] Donges, J. F., Zou, Y., Marwan, N. and Kurths, J. (2009). The backbone of the climate network. *EPL*, 87, 48007.

[103] Donges, J. F., Zou, Y., Marwan, N. and Kurths, J. (2009). Complex networks in climate dynamics. *Eur. Phys. Jour.-Spec. Top.*, 174, 157.

[104] Dornic, I., Chaté, H., Chave, J. and Hinrichsen, H. (2001). Critical coarsening without surface tension: the universality class of the voter model. *Phys. Rev. Lett.*, 87, 045701. https://doi.org/10.1103/PhysRevLett.87.045701.

[105] Dorogovtsev, S. N. (2010). *Lectures on Complex Networks*. Volume 24. Oxford University Press, Oxford.

[106] Dorogovtsev, S. N., Goltsev, A. V. and Mendes, J. F. F. (2008). Critical phenomena in complex networks. *Rev. Mod. Phys.*, 80, 1275–1335.

[107] Dorogovtsev, S. N. and Mendes, J. F. F. (2003). *Evolution of Networks: From Biological Nets to the Internet and WWW*. Oxford University Press, Oxford.

[108] Dorogovtsev, S. N., Mendes, J. F. F. and Samukhin, A. N. (2000). Structure of growing networks with preferential linking. *Phys. Rev. Lett.*, 85, 4633.

[109] Dorogovtsev, S. N., Mendes, J. F. F. and Samukhin, A. N. (2001). Size-dependent degree distribution of a scale-free growing network. *Phys. Rev. E*, 63, 062101.

[110] Ducruet, C. (2013). Network diversity and maritime flows. *Journal of Transport Geography*, 30, 77.

[111] Dunne, J. A., Williams, R. J. and Martinez, N. D. (2002). Network structure and biodiversity loss in food webs: robustness increases with connectance. *Ecology letters*, 5, 558.

[112] Eagle, N. and Pentland, A. S. (2006). Reality mining: sensing complex social systems. *Personal and Ubiquitous Computing*, 10, 255.

[113] Echenique, P., Gómez-Gardeñes, J. and Moreno, Y. (2005). Dynamics of jamming transitions in complex networks. *Eurphys. Lett.*, 71, 325.

[114] Erdös, P. and Rényi, A. (1960). On the evolution of random graphs. *Publ. Math. Inst. Hung. Acad. Sci*, 5, 17.

[115] Estrada, E. and Gómez-Gardeñes, J. (2013). Communicability reveals a transition to coordinated behavior in multiplex networks. *arXiv e-prints*.

[116] Estrada, E. and Hatano, N. (2008). Communicability in complex networks. *Phys. Rev. E*, 77, 036111.

[117] Estrada, E., Hatano, N. and Benzi, M. (2012). The physics of communicability in complex networks. *Phys. Rep.*, 514, 89.

[118] Estrada, E. and Rodriguez-Velazquez, J. A. (2005). Subgraph centrality in complex networks. *Phys. Rev. E*, 71, 056103.

[119] Evans, T. S. and Lambiotte, R. (2009). Line graphs, link partitions, and overlapping communities. *Phys. Rev. E*, 80, 016105.

[120] Fienberg, S. E., Meyer, M. M. and Wasserman, S. S. (1985). Statistical analysis of multiple sociometric relations. *J. Amer. Statist. Assoc.*, 80, 51.

[121] Fortunato, S. (2010). Community detection in graphs. *Phys. Rep.*, 486, 75.

[122] Freeman, L. C. (1977). A set of measures of centrality based on betweenness. *Sociometry*, 35.

[123] Gao, J., Buldyrev, S. V., Havlin, S. and Stanley, H. E. (2011). Robustness of a network of networks. *PRL*, 107, 195701.

[124] Gao, J., Buldyrev, S. V., Havlin, S. and Stanley, H. E. (2012). Robustness of a network formed by n interdependent networks with a one-to-one correspondence of dependent nodes. *Phys. Rev. E*, 85, 066134.

[125] Gao, J., Buldyrev, S. V., Stanley, H. E. and Havlin, S. (2012). Networks formed from interdependent networks. *Nature Phys.*, 8, 40.

[126] Gao, Z.-K., Small, M. and Kurths, J. (2017). Complex network analysis of time series. *Eurphys. Lett.*, 116, 50001.

[127] García-Pérez, G., Boguñá, M., Allard, A. and Serrano, M. A. (2015). Rethinking distance in international trade: world trade atlas 1870–2013. *arXiv preprint arXiv:1512.02233*.

[128] Garlaschelli, D. and Loffredo, M. I. (2004). Fitness-dependent topological properties of the world trade web. *Phys. Rev. Lett.*, 93, 188701.

[129] Garlaschelli, D. and Loffredo, M. I. (2005). Structure and evolution of the world trade network. *Physica A*, 355, 138.

[130] Gauvin, L., Panisson, A. and Cattuto, C. (2014). Detecting the community structure and activity patterns of temporal networks: a non-negative tensor factorization approach. *PLoS ONE*, 9, e86028.

[131] Gibbons, R. (1992). *A Primer in Game Theory*. Harvester Wheatsheaf.

[132] Girvan, M. and Newman, M. E. J. (2002). Community structure in social and biological networks. *PNAS*, **99**, 7821.

[133] Gómez, S., Díaz-Guilera, A., Gómez-Gardeñes, J., Pérez-Vicente, C. J., Moreno, Y. and Arenas, A. (2013). Diffusion dynamics on multiplex networks. *Phys. Rev. Lett.*, **110**, 028701.

[134] Gómez-Gardeñes, J., Campillo, M., Floria, L. M. and Moreno, Y. (2007). Dynamical organization of cooperation in complex topologies. *Phys. Rev. Lett.*, **98**, 108103.

[135] Gómez-Gardeñes, J., Reinares, I., Arenas, A. and Floria, M. (2012). Evolution of cooperation in multiplex networks. *Sci. Rep.*, **2**, 620. https://doi.org/10.1038/srep00620.

[136] Gómez-Gardeñes, J., Gómez, S., Arenas, A. and Moreno, Y. (2011). Explosive synchronization transitions in scale-free networks. *Phys. Rev. Lett.*, **106**, 128701.

[137] Gómez-Gardenes, J., Moreno, Y. and Arenas, A. (2007). Paths to synchronization on complex networks. *Phys. Rev. Lett.*, **98**, 034101.

[138] Gonzalez, M. C., Hidalgo, C. A. and Barabási, A.-L. (2008). Understanding individual human mobility patterns. *Nature*, **453**, 779.

[139] Granell, C., Gómez, S. and Arenas, A. Competing spreading processes on multiplex networks: awareness and epidemics. *Phys. Rev. E*, **90**, 012808.

[140] Granell, C., Gómez, S. and Arenas, A. (2013). Dynamical interplay between awareness and epidemic spreading in multiplex networks. *Phys. Rev. Lett.*, **111**, 128701.

[141] Grindrod, P. and Higham, D. J. (2014). A dynamical systems view of network centrality. In *Proc. R. Soc. A*, Volume 470, p. 20130835. The Royal Society.

[142] Guimerá, R., Díaz-Guilera, A., Vega-Redondo, F., Cabrales, A. and Arenas, A. (2002). Optimal network topologies for local search with congestion. *PRL*, **89**, 248701.

[143] Guimerá, R., Díaz-Guilera, A., Vega-Redondo, F., Cabrales, A. and Arenas, A. (2002). Optimal network topologies for local search with congestion. *Phys. Rev. Lett.*, **89**, 248701.

[144] Guleva, V. Y., Skvorcova, M. V. and Boukhanovsky, A. V. (2015). Using multiplex networks for banking systems dynamics modelling. *Procedia Computer Science*, **66**, 257.

[145] Guo, Q., Cozzo, E., Zheng, Z. and Moreno, Y. (2016). Levy random walks on multiplex networks. *Sci. Rep.*, **6**, 37641.

[146] Hackett, A., Cellai, D., Gómez, S., Arenas, A. and Gleeson, J. P. (2016). Bond percolation on multiplex networks. *Phys. Rev. X*, **6**, 021002.

[147] Halu, A., Mondragón, R. J., Panzarasa, P. and Bianconi, G. (2013). Multiplex PageRank. *PLoS ONE*, 8, e78293.

[148] Halu, A., Mukherjee, S. and Bianconi, G. (2014). Emergence of overlap in ensembles of spatial multiplexes and statistical mechanics of spatial interacting network ensembles. *Phys. Rev. E*, **89**, 012806.

[149] Halu, A., Zhao, K., Baronchelli, A. and Bianconi, G. (2013). Connect and win: the role of social networks in political elections. *Eurphys. Lett.*, **102**, 16002.

[150] Hanneke, S. *et al.* (2010). Discrete temporal models of social networks. *Electronic Journal of Statistics*, 4, 585–605.

[151] Hanneke, S. and Xing, E. P. (2007). Discrete temporal models of social networks. *Lecture Notes in Computer Science*, **4503**, 115.

[152] Hashmi, A., Zaidi, F., Sallaberry, A. and Mehmood, T. (2012). Are all social networks structurally similar? In *Advances in Social Networks Analysis and Mining (ASONAM), 2012 IEEE/ACM International Conference on*, p. 310. IEEE.

[153] Hausmann, R., Hidalgo, C. A., Bustos, S., Coscia, M., Simoes, A. and Yildirim, M. A. (2014). *The Atlas of Economic Complexity: Mapping Paths to Prosperity.* MIT Press.

[154] Heaney, M. T. (2014). Multiplex networks and interest group influence reputation: an exponential random graph model. *Social Networks*, **36**, 6681.

[155] Hofbauer, J. and Sigmund, K. (1998). *Evolutionary Games and Population Dynamics*. Cambridge University Press, Cambridge.

[156] Holme, P. and Saramäki, J. (2012). Temporal networks. *Phys. Rep.*, **519**, 97.

[157] Holme, P. and Saramäki, J. (2013). *Temporal Networks*. Springer.

[158] Horvát, E.-Á. and Zweig, K. A. (2012). One-mode projection of multiplex bipartite graphs. In *ASONAM*, pp. 599–606. IEEE Computer Society.

[159] Horvát, E.-Á. and Zweig, K. A. (2013). A fixed degree sequence model for the one-mode projection of multiplex bipartite graphs. *Social Network Analysis and Mining*, 3(4), 1209–24.

[160] Hu, Y., Ksherim, B., Cohen, R. and Havlin, S. (2011). Percolation in interdependent and interconnected networks: abrupt change from second- to first-order transitions. *Phys. Rev. E*, **84**, 066116.

[161] Hu, Y., Zhou, D., Zhang, R., Han, Z., Rozenblat, C. and Havlin, S. (2013). Percolation of interdependent networks with intersimilarity. *Phys. Rev. E*, **88**, 052805.

[162] Huang, X., Shao, S., Wang, H., Buldyrev, S. V., Stanley, H. E. and Havlin, S. (2013). The robustness of interdependent clustered networks. *Eurphys. Lett.*, **101**, 18002.

[163] Iacovacci, J. and Bianconi, G. (2016). Extracting information from multiplex networks. *Chaos*, **26**, 065306.

[164] Iacovacci, J., Rahmede, C., Arenas, A. and Bianconi, G. (2016). Functional Multiplex PageRank. *EPL*, **116**, 28004.

[165] Iacovacci, J., Wu, Z. and Bianconi, G. (2015). Mesoscopic structures reveal the network between the layers of multiplex data sets. *Phys. Rev. E*, **92**, 042806.

[166] Kalman, R. E. (1963). Mathematical description of linear dynamical systems. *Journal of the Society for Industrial and Applied Mathematics, Series A: Control*, 1(2), 152–92.

[167] Karrer, B. and Newman, M. E. J. (2011). Competing epidemics on complex networks. *Phys. Rev. E*, **84**, 036106.

[168] Karrer, B., Newman, M. E. J. and Zdeborová, L. (2014). Percolation on sparse networks. *Phys. Rev. Lett.*, **113**, 208702.

[169] Karsai, M., Perra, N. and Vespignani, A. (2014). Time varying networks and the weakness of strong ties. *Sci. Rep.*, **4**, 4001.

[170] Katz, L. (1953). A new status index derived from sociometric analysis. *Psychometrika*, **18**, 39.

[171] Kazienko, P., Musial, K. and Kajdanowicz, T. (2011). Multidimensional social network in the social recommender system. *Systems, Man and Cybernetics, Part A: Systems and Humans, IEEE Transactions on*, 41(4), 746.

[172] Kéfi, S., Miele, V., Wieters, E. A., Navarrete, S. A. and Berlow, E. L. (2016). How structured is the entangled bank? The surprisingly simple organization of multiplex ecological networks leads to increased persistence and resilience. *PLoS Bio.*, **14**, 1002527.

[173] Kim, J. Y. and Goh, K. I. (2013). Coevolution and correlated multiplexity in multiplex networks. *Phys. Rev. Lett.*, **111**, 058702.

[174] Klages, R., Radons, G. and Sokolov, I. M. (2008). *Anomalous Transport: Foundations and Applications*. John Wiley & Sons.

[175] Kleineberg, K.-K., Boguñá, M., Serrano, M. A. and Papadopoulos, F. (2016). Hidden geometric correlations in real multiplex networks. *Nature Phys.*, **12**, 1076.

[176] Kouvaris, N. E., Hata, S. and Díaz-Guilera, A. (2015). Pattern formation in multiplex networks. *Sci. Rep.*, **5**.

[177] Krapivsky, P. L., Redner, S. and Ben-Naim, E. (2010). *A Kinetic View of Statistical Physics*. Cambridge University Press, Cambridge.

[178] Krapivsky, P. L., Redner, S. and Leyvraz, F. (2000). Connectivity of growing random networks. *Phys. Rev. Lett.*, **85**, 4629.

[179] Lacasa, L., Luque, B., Ballesteros, F., Luque, J. and Nuno, J. C. (2008). From time series to complex networks: the visibility graph. *PNAS*, **105**, 4972.

[180] Lacasa, L., Nicosia, V. and Latora, V. (2015). Network structure of multivariate time series. *Sci. Rep.*.

[181] Lambiotte, R., Salnikov, V. and Rosvall, M., (2014). Effect of memory on the dynamics of random walks on networks. *Journal of Complex Networks*, **3**, 177.

[182] Lancichinetti, A. and Fortunato, S. (2012). Consensus clustering in complex networks. *Sci. Rep.*, **2**. https://doi.org/10.1038/srep00336.

[183] Latora, V. and Marchiori, M. (2001). Efficient behavior of small-world networks. *Phys. Rev. Lett.*, **87**, 198701.

[184] Latora, V., Nicosia, V. and Russo, G. (2017). *Complex Networks: Principles, Methods and Applications*. Cambridge University Press, Cambridge.

[185] Lazega, E. and Pattison, P. E. (1999). Multiplexity, generalized exchange and cooperation in organizations: a case study. *Social Networks*, **21**, 6790.

[186] Lee, K.-M., Brummitt, C. D. and Goh, K.-I. (2014). Threshold cascades with response heterogeneity in multiplex networks. *Phys. Rev. E*, **90**, 062816.

[187] Lee, K.-M., Goh, K.-I. and Kim, I.-M. (2012). Sandpiles on multiplex networks. *J. Korean Phys. Soc.*, **60**, 641.

[188] Leicht, E. A. and D'Souza, R. M. (2009). Percolation on interacting networks. *arXiv e-prints*.

[189] Lewis, K., Gonzalez, M. and Kaufman, J. (2011). Social selection and peer influence in an online social network. *PNAS*, **109**(1), 68–72.

[190] Lewis, K., Kaufman, J., Gonzalez, M., Wimmer, A. and Christakis, N. (2008). Tastes, ties, and time: a new social network dataset using facebook.com. *Social Networks*, **30**, 330.

[191] Li, C., Sun, W. and Kurths, J. (2007). Synchronization between two coupled complex networks. *Phys. Rev. E*, **76**, 046204.

[192] Li, C., Xu, C., Sun, W., Xu, J. and Kurths, J. (2009). Outer synchronization of coupled discrete-time networks. *Chaos*, **19**, 013106.

[193] Li, W., Bashan, A., Buldyrev, S. V., Stanley, H. E. and Havlin, S. (2012). Cascading failures in interdependent lattice networks: the critical role of the length of dependency links. *Phys. Rev. Lett.*, **108**, 228702.

[194] Li, W., Dai, C. and Zhou, X. J. (2015). Integrative analysis of many biological networks to study gene regulation. *Integrating Omics Data*, 68.

[195] Li, W., Liu, C.-C., Zhang, T., Li, H., Waterman, M. S. and Zhou, X. J. (2011). Integrative analysis of many weighted co-expression networks using tensor computation. *PLoS Comp. Bio.*, **7**, e1001106.

[196] Liu, Y.-Y. and Barabási, A.-L. (2016). Control principles of complex systems. *Rev. Mod. Phys*, **88**, 035006.

[197] Liu, Y.-Y., Slotine, J.-J. and Barabási, A.-L. (2011). Controllability of complex networks. *Nature*, **473**, 167.

[198] Louzada, V. H. P., Araújo, N. A. M., Andrade Jr, J. S. and Herrmann, H.J. (2013). Breathing synchronization in interconnected networks. *Sci. Rep.*, **3**, 3289.

[199] Magnani, M., Micenkova, B. and Rossi, L. (2013). Combinatorial analysis of multiple networks. *arXiv preprint arXiv:1303.4986*.

[200] Mahutga, M. C. (2006). The persistence of structural inequality? A network analysis of international trade, 19652000. *Social Forces*, **84**, 1863.

[201] Maslov, S. and Sneppen, K. (2002). Specificity and stability in topology of protein networks. *Science*, **296**, 910.

[202] Mastrandrea, R., Squartini, T., Fagiolo, G. and Garlaschelli, D. (2014). Reconstructing the world trade multiplex: the role of intensive and extensive biases. *Phys. Rev. E*, **90**, 062804.

[203] Masuda, N. (2014). Voter model on the two-clique graph. *Phys. Rev. E*, **90**, 012802.

[204] Masuda, N. and Lambiotte, R. (2016). *A Guide to Temporal Networks*. Volume 4. World Scientific.

[205] Melián, C. J., Bascompte, J., Jordano, P. and Krivan, V. (2009). Diversity in a complex ecological network with two interaction types. *Oikos*, **118**, 122.

[206] Menche, J., Sharma, A., Kitsak, M., Ghiassian, S. D., Vidal, M., Loscalzo, J. and Barabási, A.-L. (2015). Uncovering disease-disease relationships through the incomplete interactome. *Science*, **347**, 1257601.

[207] Menichetti, G., Dall'Asta, L. and Bianconi, G. (2014). Network controllability is determined by the density of low in-degree and out-degree nodes. *Phys. Rev. Lett.*, **113**, 078701.

[208] Menichetti, G., Dall'Asta, L. and Bianconi, G. (2016). Control of multilayer networks. *Sci. Rep.*, **6**, 20706. https://doi.org/10.1038/srep20706.

[209] Menichetti, G., Remondini, D., Panzarasa, P., Mondragón, R. J. and Bianconi, G. (2014). Weighted multiplex networks. *PLoS ONE*, **9**, e97857. https://doi.org/10.1371/journal.pone.0097857.

[210] Michoel, T. and Nachtergaele, B. (2012). Alignment and integration of complex networks by hypergraph-based spectral clustering. *Phys. Rev. E*, **86**, 056111.

[211] Min, B. and Goh, K.-I. (2014). Multiple resource demands and viability in multiplex networks. *Phys. Rev. E*, **89**, 040802.

[212] Min, B., Lee, S., Lee, K.-M. and Goh, K.-I. (2015). Link overlap, viability, and mutual percolation in multiplex networks. *Chaos, Solitons & Fractals*, **72**, 49.

[213] Min, B., Yi, S. D., Lee, K.-M. and Goh, K.-I. (2014). Network robustness of multiplex networks with interlayer degree correlations. *Phys. Rev. E*, **89**, 042811.

[214] Momeni, N. and Fotouhi, B. (2015). Growing multiplex networks with arbitrary number of layers. *Phys. Rev. E*, **92**, 062812.

[215] Mondragon, R. J., Iacovacci, J. and Bianconi, G. (2017). Multilink communities of multiplex networks. *arXiv preprint arXiv:1706.09011*.

[216] Montagna, M. and Kok, C. (2016). Multi-layered interbank model for assessing systemic risk. *ECB Working Paper No. 1944*.

[217] Morris, R. G. and Barthélemy, M. (2012). Transport on coupled spatial networks. *Phys. Rev. Lett.*, **109**(12), 128703.

[218] Motter, A. E. (2004). Cascade control and defense in complex networks. *Phys. Rev. Lett.*, **93**, 098701.

[219] Mucha, P. J., Richardson, T., Macon, K., Porter, M. A. and Onnela, J. P. (2010). Community structure in time-dependent, multiscale, and multiplex networks. *Science*, **328**, 876.

[220] Musiał, K., Kazienko, P. and Kajdanowicz, T. (2008). Multirelational social networks in multimedia sharing systems. In *Knowledge Processing and Reasoning for Information Society*, Warsaw, p. 275. Academic Publishing House EXIT.

[221] Musmeci, N., Nicosia, V., Aste, T., Di Matteo, T. and Latora, V. (2016). The multiplex dependency structure of financial markets. *arXiv preprint arXiv:1606.04872*.

[222] Newman, M. E. J. (2002). Assortative mixing in networks. *Phys. Rev. Lett.*, **89**, 208701.

[223] Newman, M. E. J. (2002). Spread of epidemic disease on networks. *Phys. Rev. E*, **66**, 016128.

[224] Newman, M. E. J. (2006). Modularity and community structure in networks. *PNAS*, **103**, 8577.

[225] Newman, M. E. J. (2010). *Networks: An Introduction*. Oxford University Press, New York.

[226] Newman, M. E. J. and Girvan, M. (2004). Finding and evaluating community structure in networks. *Phys. Rev. E*, **69**, 026113.

[227] Newman, M. E. J., Strogatz, S. H. and Watts, D. J. (2001). Random graphs with arbitrary degree distributions and their applications. *Phys. Rev. E*, **64**, 026118.

[228] Nicosia, V., Bianconi, G., Latora, V. and Barthélemy, M. (2013). Growing multiplex networks. *Phys. Rev. Lett.*, **111**, 058701.

[229] Nicosia, V., Bianconi, G., Latora, V. and Barthélemy, M. (2014). Non-linear growth and condensation in multiplex networks. *Phys. Rev. E*, **90**, 042807.

[230] Nicosia, V. and Latora, V. (2015). Measuring and modelling correlations in multiplex networks. *Phys. Rev. E*, **92**, 032805. https://doi.org/10.1103/PhysRevE.92.032805.

[231] Nicosia, V., Skardal, P. S., Arenas, A. and Latora, V. (2017). Collective phenomena emerging from the interactions between dynamical processes in multiplex networks. *Phys. Rev. Lett.*, **118**, 138302.

[232] Nowak, M. A. (2006). *Evolutionary Dynamics*. Harvard University Press, Cambridge.

[233] Nowak, M. A. and May, R. M. (1992). Evolutionary games and spatial chaos. *Nature*, **359**, 826.

[234] Olesen, J. M., Stefanescu, C. and Traveset, A. Strong, long-term temporal dynamics of an ecological network. *PLoS ONE*, **6**(11), e26455.

[235] Omodei, E., De Domenico, M. and Arenas, A. (2015). Characterizing interactions in online social networks during exceptional events, *Frontiers in Physics*, **3**(59).

[236] Padgett, J. F. and Ansell, C. K. (1993). Robust action and the rise of the Medici. *Am. J. Sociol.*, **98**, 1259.

[237] Palla, G., Derényi, I., Farkas, I. and Vicsek, T. (2005). Uncovering the overlapping community structure of complex networks in nature and society. *Nature*, **435**, 814.

[238] Parshani, R., Buldyrev, S. V. and Havlin, S. (2010). Interdependent networks: reducing the coupling strength leads to a change from a first to second order percolation transition. *Phys. Rev. Lett.*, **105**, 048701.

[239] Parshani, R., Buldyrev, S. V. and Havlin, S. (2011). Critical effect of dependency groups on the function of networks. *PNAS*, **108**, 1007.

[240] Parshani, R., Rozenblat, C., Ietri, D., Ducruet, C. and Havlin, S. (2010). Inter-similarity between coupled networks. *Eurphys. Lett.*, **92**, 68002.

[241] Pastor-Satorras, R., Castellano, C., Van Mieghem, P. and Vespignani, A. (2015). Epidemic processes in complex networks. *Rev. Mod. Phys*, **87**, 925.

[242] Pastor-Satorras, R., Vázquez, A. and Vespignani, A. (2001). Dynamical and correlation properties of the internet. *Phys. Rev. Lett.*, **87**, 258701.

[243] Pastor-Satorras, R. and Vespignani, A. (2001). Epidemic spreading in scale-free networks. *Phys. Rev. Lett.*, **86**(14), 3200–03.

[244] Pastor-Satorras, R. and Vespignani, A. (2002). Immunization of complex networks. *Phys. Rev. E*, **65**(3), 036104.

[245] Pattison, P. and Wasserman, S. (1999). Logit models and logistic regressions for social networks, ii: Multivariate relationships. *Br. J. Math. Stat. Psychol.*, **52**, 169.

[246] Pecora, L. M. and Carroll, T. L. (1998). Master stability functions for synchronized coupled systems. *Phys. Rev. Lett.*, **80**, 2109.

[247] Peixoto, T. P. (2015). Inferring the mesoscale structure of layered, edge-valued, and time-varying networks. *Phys. Rev. E*, **92**, 042807. https://doi.org/10.1103/PhysRevE.92.042807.

[248] Perc, M., Gómez-Gardeñes, J., Szolnoki, A., Floría, L. M. and Moreno, Y. (2013). Evolutionary dynamics of group interactions on structured populations: a review. *Jour. Roy. Soc. Interface*, **10**, 20120997.

[249] Perra, N., Gonçalves, B., Pastor-Satorras, R. and Vespignani, A. (2012). Activity driven modeling of time varying networks. *Sci. Rep.*, **2**, 469. http://doi.org/10.1038/srep00469.

[250] Pilosof, S., Porter, M. A., Pascual, M. and Kéfi, S. (2017). The multilayer nature of ecological networks. *Nature Ecology and Evolution*, **1**, 0101.

[251] Poisot, T., Canard, E., Mouillot, D., Mouquet, N. and Gravel, D. (2012). The dissimilarity of species interaction networks. *Eco. Lett.*, **15**, 1353.

[252] Poledna, S., Molina-Borboa, J. L., Martínez-Jaramillo, S., van der Leij, M. and Thurner, S. (2015). The multi-layer network nature of systemic risk and its implications for the costs of financial crises. *Journal of Financial Stability*, **20**, 70.

[253] Porter, M. A. and Gleeson, J. P. (2016). *Dynamical Systems on Networks*. Springer.

[254] Pósfai, M., Gao, J., Cornelius, S. P., Barabási, A.-L. and D'Souza, R. M. (2016). Controllability of multiplex, multi-time-scale networks. *Phys. Rev. E*, **94**, 032316.

[255] Radicchi, F. and Arenas, A. (2013). Abrupt transition in the structural formation of interconnected networks. *Nature Phys.*, **9**, 717.

[256] Radicchi, F. and Baronchelli, A. (2012). Evolution of optimal Lévy-flight strategies in human mental searches. *Phys. Rev. E*, **85**, 061121.

[257] Radicchi, F. and Bianconi, G. (2017). Redundant interdependencies boost the robustness of multiplex networks. *Phys. Rev. X*, **7**, 011013. https://doi.org/10.1103/PhysRevX.7.011013.

[258] Rahmede, C., Iacovacci, J., Arenas, A. and Bianconi, G. (2017). Centralities of nodes and influences of layers in large multiplex networks. *arXiv preprint arXiv:1703.05833*.

[259] Rapoport, A., Kugler, T., Dugar, S. and Gisches, E. J. (2009). Choice of routes in congested traffic networks: experimental tests of the braess paradox. *Games and Economic Behavior*, **65**, 538.

[260] Ravasz, E., Somera, A. L., Mongru, D. A., Oltvai, Z. N. and Barabási, A.-L. (2002). Hierarchical organization of modularity in metabolic networks. *Science*, **297**, 1551.

[261] Reis, S. D. S., Hu, Y., Babino, A. Andrade Jr, J. S., Canals, S., Sigman, M. and Makse, H. A. (2014). Avoiding catastrophic failure in correlated networks of networks. *Nature Phys.*, **10**, 762.

[262] Riascos, A. P. and Mateos, J. L. (2012). Long-range navigation on complex networks using Lévy random walks. *Phys. Rev. E*, **86**, 056110.

[263] Ribeiro, B., Perra, N. and Baronchelli, A. (2013). Quantifying the effect of temporal resolution on time-varying networks. *Sci. Rep.*, **3**, 3006.

[264] Robins, G. and Pattison, P. (2001). Random graph models for temporal processes in social networks. *Journal of Mathematical Sociology*, **25**, 5.

[265] Robins, G., Pattison, P., Kalish, Y. and Lusher, D. (2007). An introduction to exponential random graph ($p^\star$) models for social networks. *Social Networks*, **29**, 173.

[266] Rodrigues, F. A., Peron, T. K. D. M., Ji, P. and Kurths, J. (2016). The kuramoto model in complex networks. *Phys. Rep.*, **610**, 1.

[267] Rosvall, M. and Bergstrom, C. T. (2007). An information-theoretic framework for resolving community structure in complex networks. *PNAS*, **104**, 7327.

[268] Sahneh, F. D. and Scoglio, C. (2014). Competitive epidemic spreading over arbitrary multilayer networks. *Phys. Rev. E*, **89**, 062817. https://doi.org/10.1103/PhysRevE.89.062817.

[269] Salehi, M., Sharma, R., Marzolla, M., Magnani, M., Siyari, P. and Montesi, D. (2015). Spreading processes in multilayer networks. *IEEE Transactions on Network Science and Engineering*, **2**, 65.

[270] Santos, F. C. and Pacheco, J. M. (2005). Scale-free networks provide a unifying framework for the emergence of cooperation. *Phys. Rev. Lett.*, **95**, 098104.

[271] Saumell-Mendiola, A., Serrano, M. A. and Boguna, M. (2012). Epidemic spreading on interconnected networks. *Phys. Rev. E*, **86**(2 Pt 2), 026106.

[272] Scholtes, I., Wider, N., Pfitzner, R., Garas, A., Tessone, C.J. and Schweitzer, F.(2014). Causality-driven slow-down and speed-up of diffusion in non-Markovian temporal networks. *Nature communications* **5**, 5024.

[273] Sigmund, K. (2010). *The Calculus of Selfishness*. Princeton University Press, Princeton, New Jersey.

[274] Slotine, J.-J. E. and Li, W. (1991). *Applied Nonlinear Control*. Volume 199. Prentice Hall Englewood Cliffs, New Jersey.

[275] Söderberg, B. (2003). Properties of random graphs with hidden color. *Phys. Rev. E*, **68**, 026107.

[276] Söderberg, B. (2003). Random graph models with hidden color. *Acta Phys. Pol. B*, 5085.

[277] Söderberg, B. (2003). Random graphs with hidden color. *Phys. Rev. E*, **68**, 015102.

[278] Solá, L., Romance, M., Criado, R., Flores, J., García del Amo, A. and Boccaletti, S. (2013). Eigenvector centrality of nodes in multiplex networks. *Chaos*, **23**, 033131.

[279] Solé-Ribalta, A., De Domenico, M., Kouvaris, N. E., Díaz-Guilera, A., Gómez, S. and Arenas, A. (2013). Spectral properties of the laplacian of multiplex networks. *Phys. Rev. E*, **88**, 032807.

[280] Solé-Ribalta, A., Gómez, S. and Arenas, A. (2016). Congestion induced by the structure of multiplex networks. *Phys. Rev. Lett.*, **116**, 108701.

[281] Son, S.-W., Bizhani, G., Christensen, C., Grassberger, P. and Paczuski, M. (2012). Percolation theory on interdependent networks based on epidemic spreading. *Eurphys. Lett.*, **97**, 16006.

[282] Son, S.-W., Grassberger, P. and Paczuski, M. (2011). Percolation transitions are not always sharpened by making networks interdependent. *Phys. Rev. Lett.*, **107**, 195702.

[283] Song, C., Qu, Z., Blumm, N. and Barabási, A.-L. (2010). Limits of predictability in human mobility. *Science*, **327**, 1018.

[284] Sood, V., Antal, T. and Redner, S. (2008). Voter models on heterogeneous networks. *Phys. Rev. E*, **77**, 041121.

[285] Sood, V. and Redner, S. (2005). Voter model on heterogeneous graphs. *Phys. Rev. Lett.*, **94**, 178701.

[286] Sorrentino, F. (2012). Synchronization of hypernetworks of coupled dynamical systems. *New J. Phys.*, **14**, 033035.

[287] Starnini, M., Baronchelli, A., Barrat, A. and Pastor-Satorras, R. (2012). Random walks on temporal networks. *Phys. Rev. E*, **85**, 056115.

[288] Stehlé, J., Barrat, A. and Bianconi, G. (2010). Dynamical and bursty interactions in social networks. *Phys. Rev. E*, **81**, 035101.

[289] Stehlé, J. *et al.* (2011). High-resolution measurements of face-to-face contact patterns in a primary school. *PLoS ONE*, **6**, 23176.

[290] Szell, M., Lambiotte, R. and Thurner, S. (2010). Multirelational organization of large-scale social networks in an online world. *PNAS*, **107**, 13636.

[291] Tacchella, A., Cristelli, M., Caldarelli, G., Gabrielli, A. and Pietronero, L. (2012). A new metrics for countries' fitness and products' complexity. *Sci. Rep.*, **2**, 723.

[292] Tan, F., Wu, J., Xia, Y. and Tse, C. K. (2014). Traffic congestion in interconnected complex networks. *Phys. Rev. E*, **89**, 062813.

[293] Tang, L., Wu, X., Lü, J., Lu, J.-a. and D'Souza, R. M. (2016). Master stability functions for multiplex networks. *arXiv preprint*, arXiv:1611.09110.

[294] Travers, J. and Milgram, S. (1967). The small world problem. *Psychology Today*, 1, 61.

[295] Ubaldi, E., Vezzani, A., Karsai, M., Perra, N. and Burioni, R. (2017). Burstiness and tie activation strategies in time-varying social networks. *Scientific Reports*, 7, 46225.

[296] Valdano, E., Ferreri, L., Poletto, C. and Colizza, V. (2015). Analytical computation of the epidemic threshold on temporal networks. *Phys. Rev. X*, 5, 021005.

[297] Valdez, L. D., Macri, P. A., Stanley, H. E. and Braunstein, L. A. (2013). Triple point in correlated interdependent networks. *Phys. Rev. E*, 88, 050803.

[298] Valles-Catala, T., Massucci, F. A., Guimera, R. and Sales-Pardo, M. (2016). Multilayer stochastic block models reveal the multilayer structure of complex networks. *Phys. Rev. X*, 6, 011036.

[299] Valverde, S. and Solé, R. V. (2002). Self-organized critical traffic in parallel computer networks. *Physica A*, 312, 636.

[300] Vazquez, F., Eguíluz, V. M. and San Miguel, M. (2008). Generic absorbing transition in coevolution dynamics. *Phys. Rev. Lett.*, 100, 108702.

[301] Vilone, D. and Castellano, C. (2004). Solution of voter model dynamics on annealed small-world networks. *Phys. Rev. E*, 69, 016109.

[302] Viswanathan, G. M., Afanasyev, V., Buldyrev, S. V., Murphy, E. J., Prince, P. A. and Stanley, H. E. (1996). Lévy flight search patterns of wandering albatrosses. *Nature*, 381, 413.

[303] Wang, H., Li, Q., D'Agostino, G., Havlin, S., Stanley, H. E. and Van Mieghem, P. (2013). Effect of the interconnected network structure on the epidemic threshold. *Phys. Rev. E*, 88(2), 022801.

[304] Wang, P., Robins, G., Pattison, P. and Lazega, E. (2013). Exponential random graph models for multilevel networks. *Social Networks*, 35, 96.

[305] Wang, Z., Andrews, M. A., Wu, Z.-X., Wang, L. and Bauch, C. T. (2015). Coupled disease–behavior dynamics on complex networks: a review. *Phys. Life Rev.*, 15, 1.

[306] Wang, Z., Szolnoki, A. and Perc, M. (2013). Optimal interdependence between networks for the evolution of cooperation. *Sci. Rep.*, 3, 2470.

[307] Wang, Z., Wang, L., Szolnoki, A. and Perc, M. (2015). Evolutionary games on multilayer networks: a colloquium. *Eur. Phys. J. B*, 88, 124.

[308] Wasserman, S. and Faust, K. (1994). *Social Network Analysis*. Cambridge University Press, Cambridge.

[309] Watts, D. J. (2002). A simple model of global cascades on random networks. *PNAS*, 99, 5766.

[310] Watts, D. J. and Strogatz, S. H. (1998). Collective dynamics of 'small-world' networks. *Nature*, 393, 440.

[311] Williams, R. J. and Martinez, N. D. (2000). Simple rules yield complex food webs. *Nature*, 404, 180.

[312] Yamasaki, K., Gozolchiani, A. and Havlin, S. (2008). Climate networks around the globe are significantly affected by El Niño. *Phys. Rev. Lett.*, 100, 228501.

[313] Zhang, X., Boccaletti, S., Guan, S. and Liu, Z. (2015). Explosive synchronization in adaptive and multilayer networks. *Phys. Rev. Lett.*, 114, 038701.

[314] Zhang, Y., Garas, A. and Schweitzer, F. (2016). Value of peripheral nodes in controlling multilayer scale-free networks. *Phys. Rev. E*, 93, 012309.

[315] Zhao, D., Wang, L., Li, S., Wang, Z., Wang, L. and Gao, B. (2014). Immunization of epidemics in multiplex networks. *PLoS ONE*, 9, e112018.

[316] Zhao, K. and Bianconi, G. (2013). Percolation on interacting, antagonistic networks. *JSTAT*, P05005.

[317] Zhao, K., Karsai, M. and Bianconi, G. (2011). Entropy of dynamical social networks. *PLoS ONE*, **6**, e28116.

[318] Zhao, K., Stehlé, J., Bianconi, G. and Barrat, A. (2011). Social network dynamics of face-to-face interactions. *Phys. Rev. E*, **83**, 056109.

[319] Zhou, D., Gao, J., Stanley, H. E. and Havlin, S. (2013). Percolation of partially interdependent scale-free networks. *Phys. Rev. E*, **87**, 052812.

[320] Zhou, D., Stanley, H. E., D'Agostino, G. and Scala, A. (2012). Assortativity decreases the robustness of interdependent networks. *Phys. Rev. E*, **86**, 066103.

[321] Zhou, J., Yan, G. and Lai, C.-H. (2013). Efficient routing on multilayered communication networks. *Eurphys. Lett.*, **102**, 28002.

[322] Zhuang, Y. and Yağan, O. (2016). Information propagation in clustered multilayer networks. *IEEE Transactions on Network Science and Engineering*, **3**, 211.

[323] Zhuo, Y., Peng, Y., Liu, Ch., Liu, Y. and Long, K. (2011). Traffic dynamics on layered complex networks. *Physica A*, **391**, 2401.

Index